# Lecture Notes in Economics and Mathematical Systems

(Vol. 1–15: Lecture Notes in Operations Research and Mathematical Economics, Vol. 16–59: Lecture Notes in Operations Research and Mathematical Systems)

Vol. 1: H. Bühlmann, H. Loeffel, E. Nievergelt, Einführung in die Theorie und Praxis der Entscheidung bei Unsicherheit. 2. Auflage, IV, 125 Seiten. 1969. DM 16,–

Vol. 2: U. N. Bhat, A Study of the Queueing Systems M/G/1 and GI/M1. VIII, 78 pages. 1968. DM 16,–

Vol. 3: A. Strauss, An Introduction to Optimal Control Theory. VI, 153 pages. 1968. DM 16,–

Vol. 4: Branch and Bound: Eine Einführung. 2., geänderte Auflage. Herausgegeben von F. Weinberg. VII, 174 Seiten. 1972. DM 18,–

Vol. 5: Hyvärinen, Information Theory for Systems Engineers. VIII, 205 pages. 1968. DM 16,–

Vol. 6: H. P. Künzi, O. Müller, E. Nievergelt, Einführungskursus in die dynamische Programmierung. IV, 103 Seiten. 1968. DM 16,–

Vol. 7: W. Popp, Einführung in die Theorie der Lagerhaltung. VI, 173 Seiten. 1968. DM 16,–

Vol. 8: J. Teghem, J. Loris-Teghem, J. P. Lambotte, Modèles d'Attente M/G/1 et GI/M/1 à Arrivées et Services en Groupes. IV, 53 pages. 1969. DM 16,–

Vol. 9: E. Schultze, Einführung in die mathematischen Grundlagen der Informationstheorie. VI, 116 Seiten. 1969. DM 16,–

Vol. 10: D. Hochstädter, Stochastische Lagerhaltungsmodelle. VI, 269 Seiten. 1969. DM 18,–

Vol. 11/12: Mathematical Systems Theory and Economics. Edited by H. W. Kuhn and G. P. Szegö. VIII, IV, 486 pages. 1969. DM 34,–

Vol. 13: Heuristische Planungsmethoden. Herausgegeben von F. Weinberg und C. A. Zehnder. II, 93 Seiten. 1969. DM 16,–

Vol. 14: Computing Methods in Optimization Problems. Edited by A. V. Balakrishnan. V, 191 pages. 1969. DM 16,–

Vol. 15: Economic Models, Estimation and Risk Programming: Essays in Honor of Gerhard Tintner. Edited by K. A. Fox, G. V. L. Narasimham and J. K. Sengupta. VIII, 461 pages. 1969. DM 24,–

Vol. 16: H. P. Künzi und W. Oettli, Nichtlineare Optimierung: Neuere Verfahren, Bibliographie. IV, 180 Seiten. 1969. DM 16,–

Vol. 17: H. Bauer und K. Neumann, Berechnung optimaler Steuerungen, Maximumprinzip und dynamische Optimierung. VIII, 188 Seiten. 1969. DM 16,–

Vol. 18: M. Wolff, Optimale Instandhaltungspolitiken in einfachen Systemen. V, 143 Seiten. 1970. DM 16,–

Vol. 19: L. Hyvärinen Mathematical Modeling for Industrial Processes. VI, 122 pages. 1970. DM 16,–

Vol. 20: G. Uebe, Optimale Fahrpläne. IX, 161 Seiten. 1970. DM 16,–

Vol. 21: Th. Liebling, Graphentheorie in Planungs- und Tourenproblemen am Beispiel des städtischen Straßendienstes. IX, 118 Seiten. 1970. DM 16,–

Vol. 22: W. Eichhorn, Theorie der homogenen Produktionsfunktion. VIII, 119 Seiten. 1970. DM 16,–

Vol. 23: A. Ghosal, Some Aspects of Queueing and Storage Systems. IV, 93 pages. 1970. DM 16,–

Vol. 24: Feichtinger Lernprozesse in stochastischen Automaten. V, 66 Seiten. 1970. DM 16,–

Vol. 25: R. Henn und O. Opitz, Konsum- und Produktionstheorie. I. II, 124 Seiten. 1970. DM 16 -

Vol. 26: D. Hochstädter und G Jebe, Ökonometrische Methoden. XII, 250 Seiten. 1970. DM 18,–

Vol. 27: I. H. Mufti, Computational Methods in Optimal Control Problems. IV, 45 pages. 1970. DM 16,–

Vol. 28: Theoretical Approaches to Non-Numerical Problem Solving. Edited by R. B. Banerji and M. D. Mesarovic. VI, 466 pages. 1970. DM 24,–

Vol. 29: S. E. Elmaghraby, Some Network Models in Management Science. III, 177 pages. 1970. DM 16,–

Vol. 30: H. Noltemeier, Sensitivitätsanalyse bei diskreten linearen Optimierungsproblemen. VI, 102 Seiten. 1970. DM 16,–

Vol. 31: M. Kühlmeyer, Die nichtzentrale t-Verteilung. II, 106 Seiten. 1970. DM 16,–

Vol. 32: F. Bartholomes und G. Hotz, Homomorphismen und Reduktionen linearer Sprachen. XII, 143 Seiten. 1970. DM 16,–

Vol. 33: K. Hinderer, Foundations of Non-stationary Dynamic Programming with Discrete Time Parameter. VI, 160 pages. 1970. DM 16,–

Vol. 34: H. Störmer, Semi-Markoff-Prozesse mit endlich vielen Zuständen. Theorie und Anwendungen. VII, 128 Seiten. 1970. DM 16,–

Vol. 35: F. Ferschl, Markovketten. VI, 168 Seiten. 1970. DM 16,–

Vol. 36: M. P. J. Magill, On a General Economic Theory of Motion. VI, 95 pages. 1970. DM 16,–

Vol. 37: H. Müller-Merbach, On Round-Off Errors in Linear Programming. VI, 48 pages. 1970. DM 16,–

Vol. 38: Statistische Methoden I. Herausgegeben von E. Walter. VIII, 338 Seiten. 1970. DM 22,–

Vol. 39: Statistische Methoden II. Herausgegeben von E. Walter. IV, 155 Seiten. 1970. DM 16,–

Vol. 40: H. Drygas, The Coordinate-Free Approach to Gauss-Markov Estimation. VIII, 113 pages. 1970. DM 16,–

Vol. 41: U. Ueing, Zwei Lösungsmethoden für nichtkonvexe Programmierungsprobleme. VI, 92 Seiten. 1971. DM 16,–

Vol. 42: A. V. Balakrishnan, Introduction to Optimization Theory in a Hilbert Space. IV, 153 pages. 1971. DM 16,–

Vol. 43: J. A. Morales, Bayesian Full Information Structural Analysis. VI, 154 pages. 1971. DM 16,–

Vol. 44: G. Feichtinger, Stochastische Modelle demographischer Prozesse. XIII, 404 Seiten. 1971. DM 28,–

Vol. 45: K. Wendler, Hauptaustauschschritte (Principal Pivoting). II, 64 Seiten. 1971. DM 16,–

Vol. 46: C. Boucher, Leçons sur la théorie des automates mathématiques. VIII, 193 pages. 1971. DM 18,–

Vol. 47: H. A. Nour Eldin, Optimierung linearer Regelsysteme mit quadratischer Zielfunktion. VIII, 163 Seiten. 1971. DM 16,–

Vol. 48: M. Constam, FORTRAN für Anfänger. 2. Auflage. VI, 148 Seiten. 1973. DM 16,–

Vol. 49: Ch. Schneeweiß, Regelungstechnische stochastische Optimierungsverfahren. XI, 254 Seiten. 1971. DM 22,–

Vol. 50: Unternehmensforschung Heute – Übersichtsvorträge der Züricher Tagung von SVOR und DGU, September 1970. Herausgegeben von M. Beckmann. VI, 133 Seiten. 1971. DM 16,–

Vol. 51: Digitale Simulation. Herausgegeben von K. Bauknecht und W. Nef. IV, 207 Seiten. 1971. DM 18,–

Vol. 52: Invariant Imbedding. Proceedings of the Summer Workshop on Invariant Imbedding Held at the University of Southern California, June–August 1970. Edited by R. E. Bellman and E. D. Denman. IV, 148 pages. 1971. DM 16,–

Vol. 53: J. Rosenmüller, Kooperative Spiele und Märkte. IV, 152 Seiten. 1971. DM 16,–

Vol. 54: C. C. von Weizsäcker, Steady State Capital Theory. III, 102 pages. 1971. DM 16,–

Vol. 55: P. A. V. B. Swamy, Statistical Inference in Random Coefficient Regression Models. VIII, 209 pages. 1971. DM 20,–

Vol. 56: Mohamed A. El-Hodiri, Constrained Extrema. Introduction to the Differentiable Case with Economic Applications. III, 130 pages. 1971. DM 16,–

Vol. 57: E. Freund, Zeitvariable Mehrgrößensysteme. VII, 160 Seiten. 1971. DM 18,–

Vol. 58: P. B. Hagelschuer, Theorie der linearen Dekomposition. VII, 191 Seiten. 1971. DM 18,–

**continuation on page 426**

# Lecture Notes in Economics and Mathematical Systems

Managing Editors: M. Beckmann and H. P. Künzi

Systems Theory

105


# Optimal Control Theory and its Applications

Proceedings of the Fourteenth Biennial Seminar
of the Canadian Mathematical Congress
University of Western Ontario, August 12-25, 1973

Part I


Edited by Bruce J. Kirby



Springer-Verlag Berlin Heidelberg GmbH

**Managing Editors**
Prof. Dr. M. Beckmann
Brown University
Providence, RI 02912/USA

Prof. Dr. H. P. Künzi
Universität Zürich
8090 Zürich/Schweiz

**Editor**
Dr. Bruce J. Kirby
Mathematics Department
Queen's University
Kingston, Ontario, Canada

**Library of Congress Cataloging in Publication Data**

Canadian Mathematical Congress.
Optimal control theory and its applications.

(Lecture notes in economics and mathematical systems ; 105-106 : Systems theory)
Bibliography: p.
Includes index.
1. Control theory--Congresses. 2. Mathematical optimization--Congresses. I. Kirby, Bruce J., 1928- ed. II. Title. III. Series: Lecture notes in economics and mathematical systems ; 105-106.
QA402.3.C33 1974 629.8'312 74-28257

AMS Subject Classifications (1970): 49-XX, 90A99, 90D40, 92-02, 92A15

ISBN 978-3-540-07018-4 ISBN 978-3-662-01569-8 (eBook)
DOI 10.1007/978-3-662-01569-8



Originally published by Springer-Verlag Berlin · Heidelberg · New York in 1974

# PREFACE

This work (in two parts), Lecture Notes in Economics and Mathematical Systems, Volume 105 and 106, constitutes the Proceedings of the Fourteenth Biennual Seminar of the Canadian Mathematical Congress, which was held from August 12 to August 25, 1973 at the University of Western Ontario, London, Ontario.

The Canadian Mathematical Congress has held Biennual Seminars since 1947, and these have covered a wide range of topics. The Seminar reported in this publication was concerned with "Optimal Control Theory and its Applications", a subject chosen for its active growth and its wide implications for other fields. Both these aspects are exemplified in these Proceedings.

Some lectures provided excellent surveys of particular fields whereas others concentrated on the presentation of new results.

There were six distinguished Principal Lecturers: H.T. Banks, A.R. Dobell, H. Halkin, J.L. Lions, R.M. Thrall and W.M. Wonham, all of whom gave five to ten lectures during the two weeks of the Seminar. Except for Dr. Dobell's, these will all be found in Volume 105.

Besides the Principal Lecturers there were three Guest Lecturers: M.C. Delfour, V. Jurdjevic and S.P. Sethi, who presented substantial bodies of material in two or three lectures and which are included in Volume 106.

Many of the participants also spoke and reports of most of these have also been included (Volume 106).

A Seminar such as this one, involving over seventy participants and lecturers for an extended period, is a major undertaking.

Our gratitude for its success is due to the Programme Committee consisting of

Colin W. Clark (U.B.C.)
M.N. Oguztorelli (U. of Alberta)
L.F.S. Ritcey (U. of Western Ontario)
F. Stenger (U. de Montréal)
and W.R.S. Sutherland (Dalhousie University).

The Local Arrangements Committee, chaired by Dr. Ritcey, provided excellent academic and social facilities for us, as did the staff of the Mathematics Department of U.W.O.

Thanks should also be extended to Dr. John J. McNamee, Executive Director of the Canadian Mathematical Congress; and to Mrs. Eileen M. Wight of Queen's University for the excellent, but onerous, task of typing the manuscripts.

B.J. Kirby,
Chairman, Programme Committee.

Quee's University,
Kingston, Ontario, Canada.
October 1974

## TABLE OF CONTENTS



## Contents of Part II

(Lecture Notes in Economics and Mathematical Systems, Vol. 106)

# MODELING OF CONTROL AND DYNAMICAL SYSTEMS IN THE LIFE SCIENCES

H.T. Banks*

Lefschetz Center for Dynamical Systems
Division of Applied Mathematics
Brown University
Providence, R.I. 02912

## Introduction

The text below was the basis for a series of lectures given by the author at the 14th Biennial Seminar of the Canadian Mathematical Congress held at the University of Western Ontario, London, Ontario, August 12-24, 1973. Since the theme of the Seminar was "Optimal Control Theory and Its Applications", a fitting focus for these lectures might have been "applications of optimal control theory in the life sciences". As the title of our lecture notes indicates, we have chosen instead to broaden the scope of these lectures to also include the contributions to the life sciences of investigators who employ the techniques and ideas in control theory, systems analysis, differential equations, and stochastic processes. Some of these efforts will, of course, involve applications of optimal control theory. But it is our view that many of the interesting efforts being made encompass much more than an application of control theory, even though they quite often do entail a utilization of the tools and approaches of this discipline.

---

* This research was supported in part by the U.S. Air Force Office of Scientific Research Contract AFOSR-71-2078 and by the U.S. Army Contract DA-ARO-D-31-124-71-612.

A second reason for broadening our emphasis here is the paucity of reports of serious applications of optimal control theory to problems in the biological sciences. From our perusal of the literature, it appears that while optimal control theory has been employed extensively and fruitfully in engineering applications during the last decade, thoughtful and substantial efforts utilizing such techniques in the life sciences are just beginning. This is in part due to the nature of the problems encountered, where in many cases the modeling problem itself offers the most formidable of challenges to investigators.

# CHAPTER 1. A BRIEF REVIEW OF ENZYME KINETICS

Enzymes are proteins which catalyze chemical reactions that usually, but not always, take place within the cell. All proteins of cells, including enzymes, are synthesized by ribosomes. Ribosomes synthesize both "inner-use" proteins (those used within the cell and synthesized by ribosomes randomly distributed in the cell) and "outer-use" proteins. These latter proteins are synthesized by ribosomes which are attached to the membranes of the endoplasmic reticulum. This system of membranes collects these proteins which are eventually exported from the cell.

Enzymes are of primary importance in metabolic pathways which would otherwise require large amounts of energy (heat) to catalyze chains of chemical reactions. Enzymes (which are known to be highly specific for both substrate and reaction type) allow these reactions to take place at a rapid rate at lower temperatures. Roughly speaking, an enzyme joins with its substrate and lowers the energy requirements for activation of the reaction, the reaction occurs, and the enzyme is then released unchanged to be used again. This can be described in the so-called "lock-and-key" theory

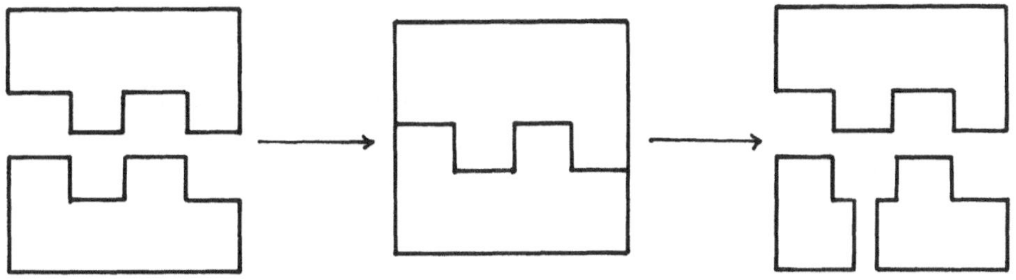

which assumes that the structure (shapes) of the enzyme and substrate molecules explain the specificity and inhibition features observed in enzymatic reactions. The schematic above is somewhat misleading since enzyme molecules are usually quite large and exceedingly complex in structure. For example, the substrate usually occupies only 10 percent of the enzyme surface during the reaction. In addition, the reactions sometimes require accessory substances which may be lightly bound to the enzyme molecule during the reaction. Another important fact that will be recalled in Chapter 3 below is that enzymes are usually present at extremely low intercellular concentrations (e.g. $10^{-7}$ Molar) and only small quantities of the enzyme are needed to catalyze the reaction. Finally, although we shall not study the control of enzyme levels within the cell in these lectures, we point out that there is a highly complex homeostatic system involving synthesis and inhibition which regulates these levels.

The kinetics (dynamics) involved in enzymatic reactions have traditionally been modeled by ordinary differential equations. A very important formulation much used (misused and abused) by modelers involves the <u>initial reaction velocity</u> expressions usually associated with the names Henri, Michaelis-Menten, and Briggs-Haldane. We develop briefly here the theory underlying these expressions (studied in the first quarter of this century) and subsequent modifications.

We consider a single substrate plus enzyme to product reaction

$$(1.1)\qquad E + S \underset{k_{-1}}{\overset{k_{+1}}{\rightleftharpoons}} ES \xrightarrow{k_{+2}} E + P$$

where it is assumed that an intermediate substrate-enzyme complex ES is formed. Further, the reaction $ES \longrightarrow E + P$ is assumed irreversible (in many cases the back reaction rate is so small that it may be ignored). Additional assumptions which are often not clearly stated are:

(i) there is an excess of substrate S in solution with the enzyme E ;

(ii) only initial reaction rates are considered, the decline of reaction velocity due to decline in substrate concentration being ignored;

(iii) the reaction $E + S \rightleftharpoons ES$ reaches equilibrium very quickly and maintains a steady state throughout the overall reaction.

Assumption (iii) is sometimes called the "quasi-equilibrium" (or "long-term stationarity" or "steady-state") assumption (see [97, 98]) and should actually be interpreted as $k_{-1} \gg k_{+2}$ . We may, under the above assumptions, write the kinetic equations

$$(1.2)\qquad \begin{aligned} \frac{d[S]}{dt} &= -k_{+1}[E][S] + k_{-1}\, ES \\ \frac{dES}{dt} &= -(k_{-1} + k_{+2})ES + k_{+1}[E][S] \\ \frac{dP}{dt} &= k_{+2}\, ES \end{aligned}$$

where $[E]$ is the concentration (in molars) of free enzyme, $[S]$ the concentration of free substrate, and $ES$ the concentration of the enzyme-substrate complex. From conservation laws we have

$$E_T = [E] + ES$$

(1.3)

$$S_T = [S] + ES$$

where $E_T$, $S_T$ represent the concentrations of the total (free and bound) amounts of enzyme and substrate present, respectively. The assumptions (i) and (ii) allow one to approximate, replacing the second equation in (1.3) by $S_T \approx [S]$. Further, the Michaelis-Menten derivation then replaces the first equation in (1.2) by $d[S]/dt = 0$ or

$$0 = -k_{+1}[E][S] + k_{-1}\, ES$$
$$= -k_{+1}\{E_T - ES\}\,[S] + k_{-1}\, ES ,$$

so that one obtains

$$ES = \frac{E_T[S]}{\frac{k_{-1}}{k_{+1}} + [S]}$$

One thus obtains the familiar expression for the initial velocity of product formation

$$dP/dt = v = \frac{k_{+2}E_T[S]}{K_M+[S]} = \frac{V_{max}[S]}{K_M+[S]}$$

or

(1.4) $$v = \frac{V_{max}S_T}{K_M+S_T}$$

where $V_{max} \equiv k_{+2}E_T$ and $K_M \equiv k_{-1}/k_{+1}$ .

In the Briggs-Haldane modification (under the same hypotheses) one uses the quasi-equilibrium assumption (iii) to write

$$dES/dt = 0$$

or

$$0 = -(k_{-1}+ k_{+2})ES + k_{+1}[E][S] ,$$

which, upon use of the first equation in (1.3) yields

(1.5) $$ES = \frac{E_T[S]}{\frac{k_{-1} + k_{+2}}{k_{+1}} + [S]} .$$

The initial velocity expression is thus found to be

(1.6) $$v = \frac{V_{max}\ S_T}{K_M + S_T}$$

where once again $V_{max} \equiv k_{+2}E_T$ , but now the "Michaelis constant" $K_m$ is given by

$$(1.7) \qquad K_M = \frac{k_{-1} + k_{+2}}{k_{+1}} \quad .$$

We point out that at maximum velocity, $V_{max}$, one has no free enzyme so that $E_T = ES$ and hence $V_{max} = k_{+2}\ ES = k_{+2}\ E_T$. Furthermore, at $v = \frac{1}{2} V_{max}$, the expressions (1.4), (1.6) yield $S_T = K_M$. In fact, one interpretation of the Michaelis constant $K_M$ (which is sometimes used as a definition in the derivation of the velocity expressions) is that $K_M$ is that value of substrate concentration $S_T$ which yields a reaction velocity one half the maximum velocity (i.e. $v = \frac{1}{2} V_{max}$).

For many investigations, the above assumptions are much too stringent and there have thus been a number of modifications proposed (see, for example, [17,30,52]). In particular, the assumptions (i) and (ii) are often objectionable. In our work on enzyme cascades detailed in Chapter 3 of these notes, both E and S are proteins with the concentrations of the substrates only ten times those of the enzymes at each stage in the cascade. Furthermore, one does have significant changes in both substrate and enzyme levels during the time course of the model. Therefore velocity expressions derived under assumptions (i) and (ii) above are inadequate for use in such instances.

We further note that in many mathematical uses of the reaction velocity expressions one wishes to ignore the formation of the intermediate ES complex and consider the reaction (1.1) as one simply of the form

$$S_T \xrightarrow{E_T} P$$

Use of the expressions (1.4), (1.6) as velocity terms then involves either very crude approximations or an implicit assumption of the form (i), which in some cases is undesirable.

For the modifications discussed here, we drop the assumptions (i) and (ii), retaining only (iii) which again implies $dES/dt = 0$ . Defining $K_M$ as in (1.7), we obtain

$$ES = \frac{[E][S]}{K_M} = \frac{\{E_T - ES\}\{S_T - ES\}}{K_M} \quad .$$

This can be written

$$(1.8) \qquad (ES)^2 - (K_M + S_T + E_T)ES + E_T S_T \doteq 0$$

which yields

$$(1.9) \qquad ES = \frac{1}{2}\left[(K_M + S_T + E_T) - \sqrt{(K_M + S_T + E_T)^2 - 4E_T S_T}\right] ,$$

where we have chosen the smaller root (minus sign) so that at $E_T = 0$ , $S_T = 0$ the expression yields $ES = 0$ . Then we obtain

$$(1.10) \qquad v = \frac{V_{max}}{2E_T}\left[(K_M + S_T + E_T) - \sqrt{(K_M + S_T + E_T)^2 - 4E_T S_T}\right] .$$

On the other hand, if $ES \ll K_M + E_T + S_T$ (which is true if

$K_M \gg E_T > ES$) , we may approximate the equation (1.8) by

$$-(K_M + S_T + E_T)ES + E_T S_T = 0$$

or

$$ES = \frac{E_T S_T}{K_M + S_T + E_T} .$$

With $V_{max}$ defined as above, this yields

$$(1.11) \qquad v = \frac{V_{max} S_T}{K_M + S_T + E_T} .$$

This expression will be used in evaluating some of the numerical approximations in the cascade model discussed in Chapter 3 below.

We return to the Briggs-Haldane formulation and indicate the changes involved if a competitive inhibitor is added to the reaction represented by equation (1.1). A competitive inhibitor is an inhibitor (chemical reagent which inhibits the catalytic action of the enzyme) whose action can be reversed by increasing the concentration of the substrate. That is, one may consider that the inhibitor and substrate "compete" for the "active site" of the enzyme, with inhibition taking place if the inhibitor occupies the site. To the equation (1.1) we must adjoin

$$(1.12) \qquad E + I \underset{k'_{-1}}{\overset{k'_{+1}}{\rightleftharpoons}} EI$$

and the first equation in (1.3) must be replaced by

$$(1.13) \qquad E_T = [E] + ES + EI ,$$

where I is the concentration of the inhibitor, EI the concentration of the enzyme-inhibitor complex. The kinetic equations (1.2) are still valid, but must be supplemented with another equation

$$\frac{d}{dt} EI = k'_{+1}[E][I] - k'_{-1} EI .$$

As before, a quasi-equilibrium assumption (iii) for both the complex formation reactions yields the approximations

$$\frac{d}{dt} ES = 0 , \quad \frac{d}{dt} EI = 0 ,$$

from which it follows that

$$(1.14) \qquad ES = \frac{[E][S]}{K_M} , \quad EI = \frac{[E][I]}{K_I}$$

where $K_I \equiv k'_{-1}/k'_{+1}$ . Using (1.13),

$$E_T = [E] + ES + \frac{[E][I]}{K_I} = ES + \{1 + \frac{[I]}{K_I}\} [E]$$

we obtain from the first equation in (1.14)

$$K_M ES = [S] \left\{ \frac{E_T - ES}{1 + [I]/K_I} \right\}$$

or

$$\left\{ K_M + \frac{K_M[I]}{K_I} + [S] \right\} ES = E_T[S] \; .$$

We thus find

$$ES = \frac{E_T[S]}{K_M + \frac{K_M[I]}{K_I} + [S]}$$

and hence

$$(1.15) \qquad v = \frac{V_{max}[S]}{K_M + \frac{K_M[I]}{K_I} + [S]}$$

If we again approximate by $S_T \approx [S]$ , we finally have

$$(1.16) \qquad v = \frac{V_{max}\, S_T}{K_M\{1 + \frac{[I]}{K_I}\} + S_T} \; .$$

In a similar manner, one may derive modified velocity expressions under relaxed assumptions (see (1.8) - (1.11) above) in the case of the presence of a competitive inhibitor. For example, if one assumes only (iii) and ignores terms $(ES)^2$ as in the derivation of (1.11) above, one obtains

$$ES = \frac{E_T S_T}{K_M\{1 + \frac{[I]}{K_I}\} + S_T + E_T}$$

which then implies

$$v = \frac{V_{max} S_T}{K_M\{1 + \frac{[I]}{K_I}\} + S_T + E_T} . \tag{1.17}$$

In the next two chapters we shall use the velocity expressions developed above to discuss two areas of modeling where optimality ideas have been fruitfully employed.

# CHAPTER 2. MODELS FOR ENZYMATICALLY ACTIVE MEMBRANES

In this chapter we discuss modeling and control problems arising in the study of enzymatically active membranes. These discussions are based on the work of Kernevez and Thomas and more detailed accounts of these investigations may be found in [63, 64, 74]. Throughout this chapter we shall, in agreement with the works cited above, use the Briggs-Haldane expressions (e.g. (1.6), (1.16)) for reaction velocities during the time course of the reaction, i.e.

$$v(t) = \frac{V_M \; S(t)}{K_M + S(t)} .$$

From Chapter 1 it should be clear to the reader exactly what assumptions and approximations are involved in so doing.

## 2.1 Modeling of the physiological phenomena using Briggs-Haldane approximations.

The basic biochemical model on which most of the mathematical models of this chapter are based involves a membrane in which enzymes are insolubly embedded by one of several means. More precisely, we assume that we have an artificial membrane $\mathfrak{m}$ of thickness $e = 50\mu$ (we recall that cell membranes are usually 75 to 100 Å or 75 to 100 $\times 10^{-4}\mu$ in thickness) separating two compartments, as depicted in Figure 2.1. The compartments I and II are 5 to 10 cm long and several centimeters in height. The compartments contain solutions of substrate (for one of the enzymes embedded in the membrane) of concentrations $S_1$, $S_2$ respectively. We are interested in how the substrate moves or is altered in the membrane, where we assume that both reaction and diffusion take place and that the membrane is initially empty of any substrate or product.

We describe several models of interest relative to these basic assumptions.

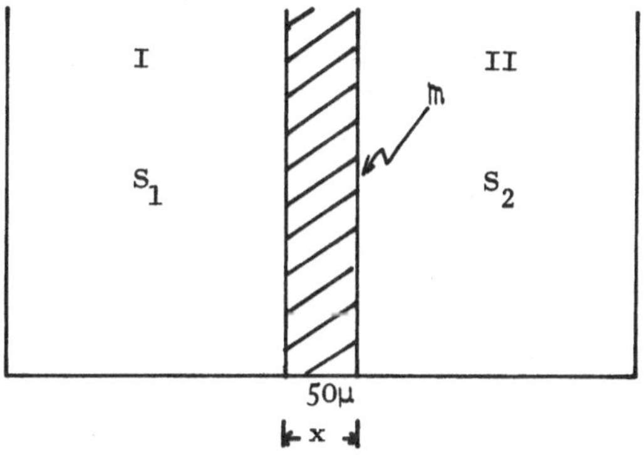


Figure 2.1

(a) Simple irreversible monoenzymatic reactions.

We assume that a simple substrate-to-product reaction as described by (1.1) takes place. Letting $S = S(x,t)$ denote the concentration of substrate at $(x,t)$ in the one-dimensional membrane, we have that the velocity due to reaction is given by

$$(2.1) \qquad \left(\frac{\partial S}{\partial t}\right)_{\text{reaction}} = -v(t) = -\frac{V_{max}\, S}{K_M + S} .$$

From the equation of continuity we find that the velocity of diffusion is related to $\bar{J}$ , the diffusion current vector, by the equation

(2.2 $$\left(\frac{\partial S}{\partial t}\right)_{\text{diffusion}} + \nabla \cdot \bar{J} = 0 \; .$$

Using Fick's law ($\bar{J} = - D_S \nabla S$ where $D_S$ is the so-called coefficient of diffusion) and assuming that $D_S$ is constant, we obtain

(2.3) $$\left(\frac{\partial S}{\partial t}\right)_{\text{diffusion}} = D_S \frac{\partial^2 S}{\partial x^2} \; .$$

Thus, the evolution of $S$ in the membrane is described by

(2.4) $$\frac{\partial S}{\partial t} - D_S \frac{\partial^2 S}{\partial x^2} + \frac{V_{max} \; S}{K_M + S} = 0 \; ,$$

or, with a rescaling of variables (including $t,x$) ,

(2.5) $$\frac{\partial s}{\partial t} - \frac{\partial^2 s}{\partial x^2} + \sigma \frac{s}{1+s} = 0 \; , \qquad x \in [0,1], \; t \in [0, T] \; .$$

Here $\sigma = V_{max} e^2/K_M D_S$ and $s = S/K_M$ . We remark that in general one finds

(2.6) $$\frac{\partial s}{\partial t} - \frac{\partial^2 s}{\partial x^2} + v = 0$$

where $v$ is a "normalized" velocity for the reaction or reactions involved.

If one assumes that the concentrations are held fixed in the compartments, one obtains boundary conditions for (2.5) given by

$$(2.7) \qquad s(0,t) = \alpha, \; s(1,t) = \beta, \; t \in [0,T] \; .$$

The initial condition is

$$(2.8) \qquad s(x,0) = 0, \quad x \in (0,1) \; .$$

In the event we assume that one or both walls of the membrane are impermeable to the substrate, the boundary conditions are modified by terms $\frac{\partial s}{\partial x}(0,t) = 0$ or $\frac{\partial s}{\partial x}(1,t) = 0$ or both. On the other hand, if the concentrations of substrate are not held fixed in compartments I or II and we have (free) flow across the membrane walls, we can easily argue that the boundary condition has the form

$$(2.9) \qquad \frac{\partial s}{\partial t} + \chi \frac{\partial s}{\partial n} = 0$$

at either $(0,t)$ or $(1,t)$ or both with initial values $s(0,0)$ or $s(1,0)$ or both specified. Here $\partial/\partial n$ is the exterior normal derivative and $\chi$ is the ratio of the volume of the membrane to the volume of solution in the adjacent compartment.

We remark that boundary conditions of the form (2.9) are appropriate if the adjacent compartment is small or if the experiment (time course) is of lengthy duration and no additional substrate is added to the compartments.

If one is also interested in the dynamics of the product $P$, one can easily see that they are given (in normalized form) by

$$\frac{\partial p}{\partial t} - \frac{D_P}{D_S}\frac{\partial^2 p}{\partial x^2} - \sigma \frac{s}{1+s} = 0 \tag{2.10}$$

where $p = P/K_M$. Assuming that the membrane walls are impermeable to the product, one has boundary conditions

$$\frac{\partial p}{\partial x}(0,t) = \frac{\partial p}{\partial x}(1,t) = 0 \tag{2.11}$$

to use with the initial condition $p(x,0) = 0$.

(b) Reactions with competitive inhibition.

We assume that along with an enzyme and substrate as in (a) above, we also have an uncontrolled competitive inhibitor present in compartments I and II which is free to diffuse throughout the membrane. Considerations similar to those detailed above yield a model with equations

$$\frac{\partial s}{\partial t} - \frac{\partial^2 s}{\partial x^2} + \sigma \frac{s}{1 + (K_M/K_I)i + s} = 0 ,$$

$$\frac{\partial i}{\partial t} - \frac{D_I}{D_S}\frac{\partial^2 i}{\partial x^2} = 0 , \tag{2.12}$$

$$\frac{\partial p}{\partial t} - \frac{D_P}{D_S}\frac{\partial^2 p}{\partial x^2} - \frac{\sigma s}{1 + (K_M/K_I)i + s} = 0 ,$$

with boundary conditions

$$s(0,t) = \alpha, \; s(1,t) = \beta \; ,$$

$$(2.13) \qquad i(0,t) = i_1, \; i(1,t) = i_2 \; ,$$

$$\frac{\partial p}{\partial x}(0,t) = \frac{\partial p}{\partial x}(1,t) = 0 \; ,$$

and initial conditions

$$s(x,0) = p(x,0) = i(x,0) = 0 \; .$$

(c) Bienzymatic double-layer membranes.

We assume that the membrane is made of two layers, each layer containing an enzyme (see Figure 2.2). In layer one, we have the reaction $S + E_1 \longrightarrow P + E_1$ which is assumed monoenzymatic irreversible with competitive inhibition by the product $P$. In layer two, $P$ is a substrate for the enzyme $E_2$ which catalyzes the irreversible reaction $P + E_2 \longrightarrow S + E_2$.

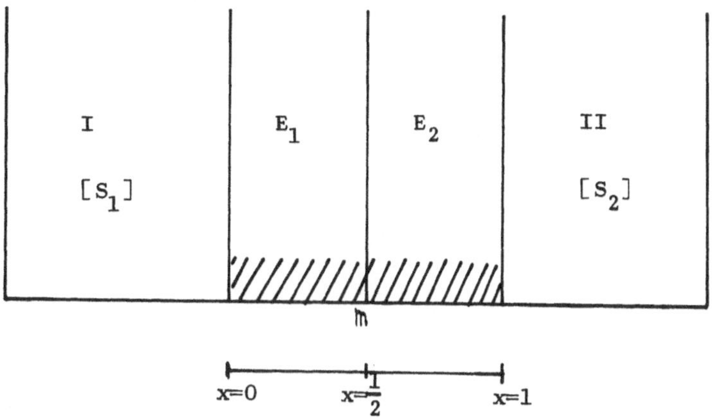


Figure 2.2

From our previous discussions we see that the normalized Briggs-Haldane reaction velocity in layer one is given by

$$(2.14) \qquad v = \frac{\sigma s}{1 + (K_M/K_P)p + s}\ , \quad 0 < x < 1/2\ ,$$

where $\sigma = V_{max}e^2/K_M D_S$ . In layer two we have

$$(2.15) \qquad v = -\frac{\sigma' p}{(K'_M/K_M) + p}\ , \quad 1/2 < x < 1\ ,$$

where $\sigma' = V'_{max}e^2/K_M D_S$ and $K'_M$ is the Michaelis constant for $E_2$ .
The model equations are then found to be

$$(2.16) \qquad \begin{aligned} &\frac{\partial s}{\partial t} - \frac{\partial^2 s}{\partial x^2} + v = 0\ , \\ &\frac{\partial p}{\partial t} - \frac{D_P}{D_S}\frac{\partial^2 p}{\partial x^2} - v = 0\ , \end{aligned} \qquad 0 < x < 1\ ,$$

where $v$ is given in (2.14) and (2.15). Appropriate boundary conditions are

$$(2.17) \qquad \frac{\partial p}{\partial x}(0,t) = \frac{\partial p}{\partial x}(1,t) = 0$$

and

$$s(0,t) = \alpha\ , \ s(1,t) = \beta$$

or

$$s(0,t) = \alpha$$

(2.19)

$$\frac{\partial s}{\partial t} + \chi \frac{\partial s}{\partial x} = 0 \quad \text{at} \quad (1,t) \quad \text{with} \quad s(1,0) = \beta \ .$$

The initial conditions are $s(x,0) = p(x,0) = 0$ .

Artificial membranes such as that described above were constructed in Laboratoire de Biochimie Médicale, Hôpital Charles Nicolle, in Rouen, France. The membranes were impregnated with ATP and the enzymes and substrates were: $E_1$ = hexokinase, $E_2$ = phosphatase, S = glucose, P = glucose-6-phosphate. The reactions involved are

$$G + ATP \xrightarrow{HK} G\text{-}6\text{-}P + ADP$$

$$G\text{-}6\text{-}P \xrightarrow{P_{ase}} G + P_i$$

and

$$ATP \longrightarrow ADP + P_i \ .$$

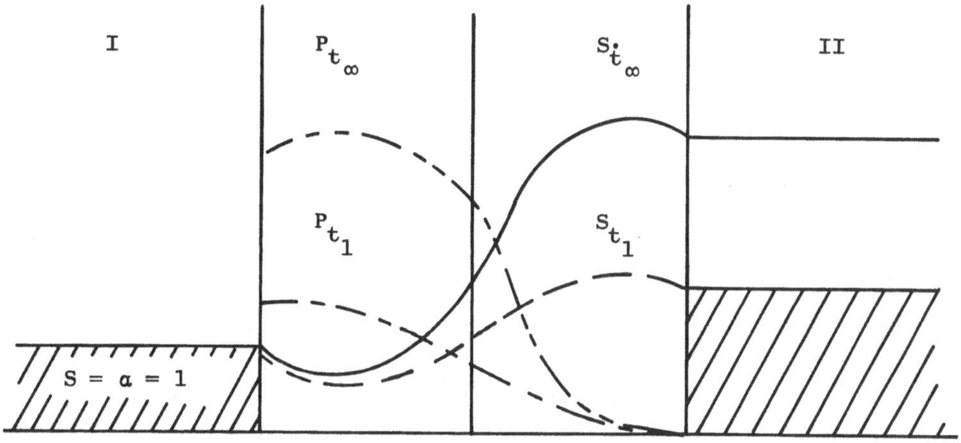


Concentration profiles with $0 < t_1 < t_\infty$ = time for steady state.

Figure 2.3

Experiments were carried out using the membranes in a configuration such as is depicted in Figure 2.2. The mathematical model (2.14) - (2.19) was used to simulate the experiments. Numerical simulation yielded the following behavior for the "system": S enters from I into layer one and is consumed (converted to P ). In layer two S is produced and leaves the membrane into compartment II where S is found at increasingly higher levels. Thus the membrane produces a "pumping action" which transports S from regions of low concentration to regions of high concentration. That is, one has flow across the membrane against the concentration gradient (difference). These numerical results for the mathematical model agree well with the experimental results. A representative set of concentration profiles is shown in Figure 2.3, where boundary conditions (2.19) were taken with $\alpha = \beta \approx 1$ .

The phenomenon modeled by the double-layer membrane described above actually occurs in vivo where it is called active transport. The mechanisms and underlying principles occurring in active transport are not yet completely understood. Theoretical and experimental work with artificial-membrane models as detailed above should aid in understanding these mechanisms.

Among the well-documented examples of active transport are the absorption of carbohydrates (glucose) across the intestinal mucosa [67, p.88] and the sodium-potassium pumps in cells [3, pp.19-28]. These pumps are shown schematically in Figure 2.4 where the solid lines represent diffusion due to permeability and the concentration gradient and the broken lines represent the pumps against the electrochemical gradients. The existence of such pumps has been demonstrated by experiments with squid axon and frog muscle fiber where in both cases Na ions are pumped out, K ions are pumped in, even though the extracellular concentrations of $Na^+$ ($K^+$) are greater (less) than the intracellular concentrations. These pumps are known to be of great importance. For example, marine bony fishes maintain a non-equilibrium level of $Na^+$ by pumping ions from the extracellular fluid (low concentrations) to their environment (the ocean - with high concentrations). Many human cells exhibit the so-called $Na^+ - K^+$ pump. In fact, the transmission of nerve impulses depends on this phenomenon.

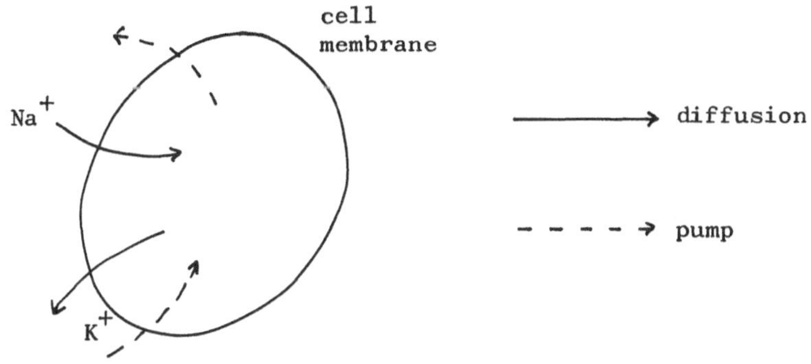


Figure 2.4

(d) <u>Tubes with diffusion, convection, and reaction</u>.

A slightly different class of problems which is nonetheless related to those discussed so far involves a column or tube in which a homogeneous, enzymatically active powder is contained, with the enzyme being fixed by covalent bonding. As shown in Figure 2.5, one

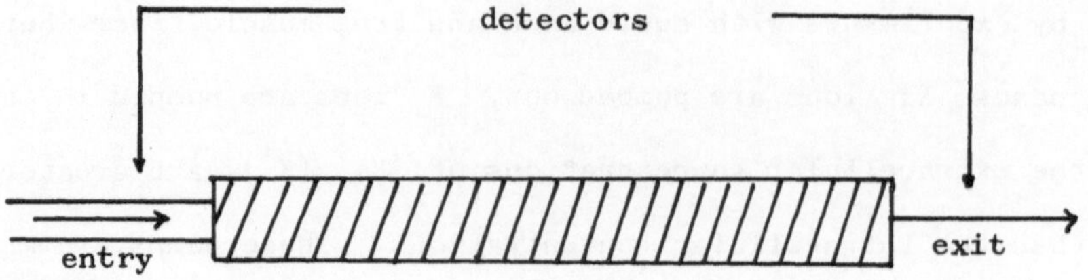


Figure 2.5

has a substrate (and possibly an inhibitor and/or activator) in a homogeneous solution injected at an entry point and flowing through the tube. A reaction $S + E \longrightarrow P + E$ takes place in the tube. Also convection is present and if the velocity of flow is sufficiently small, we may have diffusion. The model taking into account only convection and reaction terms is given by

$$\frac{\partial s}{\partial t} + c\frac{\partial s}{\partial s} + \rho\frac{s}{1+s} = 0 ,$$

(2.20)

$$\frac{\partial p}{\partial t} + c\frac{\partial p}{\partial x} - \rho\frac{s}{1+s} = 0 ,$$

where $\rho = V_{max}/K_M$ and $c$ = velocity of the flow. Appropriate boundary and initial conditions are $s(x,0) = p(x,0) = p(0,t) = 0$ and $s(0,t) = \alpha$. If diffusion is present, one must add terms $D_S(\partial^2 s/\partial x^2)$ and $D_P(\partial^2 p/\partial x^2)$ to the equations (2.20) and the boundary conditions $\partial s/\partial x = \partial p/\partial x = 0$ at the exit point.

(e) Parallel flows separated by an enzymatically active membrane.

We consider finally two flow regions separated by a

membrane $\mathfrak{m}$ as discussed in the beginning of this chapter. A solution entering the upper flow region $V_1$ contains a substrate, the solution entering the lower flow region $V_2$ contains none. The substrate diffuses and reacts $(S + E \longrightarrow P + E)$ in the membrane and passes into $V_2$ .

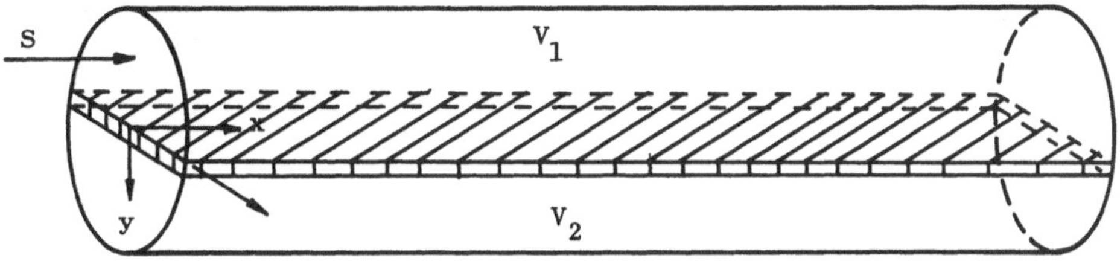


Figure 2.6

Letting $D_{\mathfrak{m}}$ and $D_V$ be the coefficients of diffusion of S and P in $\mathfrak{m}$ and $V_1$, $V_2$ respectively and c = velocity of flow in $V_1$, $V_2$ , one obtains the following equations for the normalized concentration functions $s = s(x,y,z,t)$, $p = p(x,y,z,t)$ :

In $\mathfrak{m}$ ,

$$\frac{\partial s}{\partial t} - D_{\mathfrak{m}} \quad \Delta s + F(s) = 0$$

(2.21)

$$\frac{\partial p}{\partial t} - D_{\mathfrak{m}} \quad \Delta p - F(s) = 0$$

where $F(s) = (V_{max}/K_M)(\frac{s}{1+s})$ . In $V_1$ and $V_2$ , where only convection and diffusion occur,

$$\frac{\partial s}{\partial t} + c\,\frac{\partial s}{\partial x} - D_V\,\Delta s = 0$$

(2.22)

$$\frac{\partial p}{\partial t} + c\,\frac{\partial p}{\partial x} - D_V\,\Delta p = 0\ .$$

The initial conditions, assuming $V_1, V_2, \mathfrak{m}$ are initially empty of substrate and product, are

(2.23) $$s\big|_{t=0} = p\big|_{t=0} = 0\ .$$

Appropriate boundary conditions are

(i) $\frac{\partial s}{\partial n} = \frac{\partial p}{\partial n} = 0$ on the lateral surface of the cyclinder, on the exit end of the cylinder, and on points of $\mathfrak{m}$ where $x = 0$ ;

(ii) $s = \alpha,\ p = 0$ on points of $V_1 \cap \{x = 0\}$ ;

(iii) $s = 0,\ p = 0$ on points of $V_2 \cap \{x = 0\}$ ;

(iv) $D_V \frac{\partial s}{\partial n_V} + D_{\mathfrak{m}} \frac{\partial s}{\partial n_{\mathfrak{m}}} = 0$ ; $D_V \frac{\partial p}{\partial n_V} + D_{\mathfrak{m}} \frac{\partial p}{\partial n_{\mathfrak{m}}} = 0$ on the interfaces of $V_1$ and $\mathfrak{m}$, and $V_2$ and $\mathfrak{m}$. These interface conditions are a result of conservation laws which require a balancing of the fluxes in $V_i$ and $\mathfrak{m}$ at the interface.

Questions of existence and uniqueness of solutions for classes of initial-boundary-value problems as formulated in this section have been discussed by Kernevez in [63], where he employs, among others, function space and monotone operator techniques (see also [72]).

## 2.2 Control problems arising in connection with membrane models.

The system and models described above provide a source for a number of very natrual control problems (optimal as well as other types). First, since we are already dealing with regulated or controlled phenomena, various types of optimal control problems will invariably come up in carrying out experiments. (There is in fact an entire class of biochemical mixing problems inherent here. As is well known, the area of chemical mixing has previously supplied a large number of interesting problems of a practical nature for control theorists.) Secondly, there are closely related (and overlapping) problems of the identification type which are sometimes posed as optimal "control" problems. In addition to the usual problems of "best" choice of parameters in the model, one may seek to follow a certain natural evolution by choice (optimal) of (i) concentrations of substrate, inhibitor, and/or activator on a boundary or in some interior region, (ii) flux of a substance entering or leaving a region, or (iii) a flow velocity in the case of problems with columns or tubes.

We shall discuss here two such examples. Other examples of interest are also found in the thesis of Kernevez [63].

### (a) Control of substrate flux into a membrane by concentration of inhibitors on the boundary.

We assume the model involving competitive inhibition of a irreversible monoenzymatic reaction as discussed in Section 2.1(b) and depicted in Figure 2.7.

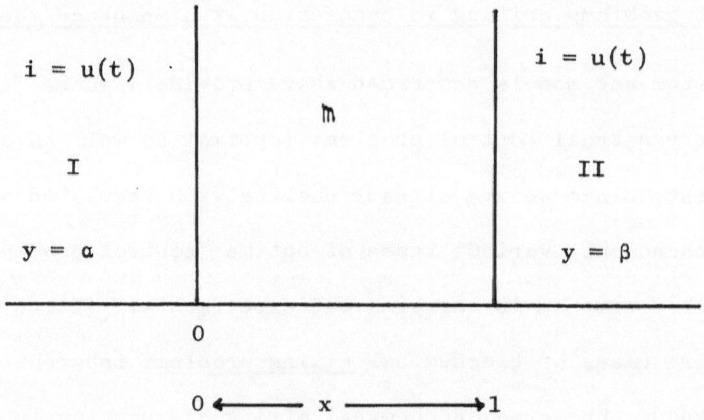


Figure 2.7

If $y$ is the "concentration" of substrate, $i$ = "concentration" of inhibitor, the normalized model can be written

$$y_t - y_{xx} + \frac{y}{1 + \lambda i + y} = 0$$

(2.24) $$(x,t) \in [0,1]\times[0,T]$$

$$i_t - i_{xx} = 0$$

with initial conditions $y(x,0) = i(x,0) = 0$ and boundary conditions $y(0,t) = \alpha$ , $y(1,t) = \beta$ , $i(0,t) = i(1,t) = u(t)$ . Here $u$ represents a control on the boundary and is to be chosen from the class of admissible controls $\mathcal{U}_{ad} \equiv \{u \in L_\infty(0,T) \mid 0 \le u(t) \le M\}$ . Experimentally, this control can be realized by addition of inhibitor to compartments I and II, or by consumption of some of the inhibitor in these compartments by chemical reactions. We assume finally that we are given an observation $z_d$ of the flux of the substrate at the boundary $x = 0$ and we wish to choose $u \in \mathcal{U}_{ad}$ to minimize

$$(2.25) \qquad J(u) = \int_{t_1}^{t_2} \{\frac{\partial y}{\partial n}(0,t) - z_d(t)\}^2 t^{2\gamma}\, dt$$

subject to (2.24) and the initial and boundary conditions. Here $0 \leq t_1 \leq t_2 \leq T$ and $\gamma > 1/2$ is a power of $t$ included as a weighting factor to insure finiteness of $J$ in (2.25).

A similar problem can be formulated if one has different controls $u_0, u_1$ and observations $z_d^0$, $z_d^1$ on the boundaries. The payoff (2.25) is then replaced by

$$J(u_0,u_1) = \int_{t_1}^{t_2} \{[\frac{\partial y}{\partial n}(0,t) - z_d^0(t)]^2 + [\frac{\partial y}{\partial n}(1,t) - z_d^1(t)]^2\} t^{2\gamma}\, dt$$

which is to be minimized subject to (2.24) with boundary conditions $i(0,t) = u_0(t)$, $i(1,t) = u_1(t)$ .

For the problem of minimizing (2.25) as formulated above, Kernevez has shown existence of an optimal control and has derived necessary conditions (in terms of a variational inequality) for optimality. In his thesis Kernevez also discusses an algorithm for the numerical solution of the problem, that work being based on a discretization of the equations (original and adjoint) and necessary conditions. We shall not go into a detailed discussion of the theoretical (optimal control) aspects of this problem or the one presented below, but only observe that they are in the spirit of the well-known work by Lions and his colleagues. We refer readers to [63,73,74].

(b) Control of the double-layer membrane.

We consider control of the transport of a substrate across a double-layer membrane as modeled in Section 2.1(c) where we assume in addition that an activator is present in layer one. If $y_1, y_2, y_3$ represent normalized concentrations of substrate, product, and activator, the model becomes:

$$
\begin{aligned}
&y_{1t} - y_{1xx} + F = 0 \\
(2.26)\qquad &y_{2t} - y_{2xx} - F = 0 \qquad\qquad (x,t) \in [0,1] \times [0,T] , \\
&y_{3t} - y_{3xx} = 0
\end{aligned}
$$

where

$$
F(x,y_1,y_2,y_3) = \begin{cases} \sigma \dfrac{y_3}{1+y_3} \dfrac{y_1}{1+y_2+y_1} & 0 < x < 1/2 \\[2ex] -\sigma \dfrac{y_2}{1+y_2} & 1/2 < x < 1 . \end{cases}
$$

For initial conditions we choose $y_i(x,0) = 0$, $i = 1,2,3$ . We assume that compartment I is very large compared to compartment II. (This is the case, for example, if II is intracellular space, the double layer is a cell membrane, and I is extracellular space.) Then natural conditions to choose on the boundary are

$$y_1(0,t) = \alpha$$

(2.27)

$$\frac{\partial y_1}{\partial t}(1,t) + \chi \frac{\partial y_1}{\partial n}(1,t) = 0 \quad \text{with} \quad y_1(1,0) = \beta ,$$

(2.28) $$\frac{\partial y_2}{\partial n}(0,t) = \frac{\partial y_2}{\partial n}(1,t) = 0$$

and

(2.29)

$$y_3(0,t) = 0$$

$$y_3(1,t) = u(t)$$

if we assume that we control the activator level at the inner membrane wall. The control $u$ is to be chosen from $\mathcal{U}_{ad} = \{u \in L_2(0,T) \,|\, 0 \leq u(t) \leq M\}$ and a payoff of interest (assuming that we wish to ensure a specified level of substrate in the intracellular space) is

(2.30) $$J(u) = \int_0^T \{y_1(1,t) - z_d(t)\}^2 \, dt .$$

Again, Kernevez has discussed in his thesis the questions of existence of optimal controls, necessary conditions for optimality, and a simple gradient algorithm for the above problem.

# CHAPTER 3. MOEDELING OF ENZYME CASCADES

It is now recognized that many of the metabolic pathways in mammals involve cascades of enzymatic reactions. In this chapter we shall discuss attempts to model such processes. After a presentation of some physiological-biochemical motivation and brief comments on modeling attempts by others, we shall turn to a discussion based on recent investigations [17] by this author and R.Miech of the Division of Bio-Medical Sciences, Brown University.

## 3.1 Physiological motivation and a review of previous modeling attempts.

Integral to certain metabolic processes are finite sequences of enzymatic reactions in which the "product" of the nth-stage enzymatic reaction acts as catalyst or "enzyme" for the "substrate" in the n+1st-stage reaction. The importance of such enzyme cascades acting as "biochemical amplifiers" by which a change in the level of a hormone circulating at extremely low concentrations (constituting a small signal) results in a substantial physiological response (large response) in the body has been accepted for some time. Recent articles in the literature (see, for example [37] have discussed the implications of the existence of such cascade systems as they relate to pharmacology and the design of drugs to inhibit or enhance the actions of these systems. We list below some of the known and/or suspected cascades which are currently of interest.

(a) Blood coagulation.

The mechanism involved in blood clotting has been studied in detail [51,52,71,77,91] and is now believed to consist of a cascade of enzymatic reactions each involving the conversion of an enzyme (factor) in inactive form to its active form as depicted in Figure 3.1, the activated form of a factor being denoted by an "a".

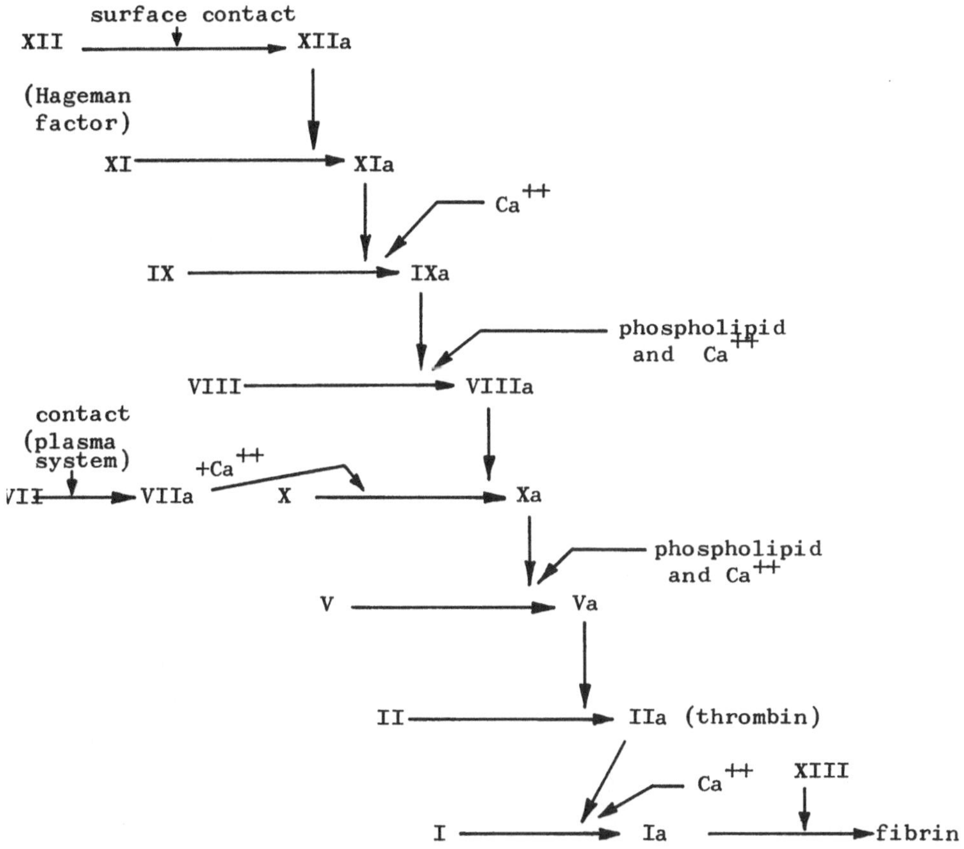


Figure 3.1

In this cascade a minute surface stimulus activates a relatively small number of molecules of Hageman factor which ultimately results in the conversion of millions of molecules of fibrinogen to fibrin.

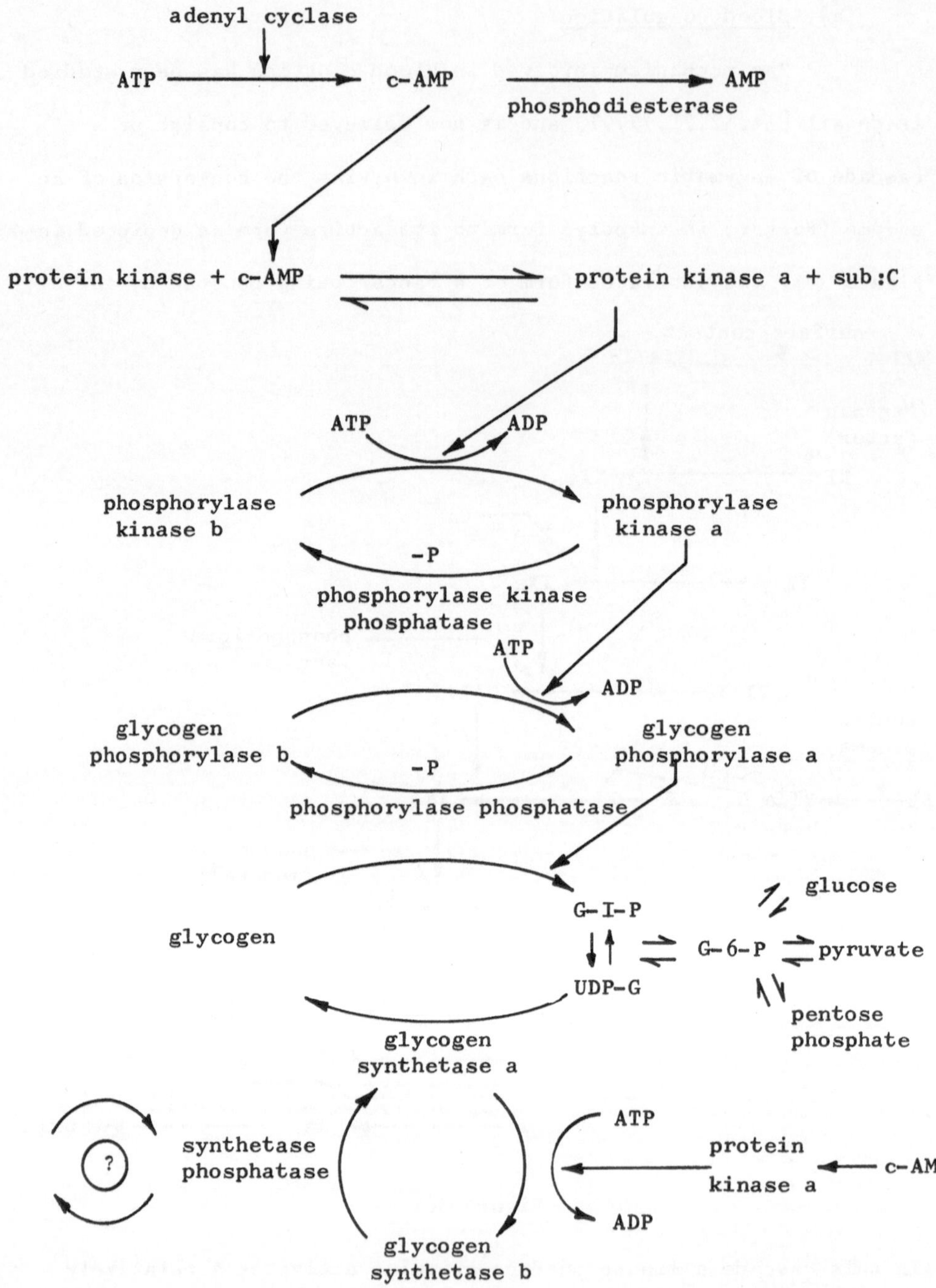
adenyl cyclase
ATP
c-AMP
AMP
phosphodiesterase
protein kinase + c-AMP
protein kinase a + sub:C
ATP
ADP
phosphorylase
kinase b
phosphorylase
kinase a
-P
phosphorylase kinase
phosphatase
ATP
ADP
glycogen
phosphorylase b
glycogen
phosphorylase a
-P
phosphorylase phosphatase
glucose
G-I-P
glycogen
G-6-P
pyruvate
UDP-G
pentose
phosphate
glycogen
synthetase a
ATP
?
synthetase
phosphatase
protein
kinase a
c-AMP
ADP
glycogen
synthetase b

Figure 3.2

(b) Glycogenolysis.

An acknowledged cascade occurs in glycogenolysis [13,22, 37,41,54,70,84,91] where an increase of cyclic AMP (present at low intracellular concentrations) initiates a sequence of enzymatic reactions which result in an increased rate of activation of glycogen phosphorylase (present at a concentration several orders of magnitude higher), the glycogen phosphorylase promoting the breakdown of glycogen to glucose-1-phosphate. More specifically, cyclic AMP-activated protein kinase converts (inactive) phosphorylase kinase b to (active) phosphorylase kinase a . Phosphorylase kinase a in turn activates glycogen phosphorylase (b to a) which, as mentioned above, promotes the breakdown of glycogen. This sequence is depicted in Figure 3.2 above. We note that cyclic AMP-activated protein kinase also affects directly glycogen synthetase (promoting conversion of a (active) to b (inactive)) and thus at the same time impedes the pathway for glycogenesis (the synthesis of glycogen).

(c) Visual excitation.

Wald [96] has suggested the possibility of the existence of a cascade in visual excitation much like that found in blood coagulation. He has proposed a cascade in which one photon stimulates a molecule of visual pigment (rhodopsin) which in turn sets off a chain of proenzyme to enzyme reactions. This chain would make the rod act as a biochemical photomultiplier, capable of yielding a large biochemical product very rapidly in return for a minimal initial input.

(d) Relaxation and contraction of smooth muscle.

The existence of an enzyme cascade is also probable in the relaxation-contraction mechanisms in certain types of smooth muscle. A typical suggested biochemical relaxation-contractile model (see, for example, [84]) is given schematically in Figure 3.3, where epinephrine induces relaxation in intestinal smooth muscle. As is shown, cyclic AMP somehow promotes binding of $Ca^{++}$ to the plasma membrane and sarcoplasmic reticulum, thus lowering the cytosol calcium level and facilitating relaxation. While the other mechanisms involved are not yet fully understood, it is quite likely that protein kinases and phosphorylase kinases such as those found in the glycogenolytic cascade are operative here also. It is known that the smooth muscle relaxation-contraction process exhibits a number of other parallels to glycogenolysis. For example, a certain amplification takes place, with a small change in c-AMP levels yielding a substantial change in the relaxation-contractile characteristics. Furthermore, certain methyl xanthines, such as caffeine and theophylline, apparently block phosphodiesterase activity, resulting in relaxation.

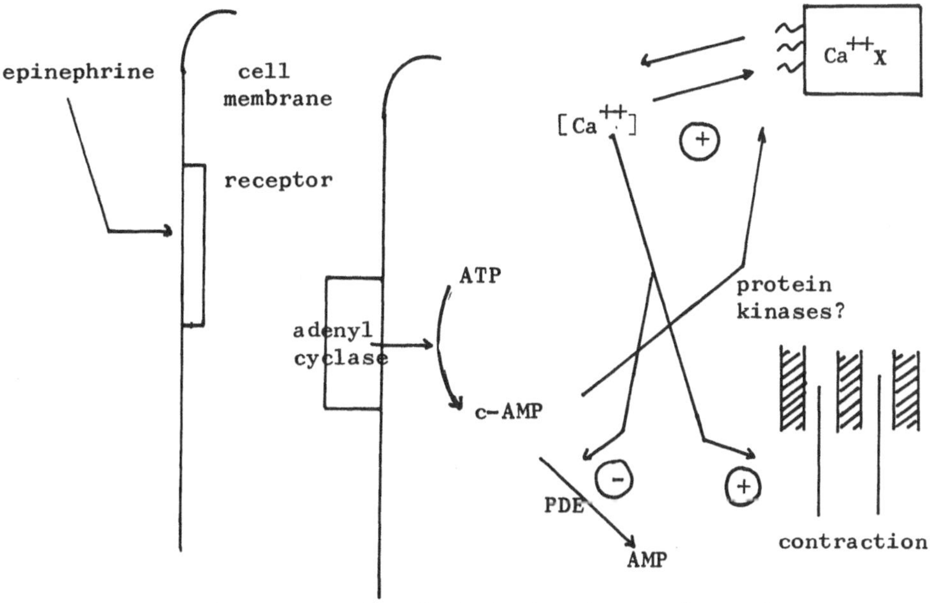


Contraction-relaxation in intestinal smooth muscle.
Figure 3.3

Previous investigations [22,37,51,52,54,71,84,92,96] have lead to the recognition of certain features that appear desirable in any mathematical model of an enzyme cascade. Two features which appear essential are "amplification" and "fast response time". By the first term we mean that a small signal stimulus or change in hormone level) at the beginning of the cascade results in a large response at the end of the sequence. The second feature refers to the relation between "time constants" of the successive stages of the cascade. Roughly, we ask that the concentration changes occur at greater speeds as we move down the cascade so that, even with delays inherent in successive steps, the end products move to their new concentration

levels faster than do the early reaction products in the cascade.
As we shall see below, such a feature is readily modeled by "stiff"
systems which often appear in biochemical differential equation
models.

While a number of people have studied the cascades discussed
above, relatively few attempts at mathematical models for these cascades have been made. Levine [71] proposed a model based on the
scheme

$$(3.1) \qquad y_1 \xrightarrow{\phantom{aa}k_i\phantom{aa}} y_{ia} \xrightarrow{\phantom{aa}K_i\phantom{aa}} y_i'$$

at the ith stage, where $y_i$, $y_{ia}$, $y_i'$ represent concentrations of
proenzyme, activated enzyme, and inactive enzyme respectively. His
model can be represented schematically by

$$
(3.2) \qquad
\begin{array}{l}
y_1 \xrightarrow{\phantom{aa}k_1\phantom{aa}} y_{1a} \xrightarrow{\phantom{aa}K_1\phantom{aa}} y_i' \\[4pt]
y_2 \xrightarrow{\phantom{aa}k_2\phantom{aa}} y_{2a} \xrightarrow{\phantom{aa}K_2\phantom{aa}} y_2' \\[4pt]
y_3 \xrightarrow{\phantom{aa}k_3\phantom{aa}} y_{3a} \xrightarrow{\phantom{aa}K_3\phantom{aa}} y_3' \\[4pt]
y_4 \xrightarrow{\phantom{aaaa}} \quad \cdots
\end{array}
\qquad ,
$$

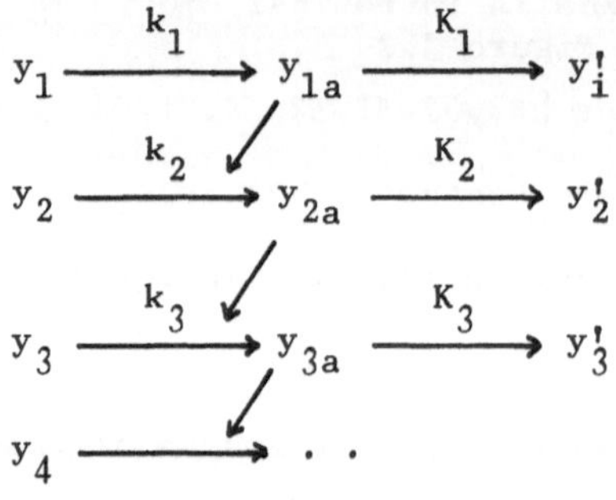

with the model equations being given by

$$(3.3) \qquad \dot{y}_{i+1a}(t) = k_{i+1} y_{ia}(t) y_{i+1}(t) - K_{i+1} y_{i+1a}(t) .$$

In the first equation $(i = 0)$, one replaces the $k_1 y_{0a} y_1$ term by a stimulus term, $k_1[U(t) - U(t-h)]y_1$, $U$ being defined by $U(z) = 1$ for $z \geq 0$, $U(z) = 0$ for $z < 0$. Levine studies this model under the assumption that the concentration of proenzyme $y_i$ remains constant at $y_1^0$, so that the model equations become

$$(3.4) \qquad \dot{y}_{i+1a}(t) = k_{i+1} y_{ia}(t) y_{i+1}(0) - K_{i+1} y_{i+1a}(t) .$$

The exact solutions of these equations can be found by standard formulae.

The assumption of constant levels of proenzyme might yield good approximations if the amount of activated proenzyme is small in comparison to that initially present, or if the proenzyme is replaced at approximately the same rate at which it is converted. This assumption may not be too crude with respect to the first stages of the cascade. But if amplification (in the sense of large amounts of enzyme activation at the later stages) is to take place, this assumption most likely yields a much too gross approximation. For example, if one has at some stage in the cascade an enzyme going from 20 percent (of the total - active plus inactive) activation to 80 percent activation (as is known to occur in glycogenolysis), the inactive (proenzyme) concentration moves from 80 percent of the total to 20 percent. We also observe that the solution curves presented by Levine do not exhibit the fast response time feature. He does not discuss this aspect of his model and it is not clear that the model will contain this feature (or amplification) at physiologically meaningful parameter values.

Hemker and Hemker [51] have also formulated mathematical models for enzyme cascades involving initial reaction velocity approximations (listed in order of increasing complexity)

$$(3.5) \qquad v = k_2 E_T$$

$$(3.6) \qquad v = k_2 \frac{E_T S}{K_M + S}$$

$$(3.7) \qquad v = \frac{k_2 E_T S}{K_M + S} (1 - e^{-bt})$$

for one-step enzymatic reactions of the type (1.1). Formula (3.5) is a very crude approximation that assumes so much substrate present that the velocity is a constant which is proportional to the total amount (fixed) of enzyme present. The expression (3.6) is the Briggs-Haldane modification of the Michaelis-Menten formulation discussed in Chapter 1 and (3.7) is a modification of (3.6) due to Gutfreund [49] which takes into account the transient time (time to reach equilibrium in $E + S \rightleftharpoons ES$ ; see the quasi-equilibrium assumption (iii) in Chapter 1). Hemker and Hemker consider both damped cascades (the type studied by Levine and represented in (3.2)) and open cascades of the form

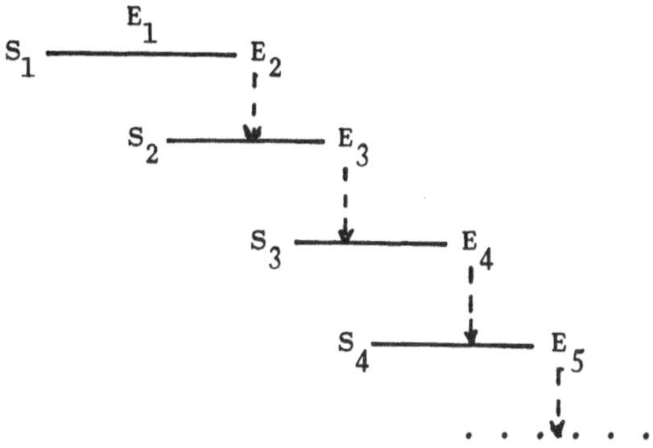


Using the approximations (3.5) - (3.7), the authors develop formulae which they claim support the argument that a "triggering effect", not amplification, is the important feature of enzyme cascades. However, the solution curves for these models are not even crude approximations to the concentration curves usually observed in cascades in vivo. One must conclude that the velocity approximations made in order to make the equations mathematically tractable (solvable in closed form) at the same time render the mathematical models less than realistic approximations to the biochemical models supported by empirical findings.

In the model discussed in detail below, we have not insisted on a model with equations that can be solved exactly, but have instead depended on numerical results from computer simulations to complement and support our (analytical) mathematical analysis.

3.2 Derivation of a model based on the cascade in glycogenolysis.

We present now a specific mathematical model based on the particular cascade found in the glycogenolytic pathway [54,84,91,92]. From our discussions of this particular model one will discern certain features and limitations involved in modeling of general cascade systems. A schematic diagram for the underlying biochemical model assumed here is given in Figure 3.4. Here AC = adenyl cyclase, PDE = phosphodiesterase, PKK' = inactive protein kinase, PKK = active protein kinase, PK' = inactive phosphorylase kinase, PK = active phosphorylase kinase, P.K.P. = phosphorylase kinase phosphatase, GP' = inactive glycogen phosphorylase, GP = active clycogen phosphorylase, P.P. = phosphorylase phosphatase, S = glycogen, P = glucose-1-phosphate. This biochemical model is based on a combination of the accepted theories on the pathways involved in glycogenolysis [54, 91, 92] in cardiac muscle along with the general models discssed in the survey paper by Rasmussen et al. [84]. We next proceed with the derivation of the mathematical model based on this biochemical model.

We assume that the stimulation involving adenyl cyclase results in a zero-order input (similar, for example, to a constant blood level of a hormone) while the degradation involving phosphodiesterase is first order (first-order removal by metabolism). Then if $c(t)$ represents the intracellular concentration of c-AMP at time $t$, we may, employing the usual order assumptions about the kinetics of the reaction with constants $k_{+1}, k_{-1}$, write

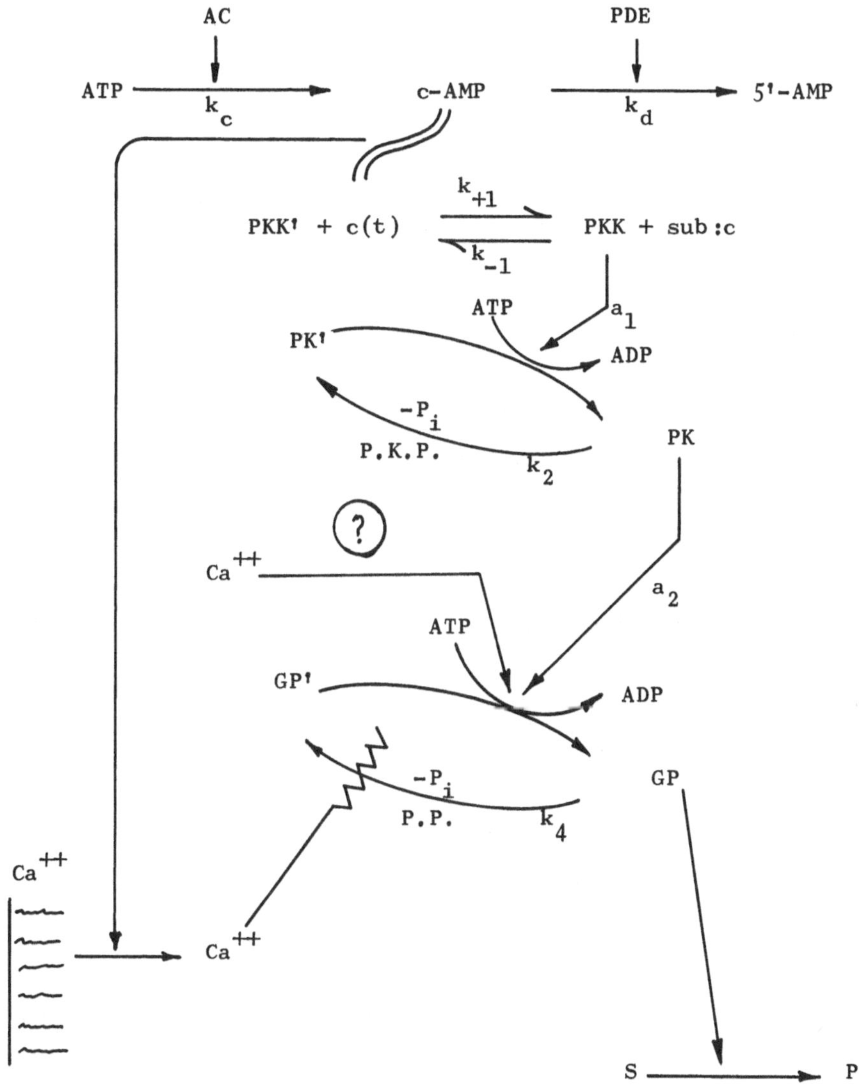


Figure 3.4

$$\dot{c}(t) = k_c - k_d c(t) - k_{+1}[PKK']c(t) + k_{-1}[PKK][sub:c] . \tag{3.8}$$

We further make the underlying assumption that the reaction $\underset{k_{-1}}{\overset{k_{+1}}{\rightleftarrows}}$ has zero equilibrium time. That is, this reaction is instantaneous relative to the time scale for the other reactions in the model involving $c(t)$, [PKK'], [PK'], [GP'], etc., so that

$$(3.9) \qquad PKK' + c(t) \rightleftharpoons PKK + sub:c$$

is always in equilibrium. Mathematically this is written as

$$(3.10) \qquad k_{+1}[PKK']c(t) - k_{-1}[PKK][sub:c] = 0 ;$$

hence equation (3.8) becomes

$$(3.11) \qquad \dot{c}(t) = k_c - k_d c(t) .$$

Schematically the reaction (3.9) can be depicted by

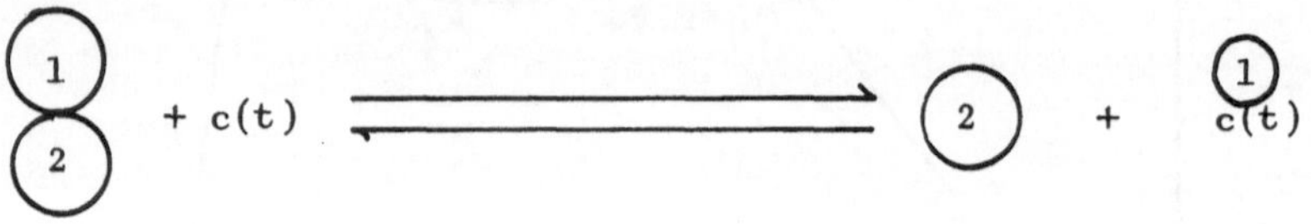


where here 1 represents an inhibitory subunit which has a high degree of affinity for c-AMP ($c(t)$) and 2 represents activated PKK (after dissociation has taken place). Letting $\alpha(t)$ denote the concentration of PKK (active) at time $t$ and $A \equiv PKK_{TOT}$ = total concentration of protein kinase (active plus inactive), we have $[PKK'](t) = A - \alpha(t)$ and equation (3.10) becomes

$$\frac{k_{+1}}{k_{-1}} = \frac{\alpha(t)[sub:c]}{(A-\alpha(t))c(t)} .$$

Defining the constant $K_{eq} \equiv k_{+1}/k_{-1}$ and noting that the above assumptions dictate [PKK] = [sub:c] (this can be verified experimentally), we find

$$K_{eq} = \frac{\alpha^2(t)}{(A-\alpha(t))c(t)}$$

or

$$\alpha(t) = \frac{1}{2}\left[\{K_{eq}^2 c^2(t) + 4\, K_{eq}\, Ac(t)\}^{1/2} - K_{eq}c(t)\right] . \tag{3.12}$$

Next we consider the reactions involving PK', PK, GP', GP and let $B \equiv PK_{TOT}$ = total concentration of phosphorylase kinase (active plus inactive), $\beta(t)$ denote the concentration of PK (active) at time t, $G \equiv GP_{TOT}$ = total concentration of glycogen phosphorylase, $\gamma(t)$ = concentration of GP (active) at time t. If we assume that P.K.P. and P.P. are present in such amounts that the back reactions PK$\longrightarrow$PK', GP$\longrightarrow$GP' can be treated with first-order kinetics with constants $k_2$, $k_4$ respectively (both in vitro and in vivo experimental support for this assumption can be found in the literature [35,36]), then the differential equation for the PK' $\rightleftharpoons$ PK reaction may be written

$$\dot{\beta}(t) = a_1\alpha(t)\{B-\beta(t)\} - k_2\beta(t) \tag{3.13}$$

where we have, of course, assumed that $\alpha$ (i.e. PKK ) is an "enzyme" with "substrate" $B-\beta$ (i.e. PK' ).

As Figure 3.4 shows, the equation for $\gamma$ is slightly more complicated. First of all, it is possible that a rise in $Ca^{++}$ must accompany the rise in [PK] in order to increase the rate of activation of GP' . But no sound data are yet available to support such a hypothesis. It is known only that in the total absence of $Ca^{++}$ no activation occurs [84] and it is quite possible that the normal cell cytosol concentration of $Ca^{++}$ is sufficient for the promotion of activation by PK . We therefore choose to represent the velocity of the forward reaction by $v(t) = a_2\beta(t)\{G-\gamma(t)\}$ . If there were no differences between the backward reaction here and that modeled in (3.13), we would use a back velocity term of $k_4\gamma(t)$ . However, it is known [84] that c-AMP releases calcium bound in the cell membrane and endoplasmic reticulum; thus an increase in intracellular c-AMP concentration leads to an increase in intracellular $Ca^{++}$ concentration. It is also known that the calcium in turn inhibits the back reaction GP $\longrightarrow$ GP' as shown in Figure 3.4. Since little is known as yet about the exact dynamics involved in this c-AMP-initiated inhibition, we have chosen to represent the backward velocity so that it is inversely proportional to $c(t)$ , the concentration of c-AMP . This no doubt rough approximation leads to the equation

$$(3.14) \qquad \dot{\gamma}(t) = a_2\beta(t)\{G-\gamma(t)\} - \frac{k_4}{c(t)}\gamma(t) .$$

We take then as our mathematical model the equations (3.11), (3.12), (3.13), (3.14). To study the model, we assume that prior to time $t = 0$ , the system is operating in steady state with levels $c_0, \alpha_0, \beta_0, \gamma_0$ . At time $t = 0$ we either perturb the system via stimulation of adenyl cyclase $(k_c)$ or inhibition of the phosphodiesterase $(k_d)$ or both. The result is a rise in the level of c-AMP to a new steady-state value $c_{ss}$ . The concentrations of $\alpha, \beta, \gamma$ also move to new steady-state levels $\alpha_{ss}, \beta_{ss}, \gamma_{ss}$ .

## 3.3 The role of optimality in model development.

The model presented above is, of course, a result of numerous modifications in our originally proposed model. One of our goals in the model was an inherent "magnification" or "amplification" feature (i.e. "small" $\Delta c$ or $\Delta\alpha$ produces "large" $\Delta\gamma$ ) when all parameters and variables are at physiologically meaningful values. We used optimality ideas throughout the stages of development to compare the limits attainable by our models with known physiological and biochemical limits on magnification.

A simplistic approach to such an endeavor would be as follows. For a given proposed model with parameters (such as the $a_i$, $k_i$, $K_{eq}$ , etc. above) and a given $\Delta\alpha$ (or $\Delta c$ ), what choices of the parameters give a maximum value to $\Delta\gamma$ ? In addition to being a nontrivial problem theoretically (one would also wish to require at the same time that the $\Delta$'s increase as we move down the cascade), this would be a rather unwieldy approach if one wished only to consider changes in certain segments of the model.

A somewhat more natural way to use the optimality ideas then is to make optimal choices at each level of the cascade. That is, given $\Delta\alpha$ , what is the maximum $\Delta\beta$ attainable?; given $\Delta\beta$ , what is $\Delta\gamma_{max}$ , etc.? These "optimal design" concepts were employed (leading to numerical optimization problems rather than optimal control problems of the Pontryagin type as in Chapter 2) keeping in mind the other features desired in the model (sigmoidicity requirements, fast response time, physiological limitations on parameters, etc.) and in this way the model presented above was developed and modified. At each stage of the cascade these ideas were used to check the feasibility of various changes and modifications with respect to limitations on numerical results produced by computer simulations for the model. For example, the inclusion of autocatalytic activation terms in certain stages of the cascade was investigated with these methods. (Conflicting evidence in the physiological literature could be found both to support and deny the existence of such mechanisms.) Furthermore, theoretical optimality considerations such as those described above led to the conjecture of the existence of feedforward inhibitory and/or stimulatory factors such as the $Ca^{++}$ inhibitory term $k_4/c(t)$ , which was then found to be supported by experimental findings.

We illustrate these ideas by describing one of the simplest of these optimality problems one encounters. Suppose one were given $\Delta\alpha$ and the equation (3.13) as the proposed model for the $\beta$-activation segment of the overall model. One finds then that the steady-state values for $\beta$ are given by

$$(3.15) \qquad \beta_0 = \frac{a_1\alpha_0 B}{k_2+a_1\alpha_0} \qquad , \qquad \beta_{ss} = \frac{a_1\alpha_{ss} B}{k_2+a_1\alpha_{ss}} \ ,$$

and, considering $\Delta\beta$ as a function of the parameters $(k_2, a_1)$, $\Delta\beta_{max}$ occurs along the curve $k_2 = a_1 \; \alpha_0\alpha_{ss}$ and is given by

$$\Delta\beta_{max} = \frac{\sqrt{\alpha_0\alpha_{ss}} \quad B \, \Delta\alpha}{(\alpha_{ss} + \sqrt{\alpha_0\alpha_{ss}})(\alpha_0 + \sqrt{\alpha_0\alpha_{ss}})} \quad .$$

The $\Delta\beta$ surface is shown in Figure 3.5. Using the information thus obtained, one can study limitations on the magnification produced at this stage in the model if one adopts the equation (3.13) for $\beta$-activation.

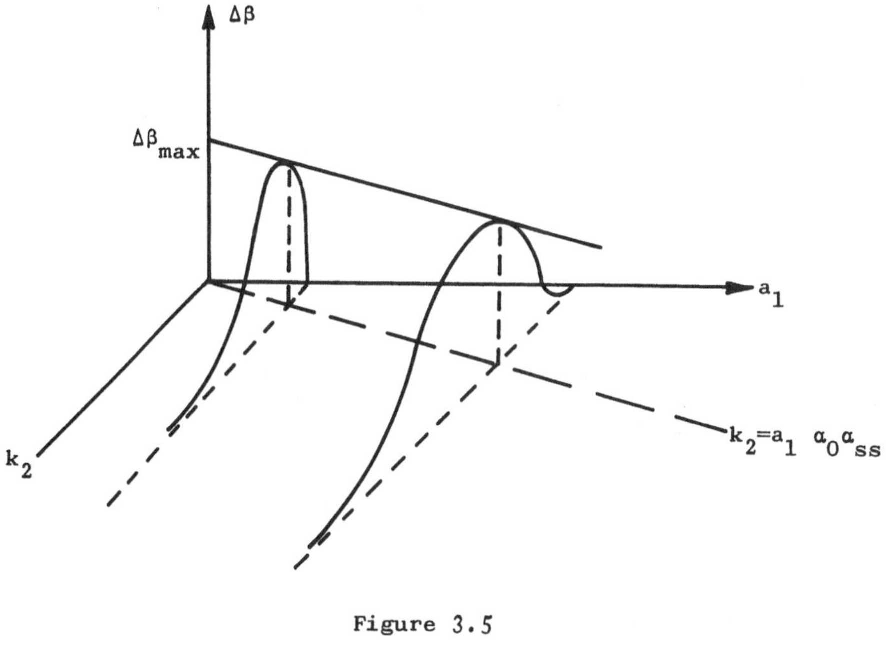


Figure 3.5

## 3.4 Qualitative and numerical results.

We have studied the qualitative behavior of the above model with respect to the parameters and modifications of the assumptions which alter the model equations (3.11) - (3.14). For example, it can be argued mathematically that the time curves for $c, \alpha, \beta, \gamma$ are nondecreasing, $\beta$ and $\gamma$ are smooth and sigmoid in shape while $\dot{c}(0^+) \neq 0 = \dot{c}(0^-)$ and $\dot{\alpha}(0^+) \neq 0 = \dot{\alpha}(0^-)$. While much of our analysis was carried out analytically, a substantial part was aided by numerous computer simulations. In these numerical runs, equation (3.11) was integrated exactly while a fourth-order Runge-Kutta integration scheme was used to obtain the solutions to equations (3.13) and (3.14). Figure 3.6 below depicts a typical solution obtained in this manner. This solution corresponds to a "stimulation" of adenyl cyclase with $k_c = 8 \times 10^{-9}$ for $t < 0$ and $k_c = 16 \times 10^{-8}$ for $t > 0$ while $k_d$ is held constant at $k_d = 8$. Values for the parameters used in this simulation are:

$$K_{eq} = .1 \qquad B = 1 \times 10^{-6}$$

$$a_1 = 1.6 \times 10^{10} \qquad G = 1 \times 10^{-5}$$

$$a_2 = 1.6 \times 10^{9} \qquad k_2 = 102.47$$

$$A = 1 \times 10^{-7} \qquad k_4 = .335 \times 10^{-5}.$$

All concentrations are in molar (M) units while the time is given in minutes. The steady-state values obtained are

$$c_0 = .1 \times 10^{-8} \qquad c_{ss} = 2 \times 10^{-8}$$

$$\alpha_0 = .031 \times 10^{-7} \qquad \alpha_{ss} = .131 \times 10^{-7}$$

$$\beta_0 = .327 \times 10^{-6} \qquad \beta_{ss} = .672 \times 10^{-6}$$

$$\gamma_0 = .134 \times 10^{-5} \qquad \gamma_{ss} = .865 \times 10^{-5} .$$

The approximate times to reach new steady-state values are $t_c = .569$, $t_\beta = .342$, $t_\gamma = .346$ . It is clear (see Figure 3.6)

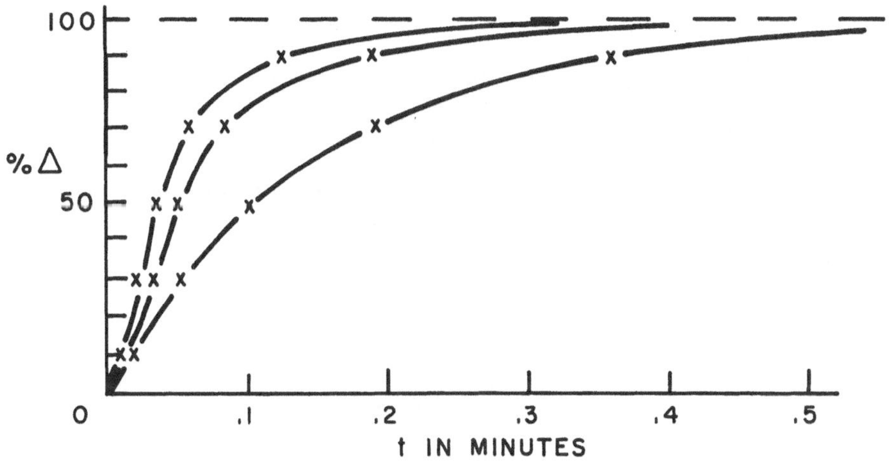


Figure 3.6

from this representative simulation that "fast response time" is a feature of the model, as is the "amplification" concept since the change in $\alpha$ ($\Delta\alpha$) is approximately 10 percent of A (the total amount - active plus inactive - of PKK ), whereas $\Delta\beta$ = 34 percent of B , $\Delta\gamma$ = 73 percent of G , so that an increase from 3 percent activation to 13 percent activation at the protein kinase stage results in a change from 13 percent activation to 86 percent activation at the glycogen phosphorylase level. These results agree qualitatively with available biochemical data.

We remark that while the magnification factor for increase in c-AMP in the above simulation is 20 (i.e., a 20-fold increase in the level of c-AMP ), our analysis has shown that the most dramatic changes in activation at the $\alpha$, $\beta$ and $\gamma$ levels when plotted against the magnification factor for c-AMP occur when the magnification factor is in the range 2-4 (see Figure 3.7). While for this range the quantitative results are not quite as dramatic as those illustrated by the above simulation, the qualitative results are the same as those for higher magnification factors. For example, for a 4-fold increase in c-AMP , the model yields changes in activation levels as follows: $\alpha$ - from 6 percent to 13 percent activation; $\beta$ - from 41 percent to 58 percent activation; $\gamma$ - from 29 percent to 70 percent activation.

Although data are not yet available to give values for the parameters $a_1$ , $a_2$ , $k_2$ , $k_4$ , one can use the velocity expressions (1.11) and (1.17) derived in Chapter 1 to argue that the values used above are in the correct numerical range. Let us consider, for example, the forward reaction velocity in the $\beta$-activation stage which is represented in the mathematical model by the $a_1 \alpha\{B-\beta\}$ term in equation (3.13). Making the quasi-equilibrium assumption of Chapter 1 and lumping the $E + S \rightleftharpoons ES$ reaction so that for our considerations we have $E_T = E + ES = \alpha$ and $S_T = S + ES = B - \beta$ at any time $t$ , we have from (1.11) that the forward velocity is approximated by

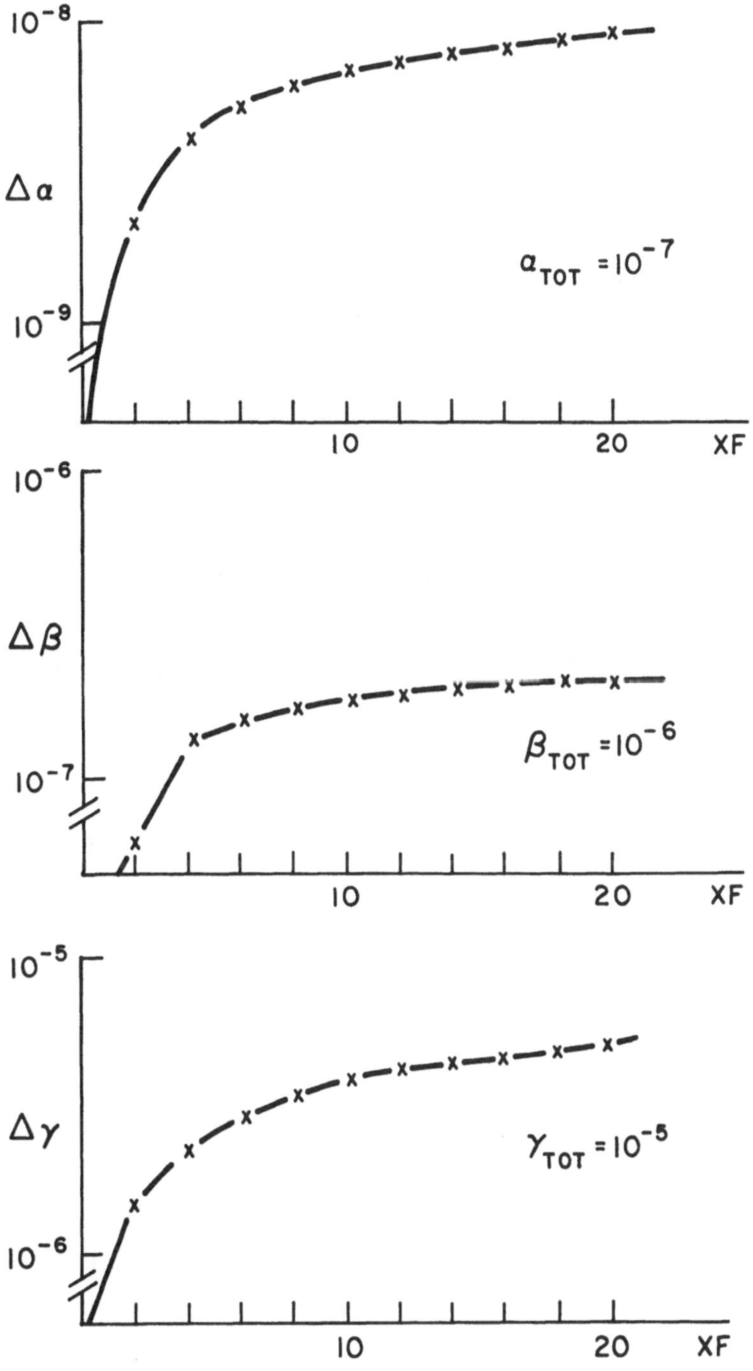


XF = magnification factor for c-AMP

Figure 3.7

$$(3.16) \qquad v(t) \sim \frac{V_{max}\{B-\beta(t)\}}{K_M + \{B-\beta(t)\} + \alpha(t)}$$

where $K_M = K_{PK}$, is the Michaelis constant for inactive phosphorylase kinase. Defining the turnover number for an enzyme as the maximum number of moles of substrate per minute converted per mole of enzyme, i.e.

$$(3.17) \qquad TN = V_{max}/E_T ,$$

one obtains

$$(3.18) \qquad v(t) \sim \frac{TN_{PKK}\alpha(t)\{B-\beta(t)\}}{K_{PK'} + \{B-\beta(t)\} + \alpha(t)} .$$

Thus we see that $a_1$ in equation (3.13) is an approximation for (or is approximated by, as the case may be) a part of the expression in (3.18), specifically

$$(3.19) \qquad a_1 \sim \frac{TN_{PKK}}{K_{PK'} + B-\beta(t) + \gamma(t)} .$$

In a similar manner, one may derive an approximation for $k_2$ given by

$$(3.20) \qquad k_2 \sim \frac{TN_{PKP}P.K.P.}{K_{PK} + P.K.P. + \beta(t)}$$

where in the model we assume that the phosphorylase kinase phosphatase is in excess and hence roughly constant. Considering the forward velocity in the equation (3.14), one finds

(3.21) $$a_2 \sim \frac{TN_{PK}}{K_{GP'} + G - \gamma(t) + \beta(t)},$$

while treating calcium as a competitive inhibitor in the back reaction and using (1.17) yields

(3.22) $$k_4/c(t) \sim \frac{TN_{PP}\ P.P.}{K_{GP}\{1 + \frac{[Ca^{++}](t)}{K_{Ca}}\} + P.P. + \gamma(t)}.$$

The best information available to date suggests that for the cascade under discussion here the Michaelis constants lie in the range $10^{-5}$ to $10^{-4}$ M while the range for turnover numbers is $10^3$ to $10^4$ $min^{-1}$. Using these values and the expressions (3.19) - (3.22), one can easily argue that the values used for $a_1$, $a_2$, $k_2$, $k_4$ in the simulation depicted in Figure 3.6 are numerically reasonable from a biochemical point of view. Furthermore, for the Michaelis constant and concentration ranges (of substrates and enzymes) used here, one can see from the work of Cha [30] that the velocity approximation represented by (1.11) and (1.17) (and used in obtaining (3.19) - (3.22)) are reasonably good approximations to the true reaction velocities.

Finally, we are able to see that the "fast response time" feature of the model is a function of the "stiffness" [34,44,83] of the system of differential equations. The curves representing the solutions of equations (3.11), (3.13), (3.14) can be approximated respectively in a qualitative sense for the above-given range of values for parameters and variables by the following functions:

$$(3.23) \qquad \hat{c}(t) = c_{ss}\{1 - (1-c_0/c_{ss})e^{-\lambda_c t}\}$$

$$(3.24) \qquad \hat{\beta}(t) = \beta_{ss}\{1 - (1-\beta_0/\beta_{ss})e^{-\lambda_\beta t}\}$$

$$(3.25) \qquad \hat{\gamma}(t) = \gamma_{ss}\{1 - (1-\gamma_0/\gamma_{ss})e^{-\lambda_\gamma t}\}$$

where $\lambda_c = 8$ , $\lambda_\beta = 1.6 \times 10^3$ , $\lambda_\gamma = 1.6 \times 10^3$ . These functions, taken together, are the solution to a stiff system of linear ordinary differential equations, one measure of the stiffness being the relative orders of magnitude of the parameters (eigenvalues) $\lambda_c$, $\lambda_\beta$, $\lambda_\gamma$.

We remark that in our investigations to date we have fortunately been spared the well-known difficulties [44] arising in the numerical integration of such systems due to the fact that we are able to solve equation (3.11) exactly. Future (and more realistic) models which include more detailed mechanisms for the stimulation of adenyl cyclase and/or the inhibition of phosphodiesterase will undoubtedly force us to face such difficulties since equation (3.11) will become much more complicated and will most likely not be exactly solvable.

# CHAPTER 4. MODELING AND CONTROL OF EPIDEMICS

One of the early attempts at mathematical modeling of the dynamics involved in the spread of disease was reported in [87]. The work of Sir Ronald Ross (1908-1911) on the spread of malaria is by now wellknown and, in fact, much of the practical work carried out later stemmed from Ross' investigations. In deriving his model consisting of two nonlinear ordinary differential equations of the form

$$(4.1) \qquad \begin{aligned} \dot{x}(t) &= ay(t)\left\{\frac{b-x(t)}{b}\right\} - rx(t) \\ \dot{y}(t) &= cx(t)\left\{\frac{d-y(t)}{d}\right\} - ey(t)\ , \end{aligned}$$

Ross ignored latency periods but subsequently pointed out that the model equations (4.1) would become differential-difference equations if the latent periods were considered. In a later paper [75], Lotka and Sharpe discussed the importance of including such delay terms in epidemic models. At about the same time Kermack and McKendrick [62] published an account of their investigations of a model based on a system involving Volterra-like integro-differential equations. Their equations were

$$(4.2) \qquad v(t) = -\dot{x}(t)$$

$$(4.3) \qquad \dot{x}(t) = -x(t)\left\{\int_0^t A(\theta)v(t-\theta)d\theta + A(t)y(0)\right\}$$

$$\dot{z}(t) = \int_0^t C(\theta)v(t-\theta)d\theta + C(t)y(0) \tag{4.4}$$

$$\dot{y}(t) = \int_0^t B(\theta)v(t-\theta)d\theta + B(t)y(0) \tag{4.5}$$

where $B(\theta) \equiv \exp[-\int_0^\theta \psi(\xi)d\xi]$, $A(\theta) \equiv \phi(\theta)B(\theta)$, $C(\theta) \equiv \psi(\theta)B(\theta)$ . Here $\psi$ is the rate of removal (recovery plus death), $\phi$ is the rate of infectivity, $x$ is the number of individuals as yet unafflicted by the disease (susceptibles), $y$ is the total number ill, and $z$ is the number removed from the population pool through death or recovery (and immunity). Although it may not be readily apparent, the equation (4.3) can be put in the form

$$\frac{d}{dt} \log x(t) = M(t) + \int_0^t N(t,\theta)x(\theta)d\theta \tag{4.6}$$

which is a Volterra-type equation. Under special circumstances including the assumptions $\psi(\theta) = \ell$, $\phi(\theta) = k$ (i.e. constant rates of infectivity and removal), the model of Kermack and McKendrick becomes

$$\begin{aligned} \dot{x}(t) &= -kx(t)y(t) \\ \dot{y}(t) &= kx(t)y(t) - \ell y(t) \\ \dot{z}(t) &= \ell y(t) . \end{aligned} \tag{4.7}$$

An analysis of the models (4.2)-(4.5) and (4.7) led to the discovery of a "threshold phenomenon" which can be roughly interpreted as follows: There is a critical value $p_c$ of the population density $p$ such that if $p \le p_c$ , introduction of one or more infecteds into the population does not result in an epidemic, while if $p > p_c$, introduction of infecteds will lead to the occurrence of a small epidemic. The value $p_c$ depends on the infectivity rate so that if either $p$ or the infectivity rate is increased slightly when $p$ is near $p_c$, a large epidemic outbreak may occur with the introduction of only several infecteds into the population. This "threshold phenomenon" theory explained certain naturally-occurring events which had long puzzled investigators. Thus, even though the model represented by equations (4.7) does not describe well the time-course of many disease outbreaks, it has been the basis of a number of models subsequently studied by other investigators.

In 1942 Wilson and Burke [101] proposed a model for the spread of infection based on differential-difference equations, i.e.

$$\dot{x}(t) = A - r(t)x(t)[A\sigma - x(t-\tau) + x(t-\tau-\sigma)] \tag{4.8}$$

where $A$ is the recruitment rate for susceptibles $x$, $\sigma$ is the latent period for exposed and infected individuals to become infectious, $\tau$ is the latent period for mosquitoes, and $r$ is a proportionality factor ("contact" rate) for exposure. Cooke [33] later discussed a very general hereditary model for disease spread which included that of Wilson and Burke as a special case. In addition to considering latency periods as in (4.8), Cooke also introduced a "resistance threshold" concept which allows repeated exposure before an individual moves from the susceptible pool to the infective pool. The model involves quite complex functional differential equations (for example, there are delays as in (4.8) above, except that they may depend on the unknown $x$ instead of being constant). Cooke pointed out a number of open mathematical questions concerning these equations which were later investigated by Hoppensteadt and Waltman [55]. In addition to a theoretical study (which provided existence, uniqueness, and continuous dependence results) of the model proposed by Cooke, they also carried out numerical simulations and considered approximate solutions to the equations. In a sequel [56] to the above paper, Hoppensteadt and Waltman modified Cooke's model so as to allow recovered individuals to later become susceptible once again. Recently Wilson [102] has shown that it is possible to solve exactly some of the model equations studied by Cooke, Hoppensteadt and Waltman.

None of the investigations reported above considers the question of control (optimal or otherwise) of epidemics. And while the above in no way constitutes a thorough review or survey (see the text by Bailey [7] and the paper by Dietz [38] for a more adequate review of the work prior to 1967 on both deterministic and stochastic models), it appears that ReVelle et al. [85, 86] were among the first to attempt to use modern control theory in the study of epidemics. These authors employed an operations-research/systems-analysis approach in studying tuberculosis in developing nations. They thus hoped to improve the decision-making process involved in tuberculosis management. The model formulated is based on nonlinear ordinary differential equation systems of the type studied by Kermack and McKendrick (see (4.7) above) and three basic types of controls are allowed in the general theoretical model: (i) drug prophylaxis (preventive treatment), (ii) drug treatment of active cases (therapy) and (iii) vaccination. However, before considering optimization problems, the authors carried out numerical simulations without applied controls in order to study trends predicted by the model and also behavior of the model with respect to parameters. Their entire investigation, and especially this phase, was hampered by the lack of any real data with which to compare their simulation results.

The controlled nonlinear model is linearized [85] and converted to an optimization problem suitable for application of linear programming (i.e. the authors obtain a constrained optimization problem with linear constraints). The authors do not really seek the "best" of all possible courses of action to reduce the number of active cases to a specified target level. Instead, their approach requires that one specify completely the "active case pattern" (trajectory of the "number of active cases" variable in the model) over a 20-year horizon and then among all control programs that yield this active case pattern, one seeks one with minimum cost. As an example, to illustrate their technique, ReVelle et al. choose four different "schedules" or active case patterns and then compare the "minimum" cost control programs for these schedules. They discuss how one might use this information to make a prediction about a true least-costly pattern (given a 20-year target). In particular, while the least "minimum cost" among their four schedules in the above example is around $7 million, they show how one might arrive at the same 20-year target of active cases (not, of course, along any of the four active case patterns) at a cost of about $4.5 million.

One of the early models which combined optimization ideas with a stochastic model was that discussed by Taylor [95] who based his theoretical model on BVD (bovine virus diarrhea) in cattle herds. He sought to arrive at a vaccination schedule that would minimize a long-term time-average cost which was formulated in such a way as to weigh the cost of herd vaccination against the cost and risks of an epidemic outbreak. Letting $X(t)$ be the number of susceptible members in a herd of population size $N$, Taylor assumes that this number increases stochastically over time until either corrective action (vaccination) is taken or an outbreak of epidemic occurs. Specifically, he assumes that $\{X(t)\}$ is a pure birth stochastic process, i.e. there are $\lambda_i$, $i = 0,1,\ldots,N-1$ such that

$$(4.9) \qquad P\{X(t + \Delta t) = i + 1 | X(t) = i\} = \lambda_i \Delta t + o(\Delta t) .$$

In addition, certain technical assumptions about the probability of exposure and outbreak are made. The costs involved are $K(i)$ if $X(t) = i$ and corrective action is taken and $C(i)$ if $X(t) = i$ and an outbreak occurs. If BVD strikes the herd, it is assumed that all of the susceptible cattle quickly become infected, some die and some recover and enjoy immunity, but eventually lose this and rejoin the susceptible pool. If corrective action is taken, then each member of the herd is provided with a period of immunity after which he is again susceptible. Taylor considers two types of corrective action timing rules (policies), i.e. 1-parameter families

$\{T(b)\}$ of rules, where $T(b)$ is the time (possibly random) that corrective action will be taken if policy $b$ is to be followed. The continuous surveillance case assumes that the number of susceptibles at any given time is an observable. Rules are specified by an integer $b$, $0 \leq b < N$ where rule $b$ calls for corrective action at the random time $T(b) = \inf\{t \mid X(t) = b\}$ (i.e. action is taken when there are $b$ or more susceptibles in the population). The second case is the fixed time (unobservable) case in which rule $b$ calls for corrective action $b$ time units after the last total immunization (i.e. $T(b) = b$).

The process begins at $t = 0$ with $X(0) = 0$ and evolves until time $S = \min\{T, T(b)\}$, where $T$ = outbreak time, when the current value of $X(t)$ is reduced to zero either by corrective action ($T(b) < T$) or an outbreak ($T(b) \geq T$). A new cycle with $X(0) = 0$ then starts. If the cycles are assumed statistically identical but independent, the long-run time average cost of operation is given by the expected cost in any one cycle divided by expected cycle time. Hence one seeks to choose $b = b^*$ so as to minimize

$$J(b) = \frac{E\{K(X(T(b)))I_{T(b)<T}\} + E\{C(X(T))I_{T\leq T(b)}\}}{E\{\min(T, T(b))\}}, \tag{4.10}$$

where $I_A$ is the indicator function for $A$ (i.e. $I_A$ is 1 if $A$ occurs, 0 otherwise). Taylor considers a number of special cases for which the cost in (4.10) can be found explicitly and which thus lead to numerical optimization problems involving the solution of

$$(4.11) \qquad \frac{\partial}{\partial b} J(b)\Big|_{b=b^*} = 0$$

for $b^*$ . He also discusses applications of these results to BVD

Jaquette [58] considered stochastic models in the same spirit as those of Taylor described above, but assumed that the stochastic process is controlled through continuous time rules of stopping type with variable intensity. That is, if $X(T) = b$ at time $T$ , then by taking control action of intensity $p$ and incurring a cost $K(b,p)$ , one obtains $X(T+0) = x$ , where $x$ is a random variable depending on the control intensity. Here $\{X(t)\}$ is a birth-death stochastic process with parameters $\{\lambda_i\}$, $\{\mu_i\}$ such that

$$P\{X(t+\Delta t) = i \,|\, X(t) = i\} = 1 - (\lambda_i + \mu_i)\Delta t + o(\Delta t)$$

$$P\{X(t+\Delta t) = i+1 \,|\, X(t) = i\} = \lambda_i \Delta t + o(\Delta t)$$

$$P\{X(t+\Delta t) = i-1 \,|\, X(t) = i\} = \mu_i \Delta t + o(\Delta t) .$$

If no action is taken, the "population" $X(t)$ grows undisturbed and other costs, associated with the state of the process, are accrued. Jaquette derives expressions for the expected cost per cycle and expected cycle length, given the control policy $\{b,p\}$ . The average cost criterion $\phi(b,p)$ (on which numerical optimization techniques - see (4.11) above - or numerical search procedures can be used) is then taken as the ratio of these expressions.

In a recent publication [60], Jaquette discusses applications of dynamic programming to control of discrete time population models. That such models may be of use in epidemiology should be obvious to readers of that paper. For a survey of some related models of interest, see also [59].

We turn next to a discussion of the work of Gupta and Rink [47, 48], which we shall present in some detail since reports of these recent efforts may not yet be readily available. The basic model itself is in the spirit of some of the models discussed earlier in this chapter involving hereditary systems, but the work is novel in that it represents one of the first efforts in epidemiology employing modern optimal control theory in the form of the celebrated maximum principle of Pontryagin. While the model is deterministic in formulation, it does include terms which recognize that the latent and infectious periods are in general nondeterministic in nature.

We begin by making the basic assumption that there are four types of individuals in our population: (i) susceptibles, (ii) exposed and infected but not yet infectious (infective) individuals, (iii) infectives, and (iv) removed individuals. In the last category are included not only those that have been removed from the population because of observable symptoms, but also those individuals who have receovered and gained immunity. The period from the time an individual is exposed and becomes infected to the time he becomes infective will be called the latent period. The infectious period will be that time that an individual is in the infectious (infective) class while the incubation period will denote the time from exposure and infectedness to the time of removal (i.e. the sum of the latent plus infectious periods). In a population of size $N$, let $X_1(t)$, $X_2(t)$, $X_3(t)$ denote respectively the number of susceptibles, infectives and removed individuals at time $t$. The basic model describing changes in the population classes is given by:

$$(4.12) \qquad \dot{X}_1(t) = -\beta_0 X_1(t) X_2(t)$$

$$(4.13) \qquad \dot{X}_2(t) = A(t) - R(t)$$

$$(4.14) \qquad \dot{X}_3(t) = R(t)$$

$$(4.15) \qquad \dot{X}_4(t) = K(t) \beta_0 X_1(t) X_2(t) .$$

Here the parameter $\beta_0$ represents an "effective contact rate" while $A(t)$ is the rate of arrival of infectives and $R(t)$ is the rate of removal of infectives. If $1 - K(t)$ represents the fraction of the population that becomes immune due to repeated exposure (and slight infection without infectiveness), then $\dot{X}_4(t)$ in (4.15) represents the rate at which infecteds (but not yet infectives) are generated. That is, we assume that not every exposure need result in an infected and hence later infective, but that some susceptibles may build up immunity through repeated exposure and "small doses" of infection.

We next normalize the model by setting $x_i = X_i/N$, $a(t) = A(t)/N$, $r(t) = R(t)/N$ and $\beta = \beta_0 N$, and thus obtain

(4.16) $$\dot{x}_1(t) = -\beta x_1(t)x_2(t)$$

(4.17) $$\dot{x}_2(t) = a(t) - r(t)$$

(4.18) $$\dot{x}_3(t) = r(t)$$

(4.19) $$\dot{x}_4(t) = K(t)\,\beta x_1(t)x_2(t)$$

The parameter $\beta$ represents (see [86]) the average number of individuals per unit time that any individual ("active" case or not) will encounter sufficiently to cause infection (if he were infectious and the other were susceptible). The value of the parameter does not depend on whether the encountered is susceptible or the individual making the encounter is infectious, nor does it depend on the population size. It is, however, characteristic of the disease and average individual's behavior in the population and thus does depend on the infectiousness of the disease, weather and meteorological conditions, and the living, working and social characteristics of the population.

We shall denote by $\theta$ and $\psi$ the latent and infectious periods respectively. If $\theta$ and $\psi$ are taken as constants, the arrival and removal rates in (4.17) and (4.18) are given by

(4.20) $$a(t) = K(t-\theta)\beta x_1(t-\theta)x_2(t-\theta)$$

(4.21) $$r(t) = K(t-\theta-\psi)\beta\, x_1(t-\theta-\psi)x_2(t-\theta-\psi)$$

amd equation (4.19) can then, of course, be omitted from the model. We note that if $K = 1$ (exposure results in infectiousness) and $\theta = \psi = 0$ , the model reduces to essentially that of Kermack and McKendrick discussed previously (the only difference being the removal rate $r(t) = \beta x_1(t)x_2(t)$ in place of $\ell z_2(t) = \ell y(t)$ in (4.7)).

It is fairly easy to make an argument for the assumption that $\theta$ and $\psi$ be random (e.g., disease-specific variability in the latent period, failure to remove infectious individuals at some constant level of symptoms, etc. - see Chapter 7 or [7]). If we then assume that $\theta$ and $\psi$ are random with probability density functions $p$ and $q$ , we find

$$(4.22) \qquad a(t) = \int_0^\infty \dot{x}_4(t-\tau)p(\tau)d\tau$$

$$(4.23) \qquad r(t) = \int_0^\infty a(t-\xi)q(\xi)d\xi \ .$$

The model then becomes

$$(4.24) \qquad \dot{x}_1(t) = -\beta_1(t)x_2(t)$$

$$(4.25) \qquad \dot{x}_2(t) = \int_0^\infty K(t-\tau)\beta x_1(t-\tau)x_2(t-\tau)p(\tau)d\tau$$

$$-\int_0^\infty q(\xi)d\xi \int_0^\infty K(t-\xi-\tau)\beta x_1(t-\xi-\tau)x_2(t-\xi-\tau)p(\tau)d\tau$$

$$(4.26) \qquad \dot{x}_3(t) = \int_0^\infty q(\xi)d\xi\int_0^\infty K(t-\xi-\tau)\beta\ x_1(t-\xi-\tau)x_2(t-\xi-\tau)p(\tau)d\tau \ .$$

If we follow Gupta and assume that $\theta$ and $\psi$ are normally distributed with means $\mu_1, \mu_2$ and variances $\sigma_1^2, \sigma_2^2$, the arrival and removal rates are then given by (4.22) and (4.23) with $p(\tau) = \mathcal{N}_1(\tau)$, $q(\xi) = \mathcal{N}_2(\xi)$ where

$$(4.27) \qquad \mathcal{N}_i(\tau) = \exp[-(\tau-\mu_i)^2/2\sigma_1^2]/\sigma_i\sqrt{2\pi} .$$

Gupta approximates these expressions by

$$(4.28) \qquad a(t) = \int_0^{2\mu_1} \dot{x}_4(t-\tau)\ \mathcal{N}_1(\tau)d\tau$$

$$(4.29) \qquad r(t) = \int_0^{2\mu_2} a(t-\xi)\ \mathcal{N}_2(\xi)d\xi ,$$

which are not bad approximations if $\mu_1, \mu_2$ are large compared to $\sigma_1, \sigma_2$. We note that the assumption that $\theta$ and $\psi$ are Gaussian actually allows a positive probability that $\theta$ or $\psi$, or both, are negative, which of course is not satisfactory. However, using $\mathcal{N}_i$ in the expressions (4.22), (4.23) ignores this part of the density curves, as does using the approximations in (4.28), (4.29). Gupta makes a further approximation for the integrals in (4.28), (4.29), obtaining

$$(4.30) \qquad a(t) \simeq \sum_{j=0}^{2\mu_1} \dot{x}_4(t-j)\ \mathcal{N}_1(j)$$

$$(4.31) \qquad r(t) \simeq \sum_{j=0}^{2\mu_2} a(t-j)\ \mathcal{N}_2(j) .$$

The resulting basic model thus obtained is

$$(4.32)\qquad \dot{x}_1(t) = -\beta x_1(t)x_2(t)$$

$$(4.33)\qquad \dot{x}_2(t) = \sum_{j=0}^{2\mu_1} K(t-j)\beta x_1(t-j)x_2(t-j)\ N_1(j)$$

$$- \sum_{j=0}^{2\mu_2} \sum_{k=0}^{2\mu_1} K(t-j-k)\beta x_1(t-j-k)x_2(t-j-k)N_1(k)N_2(j)$$

$$(4.34)\qquad \dot{x}_3(t) = \sum_{j=0}^{2\mu_2} \sum_{k=0}^{2\mu_1} K(t-j-k)\beta x_1(t-j-k)x_2(t-j-k)N_1(k)N_2(j).$$

Unfortunately, "hard" data for models of the above type apparently are not yet readily available to mathematical modelers. Nonetheless, Gupta carries out a simulation study of the model (4.32)-(4.34), choosing the parameters $\mu_1$, $\mu_2$, $\sigma_1$, $\sigma_2$ essentially at random (except that they agree roughly with values used by others in their models). He reports that the above model produces curves that agree well qualitatively with simulation curves produced by other models. In addition to exhibiting features commonly found in models of this type (e.g., "threshold" phenomena), Gupta claims that certain features are present in this model (e.g., observability of successive generations of cases) that are not found in previously proposed deterministic models.

We consider next control of the model (4.32)-(4.34) and allow two basic types of controls, one representing active immunization ("active" controls) and the other representing passive immunization ("passive" controls). The first type of immunization refers to the injection of dead or live but attenuated disease microorganisms into members of the population, resulting in the production by the vaccinated individuals of antibodies. Passive immunization refers to the direct injection of antibodies. This form of control provides immediate immunity whereas active controls involve a time delay before effectiveness. While passive immunization is more expensive, it offers the advantage of immediate effectiveness which is highly desirable if one suspects that the individual being vaccinated is already carrying the disease microorganism, since it prevents further spread of the disease by this individual.

Let $U_1(t)$ represent the rate (per day) at which members of the population are being actively vaccinated at time $t$. Since vaccines tend not to be 100 percent reliable, we assume that $U_1(t)$ has been modified by a reliability factor so that it represents the rate of "reliable" vaccinations given per day. The normalized rate is then given by $u_1(t) = U_1(t)/N$. Since we assume that the vaccine is effective only when it produces antibodies in a susceptible individual and since in any vaccination program one would not wish to test individuals to ascertain their susceptibility before vaccinating, a reasonable modeling assumption is that all individuals who present themselves will be vaccinated, but only that fraction of the vaccinations involving susceptibles will alter the population pools. Thus, the effective immunization rate for our model might be taken as $x_1(t)u_1(t)$. But if we also assume that there is a delay of h days before the antibodies become effective (i.e., a latency period for the production of antibodies), the rate of reduction of the susceptible fraction of the population should be taken as $x_1(t)u_1(t-h)$ . The model with active controls thus becomes

$$\dot{x}_1(t) = -\beta x_1(t)x_2(t) - x_1(t)u_1(t-h) \tag{4.35}$$

$$\dot{x}_2(t) = a(t) - r(t) \tag{4.36}$$

$$\dot{x}_3(t) = r(t) \tag{4.37}$$

$$\dot{x}_5(t) = u_1(t) , \tag{4.38}$$

where $x_5(t)$ represents the faction of the population vaccinated up to time t ($x_5(t)$ will be a measure of the cost of the active vaccination program). In equations (4.36), (4.37) above, we assume that $a(t)$ and $r(t)$ are as given in (4.33) and (4.34) but we will not rewrite those expressions here and below as we discuss modifications of the model due to control terms.

We turn next to the discussion of terms representing passive controls. As Gupta points out, accurate quantitative effects of passive immunization on latent and infectious periods and the duration of subsequent immunity are not yet fully known. Thus this part of the model is even more tentative and speculative than that proposed up to now, and is subject to criticism based on future findings.

There are several ways in which one might try to model passive control. If we assume that vaccinations are given randomly, removing both susceptibles and infectives from their respective pools upon inoculation, reasoning similar to that used above dictates addition of terms $-x_1(t)u_2(t)$ and $-x_2(t)u_2(t)$ to the right of equations (4.35) and (4.36) respectively, where $u_2(t) = U_2(t)/N$, $U_2(t)$ being the number of vaccinations given per day. The model then becomes

$$\dot{x}_1(t) = -\beta x_1(t)x_2(t) - x_1(t)u_1(t-h) - x_1(t)u_2(t) \tag{4.39}$$

$$\dot{x}_2(t) = a(t) - r(t) - x_2(t)u_2(t) \tag{4.40}$$

$$\dot{x}_3(t) = r(t) \tag{4.41}$$

$$\dot{x}_5(t) = u_1(t) \tag{4.42}$$

$$\dot{x}_6(t) = u_2(t) , \tag{4.43}$$

where here $x_6(t)$ denotes the fraction of the population passively vaccinated up to time $t$.

One can also formulate a model assuming that passive vaccinations are given not at random, but only to known contacts of reported cases. In this situation a term $-U_2r(t)$, where $U_2$ represents the number of infectives traced per reported case, is added to the right side of equation (4.36).

Given the model (4.39)-(4.43), one is now in a position to formulate a measure for the cost of a particular control policy. Gupta suggests that one feasible measure of the cost which takes into account the costs of untreated infectives at some final time $t_f$, the costs of the total number of reported cases, and the inoculation costs (see (4.42), (4.43)) is

$$J = c_1 x_2(t_f) + c_2 x_3(t_f) + c_3 x_7(t_f) \tag{4.44}$$

where

$$\dot{x}_7(t) = Au_1(t) + Bu_1^2(t) + Cu_2(t) + Du_2^2(t) \tag{4.45}$$

and the $c_i$ are weighting factors to be specified.

Thus the optimal control problems formulated and studied by Gupta in relation to the epidemic models discussed above are classical Mayer problems with nonlinear system dynamics of the form

$$\dot{x}(t) = f(t,x(t),x(t-\eta_1),\ldots,x(t-\eta_\nu),u(t),u(t-h)) . \tag{4.46}$$

Control problems involving such systems have also been studied in connection with models for continuous-stirred tank reactors (chemical mixing problems), gas-pressurized bipropellant rocket engines, air-traffic control models, and some population models (see, for example, the references given in [15] and [16]). In fact, it was pointed out in [9] that certain modifications in the population models of Cooke

[33] lead naturally to control problems with systems of the type (4.46). (The lags in these models, due to gestation periods, play a role analogous to the latent period delays in the above epidemic models.) A substantial amount of work on the development of a mathematical theory for such problems has been reported in the literature ([8,9,15,16,25,50,57,65,66,69,78,79,99,100]). Although there are only a few pressing mathematical questions of interest left to be explored, there is a clear need for a satisfactory treatment of numerical methods for the solution of optimal control problems of the type given above. For example, while Gupta reports on attempts to solve numerically some control problems for the above model, his methods are not based on theoretically justified techniques and algorithms.

In summary, it should be apparent that serious efforts to employ the techniques of modern optimal control theory in controlling epidemics and spread of disease are yet to be made. The discussions presented above indicate to some extent the feasibility of such attempts and the considerable difficulties (vis-a-vis model formulation and verification) that will thus be encountered.

# CHAPTER 5. MODELING OF THE CONTROL SYSTEM IN GLUCOSE HOMEOSTASIS

We turn next to an area of mathematical modeling in which, unlike that discussed in Chapter 4, there is available a substantial amount of data. Unfortunately, most of this data was not collected with mathematical modeling in mind and while it does provide some information with regard to qualitative behavior of the system, much of it is less than useful in model verification. The underlying physiological system itself is quite complex and is still not completely understood. Thus one goal of many who have attempted to model glucose homeostasis has been to provide additional understanding of and insight into the mechanisms involved.

The glucose homeostatic system, which as one might expect deals with carbohydrate metabolism, also involves in very important and fundamental ways the conversion and utilization of proteins (amino acids) and lipids (fats). In fact, one may correctly view this system as one consisting of a number of organs, glands, and sensory mechanisms interconnected in a very complex way, whose main purpose is to provide means of conversion, control, storage, and use of the various sources of energy made available through nutrients.

Among the main components which play a role in glucose homeostasis are the brain (utilization, neurogenic sensors), kidneys (utilization, filtration and reabsorption, excretion (renal thresholds)), muscle and adipose tissue (utilization, storage, conversion), the endocrine glands (pancreas, adrenals, pituitary, thyroid) and their hormones, and the liver (storage, conversion). This last organ, the liver, is considered the prime component of the system since it is the major site of action of many of the secretory agents (hormones) involved in control of the system. For example, both glucagon (secreted by the α-cells of the pancreas) and insulin (β-cells of the pancreas) are involved in glycogenesis (synthesis of glycogen, the storage form of carbohydrates), glycogenolysis (breakdown of glycogen to glucose-l-phosphate (see Chapter 3)), and gluconeogenesis through their effects on cyclic AMP which is known to inhibit synthetase and promote phosphorylation. Gluconeogenesis, which is the formation of glucose from non-carbohydrate sources such as glycerol, lactate, and amino acids, is necessarily related to the plasma levels of free fatty acids (FFA) and amino acids, and hence to lipolysis and lipogenesis (the breakdown and synthesis of lipids, e.g. triglycerides, to FFA and glycerol) in adipose tissue. Epinephrine (adrenal medulla) growth hormone (pituitary), the glucocorticoids (cortisol, etc. - adrenal cortex) and thyroxin (thyroid) are hormones which have been implicated in the mechanisms involved in lipolysis and lipogenesis in addition to others in the homeostatic system.

Even though the complexity of the system involved indicates that development of an overall mathematical model is a formidable undertaking, a number of accounts of such attempts can be found in the literature. Among those most often mentioned in survey and review articles [6,13,31] are those of Wrede [103], Ackerman, Gatewood, Rosevear, and Molnar [1,2,42,43], Antomonov et al. [4], Ceresa et al. [29], and Srinivasan, Kadish, and Sridhar [94]. A discussion of some aspects of these models may be found in [13,31]. Only one of these reports [4] mentions the use of optimal control theory, and we shall present here a brief description of that work along with a report on some investigations which represent our own involvement in the modeling of the glucose homeostatic system.

Antomonov et al. [4] propose a three-variable (glucose, insulin, epinephrine) model in which they take a black-box input-output approach and offer little physiological justification in support of their basic assumptions which result in a system of essentially linear ordinary differential equations. The homeostatic system is viewed as being made up of a number of interrelated components, the input and output of each component being described by an ordinary differential equation.

A schematic for the model is given in Figure 5.1, where $y_1$ represents the deviation of the glucose concentration in the portal vein from a "normal" or equilibrium glucose systemic concentration $y_0, y_2$ represents the deviation (from $y_0$) in the hepatic vein, and $y_3$ represents the deviation at the tissue output. The variable $y \equiv y_0 + y_2 + y_3$ is the systemic glucose concentration, and $i$ and $e$ represent the circulating levels of insulin and epinephrine respectively.

The model equations are given by

$$(5.1) \qquad \dot{y}_1 = -a_1 y_1 + k_1 r_1 \chi_{[0,\tau_a]}$$

$$(5.2) \qquad \dot{y}_2 = a_2 y_2 - a_3 y_2 - a_4(y-y_0) - b_1 i + c_1 e$$

$$(5.3) \qquad \dot{i} = -b_2 i + a_5(y-y_0)$$

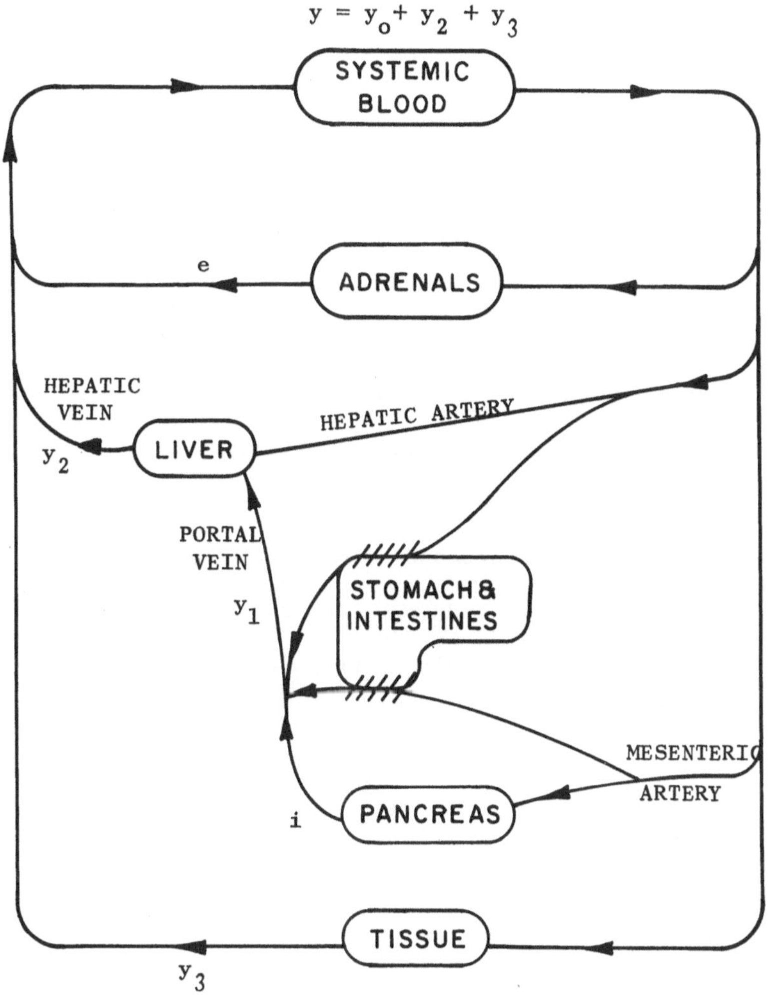


Figure 5.1

(5.4) $\dot{e} = -c_2 e + a_6(y_0 - y)$

(5.5) $\dot{y}_3 = -b_3 i - b_4 i_1 - a_7 y_3$

(5.6) $\dot{i}_1 = -b_5 i_1 + k_2 i_1^{(1)} \chi[0,\tau_i]$ , where $\chi_A$ is the characteristic function of A . Here $r_1$ is the absorption rate from the stomach and intestines into the portal vein, $i_1$ is systemic insulin received by injection into muscle tissue and $i_1^{(1)}$ is the rate of resorption due to blood flow from the area of tissue where the injection is made. Equation (5.3) is assumed to hold only when $y-y_0 > 0$ and equation (5.4) holds only when $y-y_0 < 0$ . Both equations quite obviously are based on the assumption of existence in the pancreas and adrenal glands of sensory mechanisms for systemic glucose concentration $y$ .

By fitting solutions of equations (5.1) - (5.6) to experimentally-obtained curves, Antomonov and colleagues determined values for the parameters $a_j$, $b_j$ and $c_j$ , and then carried out simulation studies of the model. Studies corresponding to various pathological situations and simulations of evolution of the homeostatic system are described in [4]. In addition, the authors mention their studies of problems involving external control variables (doses of insulin and glucose). Although details of the work are not given, they report on finding optimal controls for a minimum time criterion (minimum time to return a perturbed pathological system to equilibrium via administration of external doses) which can be approximated well by one intramuscular injection of insulin and a single oral dose of glucose.

We turn next to a description of our investigations carried out jointly with Christin A. Carter (Division of Applied Mathematics, Brown University) and H.F. Martin and J. Hologgitas (Rhode Island Hospital). In addition to providing a better understanding of the basic mechanisms and controls in glucose homeostasis, our long-term goals included a mathematical model to be used at the hospital for teaching and diagnostic purposes in work with patients suffering hormonal disorders. Many of these patients exhibit diabetic-like responses when the usual glucose-insulin data is collected during a tolerance test and it is only when data for additional variables are collected that one can uncover more specific aspects of the disorders involved.

Subjects (both patients and healthy (control) individuals) were given an oral dose of 100 gms of glucose after having fasted overnight. They were then required to remain at rest in bed for a five-hour period during which blood samples were taken every twenty minutes. Samples were large enough to permit simultaneous measurement of plasma levels of glucose ( $x_1$ in the model below), insulin $(x_2)$ , glucagon $(x_3)$ , growth hormone $(x_7)$ , and free fatty acids $(x_8)$ . Our early efforts also included checks on the plasma levels of cortisol $(x_5)$ , thyroxin $(x_6)$ , and amino acids $(x_9)$ , but these values were found to remain relatively constant during the five-hour test and were hence assumed constant in the model described below. Laboratory techniques for measuring plasma levels of epinephrine $(x_4)$ , an extremely important variable, were not yet developed at the hospital, though we were able to obtain some idea of the qualitative behavior of this variable by monitoring related variables such as blood pressure.

Attempts to use a linear model that was physiologically justifiable and that would also fit the data collected were unsuccessful.

Nonlinear components used in the version of the model presented here include

$$S(x;\ a,b) = \begin{cases} 0 & x < a \\ (x-a)/b-a) & a \leq x \leq b \\ 1 & x > b \end{cases}$$

$$H(x) = \begin{cases} 0 & x < 0 \\ 1 & x \geq 0 \end{cases}$$

$$Q(x;\ a,b) = \begin{cases} 1 & x < a \\ (b-x)/(b-a) & a \leq x \leq b \\ 0 & x > b \end{cases}$$

$$L(x;\ \alpha,A,m) = \begin{cases} (A/\alpha)x & 0 \leq x \leq \alpha \\ A + m(x-\alpha) & x > \alpha\ . \end{cases}$$

Here the piecewise linear functions $S$ and $Q$, while convenient for use in simulation studies, are meant to be only rough approximations to the sigmoid curves frequently found in saturation-limited phenomena and data. Furthermore, the functions $H$ and $L$ are no doubt only crude approximations to the much smoother functions which they represent.

The model equations are

$$(5.7)\qquad \dot{x}_1(t) = -a_{127}L(x_2(t);\alpha,A,m)S(x_1(t);\alpha_{11},\beta_{11})Q(x_7(t);$$

$$\alpha_{17},\beta_{17}) - a_{12}x_2(t) + a_{13}S(x_3(t);\alpha_{13},\beta_{13}) +$$

$$+ a_{18}S(x_8(t);\alpha_{18},\beta_{18}) + f_1(t) + r_1$$

$$(5.8)\qquad \dot{x}_2(t) = a_{21}(x_1(t)-\gamma_{21})H(x_1(t)-\gamma_{21}) + \bar{a}_{21}\dot{x}_1(t)H(\dot{x}_1(t)-\bar{\gamma}_{21}) -$$

$$- a_{22}x_2(t) + r_2$$

$$(5.9)\qquad \dot{x}_3(t) = a_{31}Q(x_1(t);\alpha_{31},\beta_{31}) - a_{33}x_3(t) + r_3$$

$$(5.10)\qquad \dot{x}_7(t) = a_{71}H(\gamma_{71}-x_1(t)) - \bar{a}_{71}\dot{x}_1(t-4/3)H(-\dot{x}_1(t-4/3))$$

$$H(\bar{\gamma}_{71}-x_1(t)) - a_{77}(x_7(t)-e_7)$$

$$(5.11)\qquad \dot{x}_8(t) = -a_{82}S(x_2(t);\alpha_{82},\beta_{82}) + a_{83}S(x_3(t);\alpha_{83},\beta_{83}) -$$

$$- a_{87}S(x_7(t);\alpha_{87},\beta_{87}) - a_{88}x_8(t) + r_8 .$$

The following paragraphs are short summaries of the assumptions underlying the above equations. A more detailed discussion, along with references to experimental support, of the assumptions may be found in [13]. Also to be found there are representative values for the parameters and a comparison of simulation of the model with actual data.

Equation (5.7) (Glucose):

i. Extrahepatic uptake is rate-limited at cell membranes; insulin promotes uptake, this effect being approximated by a Langmuir isotherm curve; plasma glucose level stimulates directly uptake while growth hormone inhibits the action of insulin at cell sites.

ii. Insulin and glucagon have opposing effects on glucose output by the liver; glucagon promotes glycogenolysis, gluconeogenesis, and inhibits glycogenesis, probably by raising the intracellular levels of cyclic AMP.

iii. Growth hormone may contribute to increased hepatic glucose output, but evidence to date does not warrant inclusion of such a mechanism.

iv. Increased levels of FFA promote increased gluconeogenesis.

v. While epinephrine is a potent hyperglycemic factor, it may be only an emergency rather than a continuous controller in homeostasis.

vi. Thyroxin and the glucocorticoids may be essential for gluconeogenesis but their effect is assumed a constant one since plasma levels of these hormones remained unchanged during normal responses.

vii. Glucose utilization by the brain is assumed constant during responses where extreme hypoglycemia is not in evidence.

viii. Renal excretion of glucose is assumed negligible during normal responses.

ix. The input function $f_1$ varies with absorption characteristics of the subject. Various forms, all having a pronounced effect during the first two to three hours of the test, are used.

Equation (5.8) (Insulin):

i. Systemic glucose levels are the major control factors in the secretion of insulin. Glucose levels above fasting level elevate a low basal secretion rate of insulin.

ii. The rate of increase of systemic glucose may also be a stimulus for insulin secretion.

iii. Both hepatic and nonhepatic degradation of insulin depend on the systemic levels of insulin.

iv. The effects of growth hormone and glucagon at physiological levels on insulin secretion are questionable and are ignored.

Equation (5.9) (Glucagon):

i. Hypoglycemia promotes secretion of glucagon while increases in blood glucose levels suppress secretion to some basal rate.

ii. Significant hepatic degradation of glucagon takes place, the rate based on circulating levels of glucagon.

iii. Other factors (such as amino acids), which do not change during normal responses, enhance glucagon secretion.

Equation (5.10) (Growth hormone):

i. The principal controllers of growth hormone levels in systemic blood appear to be the levels of glucose and growth hormone itself. Growth hormone levels are increased during hypoglycemia as well as in response to rapidly falling glucose levels. This last mechanism involves a delayed response.

ii. A negative feedback mechanism maintains an equilibrium of plasma growth hormone levels.

Equation (5.11) (Free fatty acids):

i. Insulin inhibits lipolysis while epinephrine is probably an important factor which promotes lipolysis. Growth hormone also promotes lipolysis but with a delayed effect so that its effect on lipolysis is negligible during normal responses.

ii. Growth hormone promotes uptake of fat, probably through increased oxidation in muscle. Increased plasma FFA concentration is a stimulus for FFA uptake by liver and muscle tissue for storage.

iii. Glucagon appears to enhance lipolysis, probably through its effect on cyclic AMP levels. The glucocorticoids also promote lipolysis.

The above model is by no means a final one and efforts related to this preliminary model are continuing at Rhode Island Hospital.

Related to modeling attempts such as those described above are substantial mathematical and computational questions involving parameter identification and estimation (we shall mention these questions further in Chapter 6). However, it is the opinion of this author that while "finding optimal controls" and "development of efficient methods for parameter estimation" may be important contributions of the control and system theorists, they are perhaps not the most significant benefits provided by the involvement of these workers in projects such as those detailed in these notes. Rather, we would suggest that an extremely valuable aspect of the control-theorist/ system-analyst approach is that it entails a systematic investigation of the overall system, resulting often in drastic alterations in the data collected (both in the methods and the type) by nonmathematical investigators. In some cases, development of new techniques and the recognition of factors and mechanisms heretofore ignored by biomedical researchers are promoted.

# CHAPTER 6. A SURVEY OF RECENT EFFORTS

Our discussions in previous chapters each focused on a specific topic in modeling in the life sciences and not all of the research reported on these involved the use of optimal control theory. In this chapter we present a survey of recent findings which do rely in a substantial way on the tools of modern control theory. While our literature search was not meant to be exhaustive (it did turn out to be exhausting from a personal viewpoint), the articles mentioned below do represent the results provided by a literature search of such diverse journals as IEEE Transactions on Biomedical Engineering, Journal of Theoretical Biology, Mathematical Biosciences, IEEE Transactions on Systems, Man and Cybernetics, Computers and Biomedical Research, and Computers in Biology and Medicine, as well as a number of biological and mathematical journals, and research reports from various groups in this country and abroad. The level of commitment with respect to solution of a specific biomedical modeling problem varies in the papers we found and some, while of interest to readers with a mathematical background, do not appear to involve a serious effort from the biomedical point of view.

## 6.1 Biped locomotion.

One of the more interesting projects we found described in the literature was a study of human biped locomotion carried out by Chow and Jacobson [32]. Motivated by the possibilities of programmed electrostimulation of paralyzed extremities to restore locomotion and the possible discovery of improved design procedures for artificial limbs, the authors in their theoretical study make use of a substantial body of previous work by others that involved experimentally-supported modeling efforts. Chow and Jacobson propose that normal walking obeys a certain "principle of optimality" (related to "energy"-type criteria) and while their assumptions have not been conclusively established by either experiment or theory, one can obtain a great deal of support for this concept from a number of previous studies. After deriving a mathematical model based on mechanical considerations, the authors use optimal control theory to derive optimal moment profiles that actuate locomotor elements which synthesize the patterns observed in normal gait.

The underlying feature of the biped gait on which the mechanical model is based is the "compass motion" of the lower extremities. A schematic depicting this motion is given in Figure 6.1, where only motion in the sagittal plane is considered and the solid lines

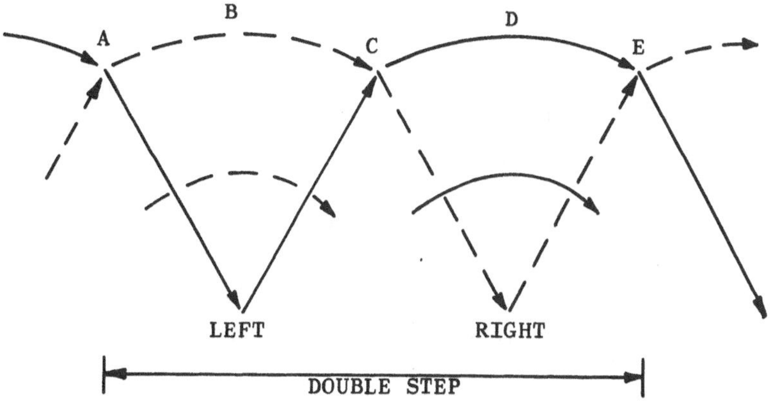


Figure 6.1

represent motion of the left leg, the broken lines represent motion of the right leg. In this figure, assume that the left leg has just completed its swing and come into the restraint position. The deploy and swing phases for the right leg force the hip to describe the arc ABC while the left leg is in a support position. The right leg then assumes a restraint position while deployment and swing of the left leg leads to the hip motion CDE. The double step can thus be described sequentially by the phases

| | | | | | |
|---|---|---|---|---|---|
| LEFT: | Restraint | Support | Deploy | Swing | Restraint |
| RIGHT: | Deploy | Swing | Restraint | Support | Deploy |

Breaking the motion into linked phases describing the stance (restraint plus support), deploy, and swing portions, one can then study the basic mechanical aspects of these motions. Carefully deriving expressions for the total kinetic energy $T$ and the total potential energy $V$ of the system, one can use Lagrange's equations

$$(6.1) \qquad \frac{d}{dt}\left(\frac{\partial T}{\partial \dot{q}_i}\right) - \frac{\partial T}{\partial q_i} + \frac{\partial V}{\partial q_i} = M_i ,$$

where the $q_i$ represent angular variables and the $M_i$ represent effective moments for the appropriate link, to derive the equations of motion. This results in a system of five nonlinear coupled second-order ordinary differential equations, a set each for the stance, deploy, and swing phases.

Chow and Jacobson then make a number of approximations and

simplifications which reduce the model to a canonical form for the sequential behavior of a single leg:

$$
\begin{aligned}
\dot{x}_1 &= x_3 \\
\dot{x}_2 &= x_4 \\
\dot{x}_3 &= R_3(x_1,x_2,x_3,x_4,u_1,u_2) \qquad (6.2)\\
\dot{x}_4 &= R_4(x_1,x_2,x_3,x_4,u_1,u_2)
\end{aligned}
$$

where $x_1$, $x_2$ are thigh and shank angles respectively and the controls $u_1$, $u_2$ are the moments generated by muscle action about the hip and knee joints respectively. The expressions $R_3$, $R_4$ are very complicated nonlinear expressions which also vary depending upon whether one is in stance, deploy, or swing portions. Included in the approximation and simplification assumptions are those which ignore certain higher-order terms and one which prescribes the hip trajectory and thus allows one to decouple the motion of the two legs. Although the expressions derived in (6.2) by Chow and Jacobson are similar to those derived by others for two-link models, their derivation is useful in that they start with an exact model and then list specifically their simplifying assumptions.

In addition to the dynamic equations (6.2), one must consider kinematic constraints which take into account foot motion. These results in equality state constraints for the stance and deploy phases of motion and inequality state constraints for the swing phase. Finally, experimental work has shown that reaction forces and ankle moments are very important, and thus to complete the basic model one must derive expressions which specify these factors.

Derivation of the performance criterion used by Chow and Jacobson in their study is based on the mechanical energy expenditure where muscles acting in agonist-antagonist pairs shorten and lengthen. The total mechanical work done by the muscle-activating system can be approximated by

$$(6.3) \qquad W = W_s + W_e = \frac{1}{2}\int_{t_0}^{t_f} r_s(t)u_s^2(t)dt - \frac{1}{2}\int_{t_0}^{t_f} r_e(t)u_e^2(t)dt$$

where $u_s$, $u_e$ are the moments generated by the shortening and lengthening muscles respectively and $r_s$, $r_e$ are functions of the moment arms $d_s$, $d_e$ and other parameters. Chow and Jacobson then approximate this quantity (6.3) by

$$(6.4) \qquad W = k\int_{t_0}^{t_f} u^2(t)dt$$

where $u = u_s - u_e$ is the net moment acting on the limbs. Thus motivated, Chow and Jacobson suggest that level locomotion is realized by programming the hip and knee moments $u_1$, $u_2$ so that the quadratic criterion

$$(6.5) \qquad J = \frac{1}{2}\int_0^{t_f} (r_1u_1^2 + r_2u_2^2)dt$$

is minimized subject to the dynamics (6.2), the kinematic constraints, and the equality and inequality state constraints. Because of the high degree of variability in the model dynamics (depending on whether one is in the deploy, swing, or stance phase), the authors actually consider three problems of the above type (each associated with one of the three phases) and argue that the suboptimal control thus obtained is a reasonable approximation to the optimal control for the original problem of minimizing J as given in (6.5). The problem of minimizing J over the stance phase essentially reduces to an algebraic problem (not involving any optimization), while the optimality problems for J over the swing and deploy phases are treated with penalty-function techniques which convert each of the constrained optimization problems into a sequence of unconstrained problems (obtained by considering an increasing sequence of weighting parameters in the penalty terms). Application of the well-known necessary conditions to the problems for the swing and deploy phase thus results as usual in two-point boundary-value problems which must be solved numerically.

By solving the above optimization problems and using the resulting sub-optimal control in numerical simulation studies, Chow and Jacobson seek to reproduce common qualitative features characteristic of non-pathological gaits. By thus testing their model and theories, they hope to establish the validity and relevance of an optimal programming approach to the study of biped locomotion. Using values for parameters obtained from experimental results reported in the literature, the authors carry out the above program. A comparison of their results with experimental findings and known "facts" reveal a good qualitative correlation. In addition to agreeing well with the findings of some previous investigators, their conclusions also shed light on some of the shortcomings of modeling attempts of others.

Chow and Jacobson close with a short discussion entailing useful ideas for practical design and quantitative study, including a "walking program via multiarc programming".

6.2 Countercurrent dialysis.

A simple countercurrent dialyzer [61,76] consists of two parallel chambers, a blood chamber and a dialyzer fluid chamber, separated by a permeable membrane. Blood containing certain substances, such as urine, to be eliminated, flows through the first chamber while a dialyzer fluid flows in the opposite direction through the other (see Figure 6.2).

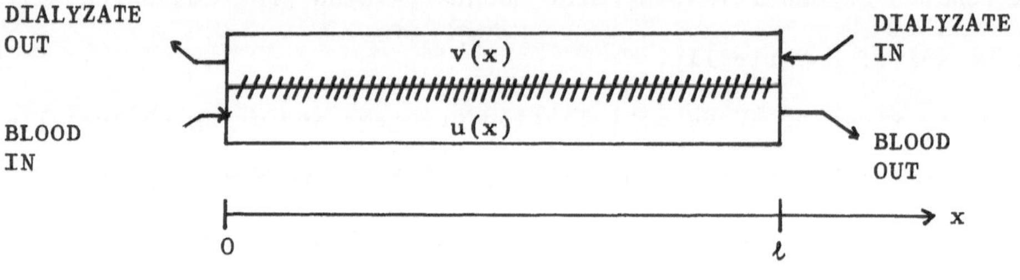


Figure 6.2

Assuming that the dialyzer is operating under quasi-static conditions (i.e., the time taken by the blood on one pass through the dialyzer is short compared to the total operating period of the dialyzer), and letting $u$ and $v$ represent the concentrations of a substance (to be eliminated) in the blood and dialyzate respectively, we have that $\partial u/\partial t = 0$, $\partial v/\partial t = 0$ . Mass balance equations which must hold for such a process dictate that

$$(6.6)\qquad \begin{aligned} du/dx &= \frac{pa}{f_b}\{v(x)-u(x)\} \\ -dv/dx &= \frac{pa}{f_d}\{u(x)-v(x)\} \end{aligned} \qquad 0 \leq x \leq \ell$$

where $f_b$, $f_d$ represent volume flow rates for blood and dialyzate respectively, $a$ is the effective transfer area of the membrane per unit length, $p$ is a permeability parameter of the membrane (in units of particles per second per $cm^2$ per unit concentration difference), and $\ell$ is the length of the dialyzer (in cm).

Assuming that the permeability parameter $p$ is spatially dependent while the transfer area and flow rates are constant, Meditch [81] considers control problems for the systems given in (6.6). Arguing that the membrane permeability should be determined so that one satisfies a "minimum complexity" criterion (which should be related to minimum cost), he poses the optimal control problem: minimize

$$(6.7)\qquad J = \frac{1}{2}\int_0^{\ell} p^2(x)dx$$

subject to (6.6) with boundary conditions $u(0) = u_0$, $v(\ell) = v_1$ (given fixed input values for $u,v$) and $u(\ell) = u_1$ (a specified desired output concentration level in the blood). The rationale behind the cost (6.7) offered by Meditch is that sharp deviations in the control values $(p(x))$ should be penalized since they represent increased complexity in design and construction of the membrane. In the event that it is not possible to construct a membrane corresponding to the optimal permeability with $u(\ell) = u_1$ , the terminal condition on $u$ may be relaxed, leading to the free-endpoint problem with cost functional

$$(6.8) \qquad J = \frac{1}{2}\{\gamma[u(\ell)-u_1]^2 + \int_0^\ell p^2(x)dx\} ,$$

where $\gamma \geq 0$ is a weighting parameter in the penalty term.

Meditch applies Pontryagin's necessary conditions to the above problems and finds that in each case the optimal permeability $p^*$ is a constant function. For the problem with cost (6.7) it is easily shown that

$$(6.9) \qquad p^*(x) = p^* = \frac{f_b f_d}{a\ell(f_d - f_b}) \ln \{\frac{f_d(u_0-v_1) - f_b(u_0-u_1)}{f_d(u_1-v_1)}\}$$

while $p^*$ must be determined by numerical solution of a transcendental equation in the case where (6.8) is to be minimized.

Two important weaknesses in his considerations are pointed out by Meditch. First, of all the design parameters in dialysis, one of the least practical with which to work is longitudinal variation of membrane permability. In fact, because of practical difficulties, one might conjecture that in design of dialysis membranes, constant membrane permeability is the usual choice. Thus the analysis by Meditch offers support for this intuitive and practical course.

A second shortcoming involves the model (6.6) itself, which entails the implicit assumption that all mass-transfer resistance is due to the membrane itself. In actual fact, effects of blood-side and dialyzate-side resistance (generally functions of length $x$ down the chamber) are present and should not be ignored. Meditch reports that investigations on improvements in this respect are in progress.

## 6.3 Drug regimens.

Assume that one has $n$ drugs to be used in therapy, either singly or in combination. Further assume that the drugs do not interact with each other and that at each time $t$ the combination of drugs present can be expressed as an equivalent amount of any one of the drugs. A three-compartment model for the ingestion,

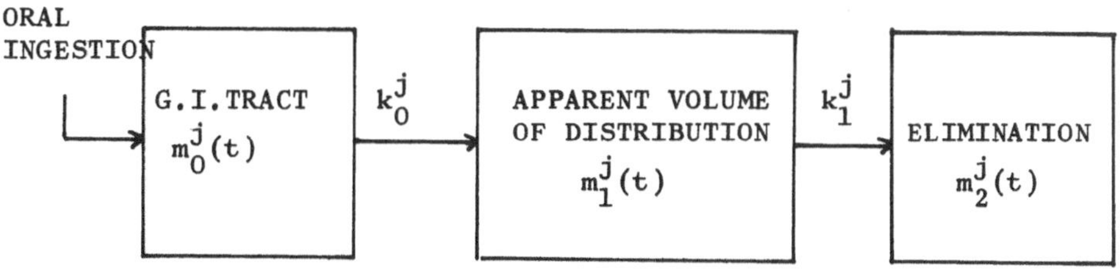


Figure 6.3

distribution, and elimination of each of the drugs is depicted schematically in Figure 6.3. If $m_0^j(t)$, $m_1^j(t)$, $m_2^j(t)$ represent the amounts at time $t$ of drug $j$ in the g.i. tract, in the apparent volume of distribution, and eliminated, then the dynamical equations [27] corresponding to Figure 6.3 are

$$(6.10)\qquad \begin{aligned} \dot m_0^j &= -k_0^j m_0^j + g^j \\ \dot m_1^j &= k_0^j m_0^j - k_1^j m_1^j \\ \dot m_2^j &= k_1^j m_1^j \quad , \end{aligned}$$

where $k_0^j(t)$, $k_1^j(t)$ are rates of absorption into the apparent volume of distribution and rates of elimination respectively, and $g^j(t)$ is the ingestion rate at time $t$. If one further assumes that drug $n$ is administered for time $t \geq 0$ while $g^j(t) \equiv 0$ for $j = 1,2,\ldots,n-1$, then the effect of all drugs can be discerned by considering the equivalent amount of drug $n$. Assuming that $m_i^j(0)$ is known for $i = 0,1$ and $j = 1,2,\ldots,n-1$, it suffices then to consider system (6.10) for $j = n$.

If drug $n$ is given orally at times $t = 0, T, 2T, \ldots, NT$, where $T$ is a fixed interval, in the amount $a_k$ at time $t = kT$, and if we assume $k_0^n$, $k_1^n$ are constant, the dynamical equations (6.10) can be replaced by a system of difference equations

$$(6.11)\qquad \begin{aligned} y_{k+1} &= ay_k + au_k \\ z_{k+1} &= cy_k + bz_k + cu_k \ . \end{aligned}$$

Here $a, b, c$ are parameters given in terms of $k_0^n$, $k_1^n$, and $y_k \equiv m_0^n(kT)$, $z_k \equiv m_1^n(kT)$, $u_k \equiv a_k$. The initial values $z_0$, $y_0$ are assumed known. If one assumes that the drug doses are available only in certain amounts, one must restrict the values $u_k = a_k$ to lie in some bounded subset $U$ (possibly discrete) of the positive real numbers.

Buell et al. [27] consider for the above model a control problem in which one seeks to maintain a therapeutically desirable level $\alpha$ of drug $n$ in the apparent volume of distribution by making a proper choice of a "control sequence" $u_0, u_1, \ldots, u_{N-1}$. The cost function chosen is

$$J = \sum_{k=1}^{N} (z_k - \beta_k)^2 \tag{6.12}$$

where $\beta_k = \alpha - w_k$, with $\{w_k\}$ a known, monotonic decreasing sequence, and the amount of drug in the apparent volume of distribution given by $v_k = z_k + w_k$. Applying dynamic programming to the resulting problem involving a linear system with quadratic cost, the authors present solutions in several special cases $(U = [0,\infty)$, $z_0 \leq \beta_0$; $z_0 > \beta_0)$. As Buell et al. point out, the above formulation requires information that is not usually available in clinical circumstances. They indicate that work is underway to extend these ideas to treat more realistic situations where the variables $z_k$ are stochastic and the rate constants vary with time and must be estimated. Some preliminary investigations on the effects of allowing for some randomness in the rate constants in systems of the above type have been reported by Soong [93].

Buell and her colleagues [26] have also used models involving systems of the type (6.10) to establish results concerning administration of drugs so as to achieve a so-called "plateau effect" (i.e., to maintain apparent volume distribution levels of drugs within certain physiologically desirable ranges).

## 6.4 Insect respiration.

In their study of the respiratory system in certain insects, Brocas and Cherruault [24] formulate certain control problems involving partial differential equations that are in the same spirit as some of those discussed in Chapter 2 above. Consider a tube (trachea - through which gas flows) of length $l$ and radius $r$ (cross-section $S = \pi r^2$) as depicted in Figure 6.4. Denote by $e$ the

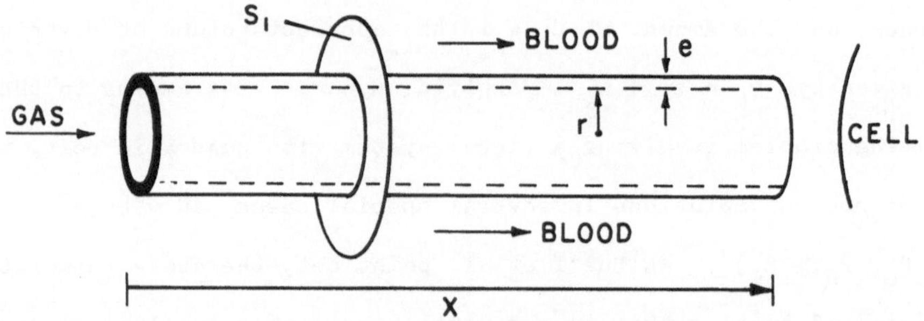


Figure 6.4

thickness of the wall of the tube and by $S_1$ the cross-section of blood surrounding the tube. Using mass balance relations and considering (i) diffusion along the axis of the tube (radial diffusion assumed negligible), (ii) convection related to movement of the gas, and (iii) diffusion across the longitudinal wall, one obtains equations which the partial pressures of oxygen, carbon dioxide, and nitrogen must satisfy. Letting $y$, $v$, $s$ denote the partial pressures in the tube of $O_2$, $CO_2$, $N_2$ respectively and $z$, $w$, $u$ denote the corresponding partial pressures in blood of these substances, one finds (the first, second, and third terms on the right in each equation represent longitudinal diffusion, convection and diffusion across the wall respectively):

(6.13)
$$\begin{aligned}
y_t &= K_{O_2} y_{xx} - \frac{1}{P_B S}(Cy)_x - D_{O_2} \frac{2}{re}(y-z) \\
v_t &= K_{CO_2} v_{xx} - \frac{1}{P_B S}(Cv)_x - D_{CO_2} \frac{2}{re}(v-w) \\
s_t &= K_{N_2} s_{xx} - \frac{1}{P_B S}(Cs)_x - D_{N_2} \frac{2}{re}(s-u) \\
z_t &= \tilde{K}_{O_2} z_{xx} - \frac{\dot{Q}}{S_1} \sigma_{O_2} z_x - D_{O_2} \frac{2\pi r}{eS_1}(z-y) \\
w_t &= \tilde{K}_{CO_2} w_{xx} - \frac{\dot{Q}}{S_1} \sigma_{CO_2} w_x - D_{CO_2} \frac{2\pi r}{eS_1}(w-v) \\
u_t &= \tilde{K}_{N_2} u_{xx} - \frac{\dot{Q}}{S_1} \sigma_{N_2} u_x - D_{N_2} \frac{2\pi r}{eS_1}(u-s)
\end{aligned}$$

to which one must add

(6.14)
$$y + v + s = P_B .$$

Here the parameters $K$, $\tilde{K}$, $D$ denote coefficients of diffusion while $C$ is the velocity of flow (convection) of the gas, $P_B$ is atmospheric pressure, $\dot{Q}$ is the blood flow rate, and $\sigma_{O_2}$, $\sigma_{CO_2}$, $\sigma_{N_2}$ denote coefficients of solubility. If one specifies appropriate boundary and initial conditions for the system (6.13), (6.14), existence of a unique solution can be guaranteed.

A natural control problem for the above system entails regulation of $CO_2$ in the blood $(w(1,t))$ by modification of ventilation

$(y(0,t),v(0,t),\dots,u(0,t))$ and the blood flow rate $\dot{Q}$. If one assumes that in the system (6.13), (6.14) one has solved for $C$ in terms of $y$, $v$, $s$, $z$, $w$, $u$, one obtains a system of six equations in six unknowns. Further assuming:

a) $\dot{Q}$ is a function of $w(1,t)$, say $\dot{Q} = \phi(w(1,t))$, where $\phi$ is specified;

b) the system regulates the entry pressures in acting on the ventilation so that $|w(1,t) - .04|$ is as small as possible;

c) the entrance and exit pressures in the blood are related by

$$\begin{aligned} &\dot{Q}\,\sigma_{CO_2}[w(0,t) - w(1,t)] = \rho_1(t) \\ (6.15)\qquad &\dot{Q}\,\sigma_{O_2}[z(0,t) - z(1,t)] = \rho_2(t) \\ &u(0,t) = u(1,t) \end{aligned}$$

where $\rho_1$, $\rho_2$ are functions depending upon the consumption of the organism;

then an obvious cost functional is

$$(6.16)\qquad J(\mathcal{U}) = \int_0^T (w(1,t) - .04)^2 dt$$

with control vector

$$\mathcal{U}(t) = (y(0,t),v(0,t)\dots,u(0,t),\dot{Q}(t)) .$$

$J$ is to be minimized subject to the reduced form of (6.13), (6.14) (i.e. six equations in six unknowns) and equations (6.15). Alternatively, one can omit (6.15) and seek to minimize

$$J_1(u) = J(u) + \int_0^T \{P_1^2(t) + P_2^2(t) + P_3^2(t)\}dt$$

where

$$P_1(t) = \dot{Q}\,\sigma_{CO_2}[w(0,t) - w(1,t)] - \rho_1(t)$$

$$P_2(t) = \dot{Q}\,\sigma_{O_2}[z(0,t) - z(1,t)] - \rho_2(t)$$

$$P_3(t) = u(0,t) - u(1,t) .$$

Brocas and Cherruault discuss existence and uniqueness of solutions of the above system (6.13), (6.14) and numerical results for special cases of problems related to those posed above.

## 6.5 Patient care and diagnostic models.

Suppose one has a patient who is ill with one of N diseases, each disease being assigned a probability of occurrence. A physician has a collection of tests (each having a certain cost) from which he may choose in trying to diagnose the patient's illness. Choosing a criterion to represent the total cost of a diagnostic procedure (which takes into account the desired goals of the diagnostic work - say minimizing average probability of death before terminal diagnosis, minimizing average losses until diagnosis, etc.), one may consider this as a control decision stochastic process. Kuznetzov and

Pchelintzev [68] discuss the formulation and exact and approximate solutions of such problems and the use of methods of linear and dynamic programming to determine optimal strategies.

An application of optimization techniques and control theory to a life-sciences-related subject that has received some attention involves operations-research-type problems of health-care delivery (patient scheduling, hospital utilization, etc.). Among the investigators who have considered problems of this nature is Esogbue [40], who formulates a discrete (in both space and time) stochastic control problem for the use of operating rooms-facilities. Decomposing the problems into a probabilistic queueing component (the numerical solution of such problems is considered in [39]) and a systems optimization component, Esogbue discusses the use of dynamic programming for the solution of these problems. Rustagi [90] suggests that dynamic programming methods will also be appropriate for solving problems in modeling of individual patient care.

## 6.6 Parameter estimation and identification.

An optimization-type problem which occurs frequently in biological modeling attempts pertains to a "best fit" of the model to experimental data. Much has been written on such identification and estimation of parameters problems [5]. One method, involving quasilinearization, has received a great deal of attention [12,14,19,20, 21,28,45,46,88] by investigators interested in biological problems. An interesting idea of Bellman and Astrom concerning the relationship between controllability and identifiability may be found in [18], where a number of fundamental questions are raised.

## REFERENCES

[1] E. Ackerman, L. Gatewood, J. Rosevear, and G. Molnar, Model studies of blood-glucose regulation, Bull. Math. Biophysics 27 (1965), 21-37.

[2] E. Ackerman, L. Gatewood, J. Rosevear, and G. Molnar, Blood glucose regulation and diabetes, Chapter 4 in Concepts and Models of Biomathematics, F. Heinmets, ed., Marcel Dekker (1969), 131-156.

[3] D.J. Aidley, The Physiology of Excitable Cells, Cambridge University Press, London, 1971.

[4] Y.G. Antomonov, N.K. Gnilitskaya, S.I. Kiforenko, and I.A. Mikulskaya, Mathematical description of the blood sugar system, Math. Biosci. 2 (1968), 435-450.

[5] K.J. Åstrom and P. Eykhoff, System identification - a survey, Automatica 7 (1971), 123-162.

[6] G.L. Atkins, Investigation of some theoretical models relating the concentrations of glucose and insulin in plasma, J. Theor. Biol. 32 (1971), 471-494.

[7] N.T.J. Bailey, The Mathematical Theory of Epidemics, Hafner Publishing Co., New York, 1957.

[8] H.T. Banks, Necessary conditions for control problems with variable time lags, SIAM J. Control 6 (1968), 9-47.

[9] H.T. Banks, A maximum principle for optimal control problems with functional differential systems, Bull. Amer. Math. Soc. 75 (1969), 158-161.

[10] H.T. Banks, Glucose homeostasis: physiological background and survey of previous mathematical models, Séminaires IRIA: analyse et controle de systemes, 1972, 15-20.

[11] H.T. Banks, Glucose homeostasis: a new mathematical model, Séminaires IRIA: analyse et controle de systemes, 1972, 21-31.

[12] H.T. Banks, On identification problems arising in biomedical modeling, Séminaires IRIA: analyse et control de systemes, 1972, 33-37.

[13] H.T. Banks and C.A. Carter, Mathematical modeling of the glucose homeostatic system in humans, CDS Lecture Notes 72-1, Brown University, July, 1972.

[14] H.T. Banks and G.M. Groome, Convergence theorems for parameter estimation by quasilinearization, J. Math. Anal. Appl. 42 (1973) 91-109.

[15] H.T. Banks and M.Q. Jacobs, The optimization of trajectories of linear functional differential equations, SIAM J. Control 8 (1970), 461-488.

[16] H.T. Banks, M.Q. Jacobs, and M.R. Latina, The synthesis of optimal controls for linear, time-optimal problems with retarded controls, J. Optimization Theory and Appl. 8 (1971), 319-366.

[17] H.T. Banks and R.P. Miech, A mathematical model for enzyme cascades, Proceedings 1973 International Conference on Cybernetics and Society, November 5-7, Boston, 236-239.

[18] R. Bellman and K.J. Åstrom, On structural identifiability, Math. Biosci. 7 (1970), 329-339.

[19] R. Bellman, J. Jacquez, R. Kalaba, and S. Schwimmer, Quasilinearization and the estimation of chemical rate constants from raw kinetic data, Math. Biosci. 1 (1967), 71-76.

[20] R. Bellman, H. Kagiwada, and R. Kalaba, Quasilinearization and the estimation of time lags, Math. Biosci. 1 (1967), 39-44.

[21] R. Bellman and R. Roth, A technique for the analysis of a broad class of biological systems, in Cybernetic Problems in Bionics, H.L. Oestreicher and D.R. Moore, eds., Gordon and Breach, New York, 1968, p. 725-737.

[22] J.M. Bowness, Epinephrine: cascade reactions and glycogenolytic effect, Science 152 (1966), 1370-1371.

[23] G.E. Briggs and J.B.S. Haldane, A note on the kinetics of enzyme action, Biochem. J. 19 (1925), 338-339.

[24] J. Brocas and Y. Cherruault, Association des systemes trachéens et circulatoires dans le transport des gaz respiratoires chez l'insecte aérien - Essai d'approche théorique par étude sur modele mathématique, in Rapport Scientifique du Laboratoire de Biologie Quantitative et Mathématique Appliquées a la Medecine, Paris, 1971-1972, p. 1-51.

[25] J. Budelis and A. Bryson, Some optimal control results for differential-difference systems, IEEE Transactions on Automatic Control AC-15 (1970), 237-241.

[26] J. Buell, R. Jelliffe, R. Kalaba, and R. Sridhar, Modern control theory and optimal drug regimens, I. The plateau effect, Math. Biosci. 5 (1969), 285-296.

[27] J. Buell, R. Jelliffe, R. Kalaba, and R. Sridhar, Modern control theory and optimal drug regimens, II. Combination therapy, Math. Biosci. 6 (1970), 67-74.

[28] J. Buell and R. Kalaba, Quasilinearization and the fitting of nonlinear models of drug metabolism to experimental kinetic data, Math. Biosci. 5 (1969), 121-132.

[29] F. Ceresa, F. Ghemi, P.F. Martini, P. Martino, G. Segre, and A. Vitelli, Control of blood glucose in normal and in diabetic subjects, Diabetes 17 (1968), 570-578.

[30] S. Cha, Magnitude of errors of Michaelis-Menten and other approximations, J. Biol. Chemistry 245 (1970), 4814-4818.

[31] W.P. Charette, A.H. Kadish, and R. Sridhar, Modeling and control aspects of glucose homeostasis, Hormonal Control Systems, Supplement I, Math. Biosci. (1969), 115-149.

[32] C.K. Chow and D.H. Jacobson, Studies of human locomotion via optimal programming, Math. Biosci. 10 (1971), 239-306.

[33] K.L. Cooke, Functional-differential equations: some models and perturbation problems, in Differential Equations and Dynamical Systems, edited by J. Hale and J. LaSalle, Academic Press, New York, 1967, 167-183.

[34] G.J. Cooper, The numerical solution of stiff differential equations, FEBS Letters 2 (1969), Supplement, p. S22-S29.

[35] C. Cori and G. Cori, The enzymatic conversion of phosphorylase a to b , J. Biol. Chemistry 158 (1945), 321-332.

[36] W. Danforth, E. Helmreich, and C.F. Cori, The effect of contraction and of epinephrine on the phosphorylase activity of frog sartorius muscle, Proc. Natl. Acad. Sci. U.S.A. 48 (1962), 1191-1199.

[37] J. Davies and P Williams, Quantitative relationship between stimulus and response in hormone action: amplification and sensitization, J. Theor. Biol. 30 (1971), 41-57.

[38] K. Dietz, Epidemics and rumours: a survey, J. Roy. Stat. Soc. A, 130 (1967), 505-527.

[39] A.M.O. Esogbue, Mathematical and computational approaches to some queuing processes arising in surgery, Math. Biosci. 4 (1969), 531-542.

[40] A.M.O. Esogbue, Dynamic programming and optimal control of variable multi-channel stochastic service systems with applications, Math. Biosci. 5 (1969), 133-142.

[41] W. Ganong, Review of Medical Physiology, 5th edition, Lange Medical Publishers, Los Altos, California, 1971.

[42] L. Gatewood, E. Ackerman, J. Rosevear, and G. Molnar, Tests of a mathematical model of the blood-glucose regulatory system, Computers Biomed. Res. 2 (1968), 1-14.

[43] L. Gatewood, E. Ackerman, J. Rosevear, and G. Molnar, Simulation studies of blood-glucose regulation: effect of intestinal glucose absorption, Computers Biomed. Res. 2 (1968), 15-27.

[44] C.W. Gear, Numerical Initial Value Problems in Ordinary Differential Equations, Prentice-Hall, Englewood Cliffs, N.J., 1971.

[45] G.M. Groome, Identification of dynamical systems, Ph.D. thesis, Brown University, June, 1972.

[46] G.M. Groome, Parameter estimation by quasilinearization: linear convergence, to appear.

[47] N.K. Gupta, Modeling and optimum control of epidemics, Ph.D. thesis, University of Alberta, Edmonton, 1972.

[48] N.K. Gupta and R.E. Rink, A model for communicable disease control, Proceedings 24th Annual Conference on Engineering in Medicine and Biology, Las Vegas (1971), p. 296.

[49] H. Gutfreund, An Introduction to the Study of Enzymes, Blackwell Scientific Publ., Oxford, 1965.

[50] A. Halanay, Optimal controls for systems with time lag, SIAM J. Control 6 (1968), 215-234.

[51] H. Hemker and P. Hemker, General kinetics of enzyme cascades, Proc. Roy. Soc. London B 173 (1969), 411-420.

[52] H. Hemker, P. Hemker, and E. Loeliger, Kinetic aspects of the interaction of blood clotting enzymes, Thromb. Diath. Haemorrh. 13 (1965), 155-175.

[53] V. Henri, Lois générales de l'action des diastases, Hermann, Paris, 1963.

[54] H. Holzer and W. Duntze, Chemical modifications of enzymes by ATP, Chapter 5 of Biochemical regulatory Mechanisms in Eukaryotic Cells, ed. by E. Kun and S. Grisolia, John Wiley, 1972, p. 115-136.

[55] F. Hoppensteadt and P. Waltman, A problem in the theory of epidemics, Math. Biosci. 9 (1970), 71-91.

[56] F. Hoppensteadt and P. Waltman, A problem in the theory of epidemics, II, Math. Biosci. 12 (1971), 133-145.

[57] D.K. Hughes, Variational and optimal control problems with delayed argument, J. Optimization Theory and Appl. 2 (1968) 1-14.

[58] D.L. Jaquette, A stochastic model for the optimal control of epidemics and pest populations, Math. Biosci. 8 (1970), 343-354.

[59] D.L. Jaquette, Mathematical models for controlling growing biological populations: a survey, Operations Res. 20 (1972), 1142-1151.

[60] D.L. Jaquette, A discrete time population control model, Math. Biosci. 15 (1972), 231-252.

[61] S. Kaplan, A. McNabb, and M.B. Wolf, Input-output relations for a counter-current dialyzer by the method of invariant imbedding, Math. Biosci. 3 (1968), 289-293.

[62] W.O. Kermack and A.G. McKendrick, A contribution to the mathematical theory of epidemics, Proc. Roy. Soc. London A 115 (1927), 700-721.

[63] J.P. Kernevez, Evolution et control de systemes bio-mathematiques, Thesis, University of Paris VI, 1972.

[64] J.P. Kernevez and D. Thomas, Diffusion-reactions in enzymatically active model-membranes, Journees d'Informatique Medicale 2 (1972), 171-182 (IRIA).

[65] G.L. Kharatishvili, A maximal principle in extremal problems with delays, in Mathematical Theory of Control, A.V. Balakrishnan and L.W. Neustadt, eds., Academic Press, New York, 1967, p. 26-34.

[66] G.L. Kharatishvili, Extremal problems in linear topological spaces, Doctoral thesis, Tbilisi State University, U.S.S.R., 1968.

[67] J.W. Kimball, Cell Biology, Addison-Wesley, Reading, Mass., 1970.

[68] P.I. Kuznetzov and L.A. Pchelintzev, The application of some mathematical methods in medical diagnostics, Math. Biosci. 5 (1969), 365-377.

[69] E.B. Lee, Variational problems for systems having delay in the control action, IEEE Transactions on Automatic Control, AC-13 (1968), 697-699.

[70] R. Levine and D. Haft, Carbohydrate homeostasis I; II, N.E.J. Medicine 283 (1970), 175-183; 237-246.

[71] S. Levine, Enzyme amplifier kinetics, Science 152 (1966), 651-653.

[72] J.L. Lions, Quelques Methodes de Resolution des Problemes aux Limites Nonlineaires, Dunrod, Paris, 1967.

[73] J.L. Lions, Optimal Control of Systems Governed by Partial Differential Equations, Springer-Verlag, New York, 1971.

[74] J.L. Lions, Some Aspects of the Optimal Control of Distributed Parameter Systems, CBMS No. 6, SIAM, Philadelphia, 1972.

[75] A. Lotka and F. Sharpe, Contributions to the analysis of malaria epidemiology, Amer. J. Hygiene 3 (1923), Jan. Supplement, 1-121.

[76] A. McNabb and M.B. Wolf, Time-dependent behavior of a countercurrent dialyzer, Math. Biosci. 3 (1968), 295-306.

[77] R. MacFarlane, An enzyme cascade in the blood clotting mechanism and its function as a biochemical amplifier, Nature 202 (1964), 498-499.

[78] A. Manitius, On the controllability conditions for systems with distributed delays in state and control, Archiwum Automat. i Telemech. 27 (1972), 363-377.

[79] A. Manitius and A. Olbrot, Controllability conditions for linear systems with delayed state and control, Archiwum Automat. i Telemech. 27 (1972), 119-131.

[80] A. Mazur and B. Harrow, Biochemistry: A Brief Course, W.B. Saunders Co., Philadelphia, 1968.

[81] J.S. Meditch, Optimal permeability character for quasi-static countercurrent dialysis, IEEE Transactions on Bio-Med. Engineering, BME-18 (1971), 181-186.

[82] L. Michaelis and M. Menten, Die Kinetik der invertinwirkung, Biochem. Z. 49 (1913), 333-369.

[83] H. Otten and L. Duysens, An extension of the steady-state approximation of the kinetics of enzyme-containing systems, J. Theor. Biol. 39 (1973), 387-396.

[84] H. Rasmussen, D. Goodman, and A. Tenenhouse, The role of cyclic AMP and calcium in cell activation, CRC Critical Reviews in Biochemistry, Feb. (1972), 95-148.

[85] C. ReVelle, F. Feldmann, and W. Lynn, An optimization model of tuberculosis epidemiology, Management Science 16 (1969), B190-B211.

[86] C. ReVelle, W. Lynn, and F. Feldmann, Mathematical models for the economic allocation of tuberculosis control activities in developing nations, Amer. Rev. of Respiratory Diseases 96 (1967), 893-909.

[87] R. Ross, The Prevention of Malaria, London, 1911.

[88] R. Roth and M. Roth, Data unscrambling and the analysis of inducible enzyme systems, Math. Biosci. 5 (1969), 57-92.

[89] S.I. Rubinow, Mathematical Problems in the Biological Sciences, CBMS No. 10, SIAM, Philadelphia, 1973.

[90] J.S. Rustagi, Dynamic programming model of patient care, Math. Biosci. 3 (1968), 141-149.

[91] H.L. Segal, Enzymatic interconversion of active and inactive forms of enzymes, Science 180 (1973), 25-32.

[92] A. Sols and C. Gaucedo, Primary regulatory enzymes and related problems, Chapter 4 in Biochemical Regulatory Mechanisms and Eukaryotic Cells, ed. by E. Kun and S. Grisolia, John Wiley, New York, 1972, p. 85-114.

[93] T.T. Soong, Pharmacokinetics with uncertainties in rate constants, II. Sensitivity analysis and optimal dosage control, Math. Biosci. 13 (1972), 391-396.

[94] R. Srinivasan, A. Kadish, and R. Sridhar, A mathematical model for the control mechanism of free fatty acid-glucose metabolism in normal humans, Computers Biomed. Res. 3 (1970), 146-166.

[95] H.M. Taylor, Some models in epidemic control, Math. Biosci. 3 (1968), 383-398.

[96] G. Wald, Visual excitation and blood clotting, Science 150 (1965), 1028-1030.

[97] C. Walter, Steady-State Applications in Enzyme Kinetics, Ronald Press, New York, 1965.

[98] C. Walter, Enzyme Kinetics, Ronald Press, New York, 1966.

[99] J. Warga, Optimal Control of Differential and Functional Equations, Academic Press, New York, 1972.

[100] J. Warga, Optimal controls with pseudo-delays, to appear.

[101] E. Wilson and M. Burke, The epidemic curve, Proc. Nat. Acad. Sci. U.S.A. 28 (1942), 361-367.

[102] L.O. Wilson, An epidemic model involving a threshold, Math. Biosci. 15 (1972), 109-121.

[103] D. Wrede, Development of a mathematical model for a biological feedback system with particular application to glucose metabolism, Ph.D. thesis, University of Cincinnati, 1963.

# NECESSARY CONDITIONS IN MATHEMATICAL PROGRAMMING AND OPTIMAL CONTROL THEORY

by

Hubert Halkin*

Department of Mathematics
University of California at Dan Diego

## Introduction

In these lectures I will present a theory of necessary conditions for nonlinear optimizations problems in infinite-dimensional spaces and I will apply the results of the theory to the optimal control of systems described by families of non-linear ordinary differential equations.

The very large set of known results related to necessary conditions in (continuous) optimization theory can be divided in two categories. For the first category of results some specific assumptions are made a priori which guarantee that the Lagrange multiplier corresponding to the cost function can be chosen to be $-1$ whereas no such assumption is made for the results of the second category, and, as a consequence, nothing is known a priori in that case concerning the Lagrange multiplier corresponding to the cost function beside the fact that it is nonpositive. In order for the second category of results to be

*This paper is the write-up of a series of lectures given at the 14th Biennial Seminar of the Canadian Mathematical Congress. This paper was written while the author was a visiting member of the Centre de Recherches Mathématiques at the Université de Montréal.

nontrivial one must prove the existence of a vector of Lagrange multipliers which is not identically zero. In the case of results of the first category the possibility of a vector of Lagrange multipliers which is identically zero is ruled out by the fact that the Lagrange multiplier corresponding to the cost function may be chosen to be -1 . The linear independence of the gradients of the constraints at the optimal point in "classical" mathematical programming, Constraints Qualifications in "modern" mathematical programming, normality conditions in calculus of variations, controlability conditions in optimal control theory, etc., are all examples of the assumptions characterizing the first category of results, (which under those assumptions are relatively easy to establish). In the second category of results one finds Carathéodory's Multiplier Rule for "classical" mathematical programming, John's Multiplier Rule for "modern" mathematical programming, McShane's Multiplier Rule in calculus of variations and Pontryagin's Maximum Principle in optimal control theory. The present paper belongs to this second category and will contain a unified treatment of necessary conditions which includes the above results of Carathéodory, John, McShane and Pontryagin.

A key element in my 1963 dissertation was a specific application of the Brouwer Fixed Point Theorem. In their proof of the Maximum Principle, Pontryagin et al. had given instead a sketch of an algebraic topological argument to deal with the same specific difficulty. This specific application of the Brouwer Fixed Point Theorem became a standard fixture of the subsequent literature of nonlinear optimal control and for a long while I remained persuaded that the recourse to the Brouwer Fixed Point Theorem (or to an equivalent result from Algebraic Topology) was strictly necessary in order to establish meaningful necessary conditions for nonlinear optimization problems. I am very pleased to report that the situation is much better than I had been thinking since 1963. In these lectures I will show that, at the cost of very minimal and reasonable assumptions, a theory of necessary condition for nonlinear optimization problems can be obtained without using the Brouwer Fixed Point Theorem, but by using instead the much simpler (and constructively more acceptable) contraction mapping theorem.

Essential guidelines for the reader.

The present paper contains three types of messages: the normal message, the bracketed message and the braced message. The reader should split his personality into two parts: Reader A and Reader B. Reader A should read the normal and the braced messages, whereas Reader B should read the normal and the bracketed messages. For instance, if the text reads: "when the function is [strongly] differentiable we shall establish our result by using the {Brouwer Fixed Point} [Contraction Mapping] Theorem", Reader A should read: "when the function is differentiable we shall establish our result by using the Brouwer Fixed Point Theorem" and Reader B should read "when the function is strongly differentiable we shall establish our result by using the Contraction Mapping Theorem". Readers A and B will be presented with the same results but the results presented to Reader A will require slightly weaker assumptions than the results presented to Reader B. However whereas Reader A will need the Brouwer Fixed Point Theorem, Reader B will be able to get along with the simpler (and more constructive) Contraction Mapping Theorem.

## §1. A preview of the Multiplier Rule

In this section we consider the following optimization problem: we are given a normed linear space $X$, a subset $L$ of $X$, non-negative integers $\mu$ and $m$, a function $\varphi = (\varphi_{-\mu},\ldots,\varphi_{-1},\varphi_0,\varphi_1,\ldots,\varphi_m)$ from $X$ into $R^{\mu+m+1}$ and we want to minimize $\varphi_0(x)$ subject to $x \in L$, $\varphi_i(x) \leq 0$ for $i = -\mu,\ldots,-1$ and $\varphi_i(x) = 0$ for $i = 1,\ldots,m$. In other words, if $A = \{x : x \in L, \varphi_i(x) \leq 0$ for $i = -\mu,\ldots,-1$ and $\varphi_i(x) = 0$ for $i = 1,\ldots,m\}$ we want to determine an element $x_0 \in A$ such that $\varphi_0(x_0) \leq \varphi_0(x)$ for all $x \in A$. Such $x_0$ is called an optimal solution. This paper is entirely devoted to the study of necessary conditions; i.e. we assume that an optimal solution $x_0$ exists and we want to state and prove some interesting properties of $x_0$. The basic necessary condition for this optimization problem takes the form of a Multiplier Rule which is stated below. We shall prove this Multiplier Rule under various sets of specific assumptions given later. However, in order to identify the elements used in the statement of the Multiplier Rule it is necessary at this point to state informally two of these assumptions: 1) there exists a set $N$ which "approximates" the set $L$ around $x_0$ and 2) there exists a function $h = (h_{-\mu},\ldots,h_m)$ which "approximates" the function $\varphi$ around $x_0$ and such that $h_i$ is convex for

$i = \mu,\ldots,-1,0$ and affine for $i = 1,\ldots,m$ .

Multiplier Rule. If $x_0$ is an optimal solution then there exists a nonzero vector $\lambda = (\lambda_{-\mu},\ldots,\lambda_m) \in R^{\mu+m+1}$ such that

(i) $\lambda \cdot h(x) \leq \lambda \cdot h(x_0)$ for all $x \in N$

(ii) $\lambda_i \leq 0$ for $i = -\mu,\ldots,0$

(iii) $\lambda_i \varphi_i(x_0) = 0$ for $i = -\mu,\ldots,-1$ .

Moreover there exists a continuous affine function $\ell = (\ell_{-\mu},\ldots,\ell_m)$ from $X$ into $R^{\mu+m+1}$ such that

(iv) $\ell(x_0) = h(x_0)$

(v) $\ell_i = h_i$ for $i = 1,\ldots,m$

(vi) $\ell_i(x) \leq h_i(x)$ if $i = -\mu,\ldots,0$ and $x \in X$

(vii) $\lambda \cdot \ell(x) \leq \lambda \cdot \ell(x_0)$ for all $x \in N$

Our proof of the Multiplier Rule will be based on standard results of convex analysis (separation of disjoint convex sets for instance) and the following type of

Interior Mapping Principle. If $N$ is a subset of a normed linear space $X$ which "approximates" the set $L \subset X$ around $x_0$ , if $\varphi$ is a mapping from $X$ into a finite-dimensional Euclidean space $Y$ which is continuous in a neighborhood of $x_0$ , if $h$ is a continuous affine mapping which "approximates" $\varphi$ around $x_0$ and if $h(x_0) \in \operatorname{int} h(N)$ then $\varphi(x_0) \in \operatorname{int} \varphi(L)$ .

## §2. Simplicial Approximation

Let $Z$ be the set of positive integers. If $k \in Z$ we shall say that a subset $N$ of a normed linear space $X$ is a [strong] simplicial approximation of order $k$ to $L \subset X$ around $x_0 \in X$ if for all $\varepsilon_1, \varepsilon_2 > 0$ and for all sets $\{x_1,\ldots,x_\ell\}$ with $\ell \leq k$ elements in general position in $N$ there exist an $\eta \in (0, \varepsilon_2]$ and a continuous function $f$ from $S_\eta = \text{co}\{x_0 + \eta(x_i - x_0) : i = 1,\ldots,\ell\}$ into $L$ such that

$$|f(x) - x| \leq \varepsilon_1 \eta \quad \text{whenever} \quad x \in S_\eta$$

[and $|f(x'') - f(x') - (x'' - x')| \leq \varepsilon_1 |x'' - x'|$ whenever $x'$ and $x'' \in S_\eta$].

For any $\ell \in Z$ let $\Lambda_\ell$ be the subset of $R^{\ell+1}$ defined by $\Lambda_\ell = \{\alpha = (\alpha_0,\alpha_1,\ldots,\alpha_\ell) : \alpha_i \geq 0$ for $i = 0,\ldots,\ell$, $\Sigma_{i=0,\ldots,\ell}\alpha_i = 1\}$. The element $(1,0,\ldots,0)$ of $\Lambda_\ell$ will be denoted by $\alpha^0$. On $\Lambda_\ell$ we shall use the norm $|\alpha'' - \alpha'| = \Sigma_{i=0,\ldots,\ell}|\alpha_i'' - \alpha_i'|$. For every $\eta \in [0,2]$ let $\Lambda_\ell^\eta = \{\alpha : \alpha \in \Lambda_\ell, |\alpha - \alpha^0| = \eta\}$. With the help of the above notations we can say that a subset $N$ of a normed linear space $X$ is a [strong] simplicial approximation of order $k$ to $L \subset X$ around $x_0 \in X$ if and only if for all $\varepsilon_1, \varepsilon_2 > 0$ and for all sets $\{x_1,\ldots,x_\ell\}$ with $\ell \leq k$ elements in general position in $N$ there exist an $\eta \in (0,\varepsilon_2]$ and a continuous function $\zeta$ from $\Lambda_\ell^\eta$ into $L$ such that

$$|\zeta(\alpha) - \Sigma_{i=0,\ldots\ell}\alpha_i x_i| \leq \varepsilon_1 \eta \quad \text{whenever} \quad \alpha \in \Lambda_\ell^\eta$$

$$\left[\begin{array}{l}\text{and} \quad |\zeta(\alpha'') - \zeta(\alpha') - \Sigma_{i=0,\ldots,\ell}(\alpha''_i - \alpha'_i)\, x_i| \le \epsilon\, |\alpha'' - \alpha'| \\ \qquad\qquad \text{whenever} \quad \alpha' \text{ and } \alpha'' \in \Lambda^{\eta}_{\ell}\end{array}\right]$$

Remark. The above definition would remain conceptually unchanged but would take a simpler appearance if a single $\epsilon > 0$ was used instead of the two $\epsilon_1$ and $\epsilon_2 > 0$. I prefer the above definition because it underlines the two different roles that a single $\epsilon$ would play and hence it makes it easier (for me at least) to follow the developments given in this paper.

We shall say that two functions f and g from a normed linear space X into a normed linear space Y are [strongly] tangent at the point $x_0 \in X$ if for all $\epsilon > 0$ there exists a $\delta > 0$ such that

$$|f(x) - g(x)| \le \epsilon\, |x - x_0| \quad \text{whenever} \quad |x - x_0| \le \delta$$

$$\left[\begin{array}{l}\text{and} \quad |(f(x_2) - g(x_2)) - (f(x_1) - g(x_1))| \le \epsilon\, |x_2 - x_1| \\ \qquad\qquad \text{whenever} \quad |x_1 - x_0| \text{ and } |x_2 - x_0| \le \delta\end{array}\right].$$

Furthermore we shall say that a function f from a normed linear space X into a normed linear space Y is [strongly] differentiable at the point $x_0 \in X$ if there exists a continuous linear mapping A from X into Y such that the function g defined by

$$g(x) = f(x_0) + A(x - x_0) \quad \text{for all} \quad x \in X$$

is [strongly] tangent to the function f at the point $x_0$.

Proposition 2.1. If $k \in Z$ , if $X$ and $Y$ are normed linear spaces, if $x_0 \in X$ , if $N \subset X$ is a [strong] simplicial approximation of order $k$ to $L \subset X$ around $x_0$ , if $\varphi$ is a mapping from $X$ into $Y$ which is continuous on a neighborhood of $x_0$ , and if $h$ is a continuous affine mapping from $X$ into $Y$ such that $\varphi$ and $h$ are [strongly] tangent around $x_0$ , then $h(N)$ is a [strong] simplicial approximation of order $k$ to $\varphi(L)$ around $h(x_0) = \varphi(x_0)$ .

Proof of Proposition 2.1. There is no loss of generality by assuming that $\varphi$ is continuous over the entire space $X$ since this can be achieved without enlarging the set $\varphi(L)$ by modifying the function $\varphi$ outside a closed ball containing $x_0$ in its interior. Let $\varepsilon_1$ and $\varepsilon_2 > 0$ and let $\{y_1,\ldots,y_\ell\}$ be a set of $\ell \le k$ elements in general position in $h(N)$ . Let $y_0 = h(x_0)$ and for every $i \in \{1,\ldots,\ell\}$ let $x_i \in N$ be such that $y_i = h(x_i)$ . Since the function $h$ is affine it follows that the elements $\{x_1,\ldots,x_\ell\}$ are in general position. There exists a real number $\tau > 0$ such that

$$|\varphi(\Sigma_{i=0,1,\ldots,\ell}\alpha_i x_i^*)-h(\Sigma_{i=0,1,\ldots,\ell}\alpha_i x_i^*| \le \frac{\varepsilon_1}{2}|\alpha - \alpha^0|$$

whenever $\alpha \in \Lambda_\ell, ||\alpha - \alpha^0| \le \tau,\ x_0^* = x_0$ and $|x_i^* - x_i| \le 1$ for $i = 1,\ldots,\ell$ .

[ and even

$$|\varphi(\Sigma_{i=0,\ldots,\ell}\alpha_i''x_i^*)-\varphi(\Sigma_{i=0,\ldots,\ell}\alpha_i'x_i^*)-(h(\Sigma_{i=0,\ldots,\ell}\alpha_i''x_i^*)-$$
$$-h(\Sigma_{i=0,\ldots,\ell}\alpha_i'x_i^*))| \le \frac{\epsilon_1}{2}|\alpha''-\alpha'|$$

whenever $\alpha',\alpha'' \in \Lambda_\ell$, $|\alpha'-\alpha^0|, |\alpha''-\alpha^0| \le \tau$, $x_0^* = x_0$ and

$|x_i^* - x_i| \le 1$ for $i = 1,\ldots,\ell$ ] .

Let $|h| = \sup_{|x-x_0|\le 1}|h(x)-h(x_0)|$ . Let $\epsilon_1^* = \min\{1, \epsilon_1/2|h|\}$ and let $\epsilon_2^* = \min\{\epsilon_2, \tau, 1\}$ . Since N is a [strong] simplicial approximation to L around $x_0$ we know that there exists an $\eta^* \in (0, \epsilon_2^*]$ and a continuous function $\zeta^*$ from $\Lambda_\ell^{\eta^*}$ into L with

$$|\zeta^*(\alpha) - \Sigma_{i=0,1,\ldots,\ell}\alpha_i x_i| \le \epsilon_1^*\eta^* \quad \text{whenever} \quad \alpha \in \Lambda_\ell^{\eta^*}$$

[ and even

$$|\zeta^*(\alpha'')-\zeta^*(\alpha')-\Sigma_{i=0,1,\ldots,\ell}(\alpha_i''-\alpha_i')x_i| \le \epsilon_1^*|\alpha'' \ \alpha'|$$

whenever $\alpha'$ and $\alpha'' \in \Lambda_\ell^{\eta^*}$ . ]

Whenever $|\alpha - \alpha^0| = \eta^*$ we may write

$$\zeta^*(\alpha) = \alpha_0x_0 + \Sigma_{i=1,\ldots,\ell}\alpha_ix_i + \eta^*z(\alpha) = \alpha_0x_0 + \Sigma_{i=1,\ldots,\ell}\alpha_ix_i^*$$

where $z(\alpha)$ is some element of X such that $|z(\alpha)| \le \epsilon_1^* \le 1$

and where $x_i^* = x_i + \dfrac{\alpha_i}{\alpha_1+\ldots+\alpha_\ell}\eta^*z(\alpha)$ for $i = 1,\ldots,\ell$ .

Since $|x_i^* - x_i| = \dfrac{\alpha_i}{\alpha_1+\ldots+\alpha_\ell}\eta^*|z(\alpha)| \le 1$ for $i = 1,\ldots,\ell$

and since $\eta^* \leq \tau$ we have then $|\varphi\zeta^*(\alpha)) - h(\zeta^*(\alpha))| \leq \frac{\varepsilon_1}{2}\eta^*$ whenever $\alpha \in \Lambda_\ell^{\eta^*}$

[and even

$$|(\varphi(\zeta^*(\alpha'')) - \varphi(\zeta^*(\alpha')) - (h(\zeta^*(\alpha'')) - h(\zeta^*(\alpha'')))| \leq \frac{\varepsilon_1}{2}|\alpha'' - \alpha'|$$

whenever $\alpha'$ and $\alpha'' \in \Lambda_\ell^{\eta^*}$].

Let $\eta = \eta^*$ and let $\zeta$ be defined by $\zeta(\alpha) = \varphi(\zeta^*(\alpha))$ for all $\alpha \in \Lambda_\ell^\eta$. Since the functions $\varphi$ and $\zeta^*$ are continuous it follows that the function $\zeta$ is continuous. We have then $\eta = \eta^* \leq \varepsilon_2^* \leq \varepsilon_2$ and whenever $|\alpha - \alpha^0| = \eta$ we have

$$\begin{aligned} |\zeta(\alpha) - \Sigma_{i=0,1,\ldots,\ell}\alpha_i y_i| &= |\varphi(\zeta^*(\alpha)) - \Sigma_{i=0,1,\ldots,\ell}\alpha_i h(x_i)| \\ &\leq |\varphi(\zeta^*(\alpha)) - h(\zeta^*(\alpha))| + |h||\zeta^*(\alpha) - \Sigma_{i=0,1,\ldots,\ell}\alpha_i x_i| \\ &\leq \frac{\varepsilon_1}{2}\eta + |h|\varepsilon_1^*\eta^* \leq \varepsilon_1\eta \end{aligned}$$

[Moreover, whenever $|\alpha' - \alpha^0|$ and $|\alpha'' - \alpha^0| = \eta$ we have

$$\begin{aligned} &|\zeta(\alpha'') - \zeta(\alpha') - \Sigma_{i=0,1,\ldots,\ell}(\alpha_i'' - \alpha_i')y_i| \\ &= |\varphi(\zeta^*(\alpha'')) - \varphi(\zeta^*(\alpha')) - \Sigma_{i=0,1,\ldots,\ell}(\alpha_i'' - \alpha_i')h(x_i)| \\ &\leq |(\varphi(\zeta^*(\alpha'')) - \varphi(\zeta^*(\alpha'))) - (h(\zeta^*(\alpha'')) - h(\zeta^*(\alpha')))| + \\ &\qquad |h||\zeta^*(\alpha'') - \zeta^*(\alpha') - \Sigma_{i=0,1,\ldots,\ell}(\alpha_i'' - \alpha_i')x_i| \\ &\leq \frac{\varepsilon_1}{2}|\alpha'' - \alpha'| + |h|\varepsilon_1^*|\alpha'' - \alpha'| \leq \varepsilon_1|\alpha'' - \alpha'| . \end{aligned}$$
]

This concludes the proof of Proposition 2.1.

Proposition 2.2. If $k \in Z$ , if $N \subset R^k$ is a [strong] simplicial approximation of order $(k + 1)$ to $L \subset R^k$ around $x_0 \in R^k$ and if $x_0 \in \text{int } N$ then $x_0 \in \text{int } L$ .

Proof of Proposition 2.2. If $x_0 \in \text{int } N$ , then there is an $\varepsilon \in (0,1]$ and a set $\{x_1,\ldots,x_{k+1}\} \subset N$ in general position such that $\text{co}\{x_1,\ldots,x_{k+1}\} \subset N$ and $x \in \text{co}\{x_1,\ldots,x_{k+1}\}$ whenever $|x - x_0| < \varepsilon$ . Let $\eta \in (0,1]$ and let $f$ be a continuous mapping from $S_\eta = \text{co}\{x_0 + \eta(x_i - x_0):i = 1,\ldots,k+1\}$ into $L$ such that

$$|f(x) - x| \leq \frac{\varepsilon}{2}\eta \quad \text{whenever} \quad x \in S_\eta$$

[and $|f(x'') - f(x') - (x'' - x')| \leq \varepsilon/2\,|x'' - x'|$ whenever $x'$ and $x'' \in S_\eta$] .

For every $x_*$ with $|x_* - x_0| \leq \varepsilon\eta/2$ let $\varphi_{x_*}$ be a function defined over $S_\eta$ by $\varphi_{x_*}(x) = x_* + x - f(x)$ . The function $\varphi_{x_*}$ is continuous and maps $S_\eta$ into itself since $|\varphi_{x_*}(x) - x_0| \leq |x_* - x_0| + |x - f(x)| \leq \varepsilon\eta/2 + \varepsilon\eta/2 = \varepsilon\eta$ . [Moreover for all $x'$ and $x'' \in S_\eta$ we have $|\varphi_{x_*}(x'') - \varphi_{x_*}(x')| \leq |f(x'')-f(x') - (x'' - x')| \leq \frac{\varepsilon}{2}|x' - x''|$] . Let $x_{**}$ be the fixed point of $\varphi_{x_*}$ . {The existence of $x_{**}$ is guaranteed by the Brouwer Fixed Point Theorem}. [The existence of $x_{**}$ is guaranteed by the Contraction Mapping Theorem since $\varepsilon/2 < 1$] . We have then $x_* + x_{**} - f(x_{**}) = x_{**}$ and hence

$f(x_{**}) = x_*$ . In other words we have proved that $x_* \in L$ whenever $|x_* - x_0| \leq \frac{\varepsilon\eta}{2}$ . This concludes the proof of Proposition 2.2.

Proposition 2.3. (Interior Mapping Theorem). If $k \in Z$ , if a subset $N$ of a normed linear space $X$ is a [strong] simplicial approximation of order $(k + 1)$ to a set $L \subset X$ around $x_0 \in X$ , if $\varphi$ is a function from $X$ into $R^k$ which is continuous in a neighborhood of $x_0$ , if $h$ is a continuous affine mapping from $X$ into $R^k$ such that $\varphi$ and $h$ are [strongly] tangent around $x_0$ , and if $h(x_0) \in \text{int } h(N)$ then $\varphi(x_0) \in \text{int } \varphi(L)$ .

Proof of Proposition 2.3. According to Proposition 2.1 we know that $h(N)$ is a [strong] simplicial approximation of order $(k + 1)$ to $\varphi(L)$ around $h(x_0) = \varphi(x_0)$ . From Proposition 2.2 it then follows that $\varphi(x_0) \in \text{int } \varphi(L)$ . This concludes the proof of Proposition 2.3.

Proposition 2.4. If $k \in Z$ , if a subset $N$ of a normed linear space $X$ is a [strong] simplicial approximation of order $k$ to $L \subset X$ around $x_0 \in X$ then $\overline{N}$ is a [strong] simplicial approximation of order $k$ to $L$ around $x_0$ .

Proof of Proposition 2.4. Let $\overline{\varepsilon_1}$ and $\overline{\varepsilon_2} > 0$ be given, let $\{\overline{x_1}, \ldots, \overline{x_\ell}\}$ be a set of $\ell \leq k$ elements in general position in

$\bar{N}$ and let $\overline{x_0} = x_0$ . For $i = 1,\dots,\ell$ let $x_i \in N$ be such that $|x_i - \overline{x_i}| \leq \frac{\overline{\varepsilon_1}}{2}$ and such that the elements $\{x_1,\dots,x_\ell\}$ are in general position. Let $\eta \in (0, \overline{\varepsilon_2}]$ and let $\zeta$ be a continuous function from $\Lambda_\ell^\eta$ into L such that

$$|\zeta(\alpha) - \Sigma_{i=0,1,\dots,\ell}\alpha_i x_i| \leq \frac{\overline{\varepsilon_1}}{2}\eta \quad \text{whenever} \quad \alpha \in \Lambda_\ell^\eta$$

$$[\text{and} \quad |\zeta(\alpha'') - \zeta(\alpha') - \Sigma_{i=0,1,\dots,\ell}(\alpha_i'' - \alpha_i')\, x_i| \leq \frac{\overline{\varepsilon_1}}{2}|\alpha''-\alpha'| \quad \text{whenever} \quad \alpha',\alpha'' \in \Lambda_\ell^\eta] .$$

We have then

$$|\zeta(\alpha) - \Sigma_{i=0,1,\dots,\ell}\alpha_i\overline{x_i}| \leq |\zeta(\alpha) - \Sigma_{i=0,1,\dots,\ell}\alpha_i x_i| + $$
$$+ |\Sigma_{i=1,\dots,\ell}\alpha_i(\overline{x_i} - x_i)| \leq \frac{\overline{\varepsilon_1}}{2}\eta + |\alpha - \alpha^0|\frac{\overline{\varepsilon_1}}{2} \leq \bar{\varepsilon}_1\eta$$
$$\text{whenever} \quad |\alpha - \alpha^0| = \eta \quad \text{i.e.} \quad \alpha \in \Lambda_\ell^\eta$$

$$\begin{bmatrix} \text{and} \quad |\zeta(\alpha'') - \zeta(\alpha') - \Sigma_{i=0,1,\dots,\ell}(\alpha_i'' - \alpha_i')\,\overline{x_i}| \\ \leq |\zeta(\alpha'') - \zeta(\alpha') - \Sigma_{i=0,1,\dots\ell}(\alpha_i'' - \alpha_i')\, x_i| + \\ + |\Sigma_{i=1,\dots,\ell}(\alpha_i'' - \alpha_i')(\overline{x_i} - x_i)| \leq \frac{\overline{\varepsilon_1}}{2}|\alpha'' - \alpha'| + \\ + |\alpha'' - \alpha'|\frac{\overline{\varepsilon_1}}{2} \leq \underline{\overline{\varepsilon_1}}|\alpha'' - \alpha'| \quad \text{whenever} \quad \alpha' \text{ and } \alpha'' \in \Lambda_\ell^\eta \end{bmatrix}$$

This concludes the proof of Proposition 2.4.

## §3. Interior Convex Approximation.

We shall say that a subset $N$ of a normed linear space $X$ is an interior convex approximation to a set $L \subset X$ around $x_0 \in X$ if (i) $N$ is open and convex, and (ii) for all $x \in N$ there exist numbers $\epsilon_1$ and $\epsilon_2 > 0$ such that for all $\eta \in (0, \epsilon_2]$ and all $y \in X$ with $|y - x| \leq \epsilon_1$ we have $x_0 + \eta(y - x_0) \in L$. If $X$ is a normed linear space, if $x_0 \in X$ and if $\varphi$ and $h$ are real-valued functions defined on $X$ then we shall say that $h$ is super tangent to $\varphi$ at $x_0$ if $h(x_0) = \varphi(x_0)$ and if for all $\epsilon > 0$ there exists a $\delta > 0$ such that $\varphi(x) - h(x) \leq \epsilon|x - x_0|$ whenever $|x - x_0| \leq \delta$.

Proposition 3.1. If $X$ is a normed linear space, if $x_0 \in X$ and if $\varphi$ and $h$ are real-valued functions on $X$ such that

(i) $h$ is continuous and convex on $X$

(ii) $h$ is super tangent to $\varphi$ at $x_0$

then the set $\{x : x \in X, h(x) < h(x_0)\}$ is an interior convex approximation to the set $\{x : x \in X, \varphi(x) < \varphi(x_0)\}$ around $x_0$.

Proof of Proposition 3.1. There is no loss of generality by assuming that $\varphi(x_0) = h(x_0) = 0$. Let $\Omega = \{x : x \in X, h(x) < 0\}$ and let $S = \{x : x \in X, \varphi(x) < 0\}$. If $\Omega = \Phi$ the Proposition 3.1 is trivially true. We shall thus assume that $\Omega \neq \Phi$. Let $x \in \Omega$ and let $4\delta = -h(x)$. We have then $\delta > 0$. Since the function $h$ is continuous then there exists an $\epsilon_1 > 0$ such that $h(y) \leq -2\delta$

for all $y \in X$ with $|y - x| \leq \varepsilon_1$. Since the function $h$ is convex and since $h(x_0) = 0$ it follows that $h(x_0 + \eta(y - x_0)) \leq$ $\leq - 2\delta\eta$ for all $\eta \in [0,1]$ and all $y \in X$ with $|y - x| \leq \varepsilon_1$. Since the function $h$ is super tangent to $\varphi$ at $x_0$ then there exist an $\varepsilon_2 > 0$ such that $\varphi(x_0 + \eta(y - x_0)) - h(x_0 + \eta(y - x_0)) \leq$ $\leq \delta\eta$ whenever $\eta \in [0, \varepsilon_2]$ and $y \in X$ with $|y - x| \leq \varepsilon_1$. For all $\eta \in (0, \varepsilon_2]$ and $y \in X$ with $|y - x| \leq \varepsilon_1$ we have then $\varphi(x_0 + \eta(y - x_0) \leq \delta\eta + h(x_0 + \eta(y - x_0)) \leq \delta\eta - 2\delta\eta \leq - \delta\eta < 0$ and hece $x_0 + \eta(y - x_0) \in S$. This concludes the proof of Proposition 3.1.

<u>Proposition 3.2.</u> If a subset $N_1$ of a normed linear space $X$ is an interior convex approximation to $L_1 \subset X$ around $x_0 \in X$, if $k \in Z$ and if $N_2 \subset X$ is a [strong] simplicial approximation of order $k$ to $L_2 \subset X$ around $x_0$ then $N_1 \cap N_2$ is a [strong] simplicial approximation of order $k$ to $L_1 \cap L_2$ around $x_0$.

<u>Proof of Proposition 3.2.</u> Let $\varepsilon_1$ and $\varepsilon_2 > 0$ and let $\{x_1,\ldots,x_\ell\}$ be a set of $\ell \leq k$ elements in general psotion in $N_1 \cap N_2$. We must show the existence of an $\eta \in (0, \varepsilon_2]$ and of a continuous function $f$ from $S_\eta = co\{x_0 + \eta(x_i - x_0) : i = 1,\ldots,\ell\}$ into $L_1 \cap L_2$ such that $|f(x) - x| \leq \varepsilon_1\eta$ whenever $x \in S_\eta$

[and $|f(x'') - f(x') - (x'' - x')| \leq \varepsilon_1 |x''-x'|$ whenever

$x'$ and $x'' \in S_\eta$].

Since $\mathrm{co}\{x_1,\ldots,x_\ell\}$ is a <u>compact</u> subset of the set $N_1$ and since $N_1$ is an interior convex approximation to the set $L_1$ around $x_0$ we know that there exist numbers $\overline{\varepsilon_1} \in (0, \varepsilon_1]$ and $\overline{\varepsilon_2} \in (0, \varepsilon_2]$ such that $x_0 + \eta(x - x_0) \in L_1$ whenever $\eta \in (0, \overline{\varepsilon_2}]$ and $|x - x^*| \leq \overline{\varepsilon_1}$ for some $x^* \in \mathrm{co}\{x_1,\ldots,x_\ell\}$, i.e. such that $y \in L_1$ whenever $|y - y^*| \leq \overline{\varepsilon_1}\eta$ for some $\eta \in (0, \overline{\varepsilon_2}]$ and some $y^* \in S_\eta$. Since $N_2$ is a [strong] simplicial approximation of order $k$ to $L_2$ around $x_0$ we know that there exist an $\bar{\eta} \in (0, \overline{\varepsilon_2}]$ and a function $\bar{f}$ from $S_{\bar{\eta}}$ into $L_2$ such that

$$|\bar{f}(x) - x| \leq \bar{\varepsilon}_1\bar{\eta} \quad \text{whenever} \quad x \in S_{\bar{\eta}}$$

[and $|\bar{f}(x'') - \bar{f}(x') - (x'' - x')| \leq \bar{\varepsilon}_1|x'' - x'|$ whenever $x', x'' \in S_{\bar{\eta}}$] .

Since $\bar{\eta} \in (0, \overline{\varepsilon_2}]$, then for all $x \in S_\eta$ we have $\bar{f}(x) \in L_1$ and hence $\bar{f}(x) \in L_1 \cap L_2$. We conclude the proof of Proposition 3.2 by letting $f = \bar{f}$ and $\eta = \bar{\eta}$.

<u>Proposition 3.3</u>. If $k \in Z$, if a subset $N$ of a normed linear space $X$ is a [strong] simplicial approximation of order $k$ to a set $L \subset X$ around $x_0 \in X$, and if $\varphi$ and $h$ are functions on $X$ such that (i) $h$ is continuous and convex and (ii) $h$ is super tangent to $\varphi$ at $x_0$, then $\tilde{N} = \{x : x \in N, h(x) < h(x_0)\}$ is a [strong] simplicial approximation of order $k$ to the set $\tilde{L} = \{x : x \in L, \varphi(x) < \varphi(x_0)\}$ around $x_0$.

Proof of Proposition 3.3. Let $\Omega = \{x : x \in X, h(x) < h(x_0)\}$ and let $S = \{x : x \in X, \varphi(x) < \varphi(x_0)\}$. According to Proposition 3.1 we know that $\Omega$ is an interior convex approximation to $S$ around $x_0$. According to Proposition 3.2 we know that $\tilde{N} = \Omega \cap N$ is a [strong] simplicial approximation of order $k$ to the set $\tilde{L} = S \cap L$ around $x_0$. This concludes the proof of Proposition 3.3.

## §4. Corollaries to the Hahn-Banach Theorem

The results of this section shall be used only in proving the second half of the multiplier rule.

Proposition 4.1. (One version of the Hahn-Banach Theorem). If $\Omega$ is a nonempty convex subset of a normed linear space $X$, if $f_1$ is a continuous concave function on $X$, if $f_2$ is a convex function on $\Omega$ such that $f_1(x) \leq f_2(x)$ for all $x \in \Omega$, then there exists a continuous affine function $\omega$ on $X$ such that $f_1(x) \leq \omega(x)$ for all $x \in X$ and $\omega(x) \leq f_2(x)$ for all $x \in \Omega$.

Proof of Proposition 4.1. The sets $K_1 = \{(x,t) : x \in X, t < f_1(x)\}$ and $K_2 = \{(x,t) : x \in \Omega, t > f_2(x)\}$ are disjoint nonempty convex sets in $X \times R$. Moreover the set $K_1$ is open since the function $f_1$ is continuous. According to the (geometric version of the) Hahn-Banach Theorem there exists a nonconstant continuous affine functional $\gamma$ on $X \times R$ such that $\gamma(x,t) < 0$ for all

$(x,t) \in K_1$ and $\gamma(x,t) \geq 0$ for all $(x,t) \in K_2$ . The functional $\gamma$ is of the form $\gamma(x,t) = \alpha(x) + \beta t + \delta$ where $\alpha$ is a continuous linear functional on $X$ and where $\beta$ and $\delta$ are real numbers. Since $\gamma$ is nonconstant on $X \times R$ we know that $\alpha$ and $\beta$ cannot be both zero. We must have $\beta \neq 0$ since otherwise we would have $\alpha(x) + \delta < 0$ for all $x \in X$ and $\alpha(x) + \delta \geq 0$ for all $x \in \Omega$ which cannot be since $\Omega$ is nonempty. Moreover we must have $\beta \geq 0$ (and hence $\beta > 0$) since for a fixed $x$ the set $K_1$ contains elements of the form $(x,t)$ for arbitrarily large negative $t$ . For every $x \in X$ we have then $\alpha(x) + \beta t + \delta \leq 0$ for all $t < f_1(x)$ , which implies $\alpha(x) + \beta\, f_1(x) + \delta \leq 0$ and hence $f_1(x) \leq -\frac{1}{\beta}(\alpha(x) + \delta)$ . Similarly for every $x \in \Omega$ we have $\alpha(x) + \beta t + \delta \geq 0$ for all $t > f_2(x)$ , which implies $\alpha(x) + \beta f_2(x) + \delta \geq 0$ and hence $f_2(x) \geq -\frac{1}{\beta}(\alpha(x) + \delta)$. We conclude the proof of Proposition 4.1 by defining the affine function $\omega$ by the relation $\omega(x) = -\frac{1}{\beta}(\alpha(x) + \delta)$ for all $x \in X$ .

<u>Proposition 4.2</u>. If $g_1,\dots,g_k$ are given continuous concave functions on a normed linear space $X$ and if $\ell$ is a given continuous affine function on $X$ such that $\Sigma_{i=1,\dots,k} g_i(x) \leq$ $\leq \ell(x)$ for all $x \in X$ then there exist continuous affine functions $\ell_1,\dots,\ell_k$ on $X$ such that $\ell(x) = \Sigma_{i=1,\dots,k}\ell_i(x)$ whenever $x \in X$ and $g_i(x) \leq \ell_i(x)$ whenever $i \in \{1,\dots,k\}$ and $x \in X$ . Moreover if for some $x_0 \in X$ we have $\ell(x_0) = \Sigma_{i=1,\dots,k} g_i(x_0)$ then $\ell_i(x_0) = g_i(x_0)$ for all $i \in \{1,\dots,k\}$ .

Proof of Proposition 4.2. Proposition 4.2 is trivial when $k = 1$. If $k > 1$ it is sufficient to prove Proposition 4.2 in the case $k = 2$. (Indeed since the sum of two continuous concave functions is a continuous concave function the general proof for $k > 1$ is obtained by repeating $k - 1$ times the proof for $k = 2$). We let $\Omega = X$, $f_1 = g_1$ and $f_2 = \ell - g_2$ and we apply Proposition 4.1. We then conclude the proof of Proposition 4.2 by letting $\ell_1 = \omega$ and $\ell_2 = \ell - \omega$.

## §5. Proof of the Multiplier Rule.

In this section we return to the optimization problem introduced in §1. We shall now give a list of assumptions:

H1. The set $N$ is a [strong] simplicial approximation of order $(m + 1)$ to $L$ around $x_0$, the set $\bar{N}$ is convex and $x_0 \in \bar{N}$.

H2. The function $\varphi_i$ is continuous for $i = 1,\dots,m$ in a neighborhood of $x_0$.

H3. The function $h$ is continuous over $X$.

H4. For every $i = -\mu,\dots,0$ the function $h_i$ is convex and super tangent to the function $\varphi_i$ at $x_0$.

H5. For every $i = 1,\dots,m$ the function $h_i$ is affine and [strongly] tangent to the function $\varphi_i$ at $x_0$.

Our proof of the Multiplier Rule (see Statement in §1) will be given in two steps. In the first step, which we call the proof of the Convex Multiplier Rule, we shall prove the first half of the Multiplier Rule (up to and including Relation (iii)). In the second step, which we call the prooof of the Affine Support Theorem, we shall prove the second half of the Multiplier Rule. The proof of the Convex Multiplier Rule will depend on the results of Sections 2 and 3 whereas the proof of the Affine Support Theorem will depend on the results of Section 4.

Proof of Convex Multiplier Rule. In this proof we assume that $m > 0$ . In a later remark we shall give the modifications and simplifications of that proof which are applicable to the case $m = 0$ . According to Proposition 2.4 we know that $\bar{N}$ is a [strong] simplicial approximation of order $(m + 1)$ to $L$ around $x_0$ . Since $\bar{N}$ is convex it follows that $\mathrm{co}N \subset \bar{N}$ , and hence that $\mathrm{co}N$ is a [strong] simplicial approximation of order $(m + 1)$ to $L$ around $x_0$ . We let $M = \mathrm{co}N$ . The set $M$ is convex and $x_0 \in \bar{M}$ . We may assume without loss of generality that $\varphi_0(x_0) = 0$ . Let $\tilde{M} = \{x : x \in M, h_i(x) < 0 \text{ for } i = -\mu,\ldots,0\}$ and let $\tilde{L} = \{x : x \in L, \varphi_i(x) < 0 \text{ for } i = -\mu,\ldots,0\}$ . From Proposition 3.3 we know that $\tilde{M}$ is a [strong] simplicial approximation of order $(m + 1)$ to $\tilde{L}$ around $x_0$ . Let $h^+ = (h_1,\ldots,h_m)$ and let $\varphi^+ = (\varphi_1,\ldots,\varphi_m)$ . We have then $h^+(x_0) = \varphi^+(x_0) = 0$ and $0 \notin \mathrm{int}\, h^+(\tilde{M})$ since otherwise Proposition 2.3 would give $0 \in \mathrm{int}\, \varphi^+(\tilde{L})$ and a fortiori $0 \in \varphi^+(\tilde{L})$ which contradicts the optimality of $x_0$ . From the fact that $h^+(\tilde{M})$ is convex and that $0$ belongs to the boundary of $h^+(\tilde{M})$ we know that there exists a nonzero $\lambda^+ = (\lambda_1^+, \ldots, \lambda_m^+)$ such that $0 \leq \Sigma_{i=1,\ldots,m} \lambda_i^+ h_i(x)$ for all $x \in \tilde{M}$ . Let $h^\Delta = (h_{-\mu}, \ldots, h_{-1}, h_0, \Sigma_{i=1,\ldots,m} \lambda_i^+ h_i)$ and $K^\Delta = \{(\alpha_{-\mu}, \ldots, \alpha_0, \alpha_{+1}) : \alpha_i < 0 \text{ for } i = -\mu, \ldots, +1\}$ . The sets $h^\Delta(M)$ and $K^\Delta$ are disjoint, hence the set $h^\Delta(M) - K^\Delta$ and $K^\Delta$ are disjoint (since $K^\Delta + K^\Delta = K^\Delta$) , and hence the sets $H^\Delta(M) - K^\Delta$ and $K^\Delta$ are separated (since $h^\Delta(M) - K^\Delta$ and $K^\Delta$ are convex), and hence the sets $H^\Delta(M)$ and $K^\Delta$ are separated (since $0 \in \overline{K^\Delta}$) . We have moreover $h^\Delta(x_0) \in \overline{h^\Delta(M)} \cap \overline{K^\Delta}$ and hence $\overline{K^\Delta}$ and $h^\Delta(M)$ are separated by some hyperplane passing through $h^\Delta(x_0)$ , i.e. there exists a nonzero $\lambda^\Delta = (\lambda_{-\mu}^\Delta, \ldots, \lambda_{+1}^\Delta)$ such that

$$\Sigma_{i=-\mu,\ldots,+1} \lambda_i^\Delta h_i^\Delta(x) \leq \Sigma_{i=-\mu,\ldots,+1} \lambda_i^\Delta h_i^\Delta(x_0) \leq \Sigma_{i=-\mu,\ldots,+1} \lambda_i^\Delta \alpha_i^\Delta$$

for all $x \in M$ and for all $\alpha \in \overline{K^\Delta}$ .

If we let $K = \{(\alpha_{-\mu}, \ldots, \alpha_m) : \alpha_i < 0 \text{ for } i = -\mu, \ldots, 0$ and $\alpha_i = 0$ for $i = 1, \ldots, m\}$ then we can state that $h(M)$ and $K$ are separated by some hyperplane passing through $h(x_0)$ . Indeed if we let $\lambda = (\lambda_{-\mu}, \ldots, \lambda_m)$ where $\lambda_i = \lambda_i^\Delta$ for $i = -\mu, \ldots, 0$ and $\lambda_i = \lambda_{+1}^\Delta \lambda_i^+$ for $i = 1, \ldots, m$ then we have $\lambda \neq 0$ and

$$\Sigma_{i=-\mu,\ldots,+m} \lambda_i h_i(x) \leq \Sigma_{i=-\mu,\ldots,+m} \lambda_i h_i(x_0) \leq \Sigma_{i=-\mu,\ldots,m} \lambda_i \alpha_i$$

for all $x \in M$ and for all $\alpha \in \bar{K}$ .

The vector $\lambda \neq 0$ satisfies conditions (i), (ii) and (iii) of the Convex Multiplier Rule as we show now. Condition (i) is a part of the above inequality. We must have Condition (ii) since otherwise (in the case of $\lambda_j > 0$ for some $j \in \{-\mu,\dots,0\}$) the expression $\Sigma_{i=-\mu,\dots,m}\lambda_i\alpha_i$ could assume arbitrarily large negative values (by taking large negative values of $\alpha_j$) . We must have Condition (iii): we know already that $\lambda_i\varphi_i(x_0) \geq 0$ for all $i \in \{-\mu,\dots,-1\}$ since $\lambda_i \leq 0$ and $\varphi_i(x_0) \leq 0$ for all $i \in \{-\mu,\dots,-1\}$ ; we cannot have $\lambda_j\varphi_j(x_0) > 0$ for some $j \in \{-\mu,\dots,-1\}$ since this would imply $\lambda_j < 0$ and $\varphi_j(x_0) < 0$ and for $\alpha^* \in \bar{K}$ defined by $\alpha_i^* = \varphi_i(x_0)$ for $i \neq j$ and $\alpha_j^* = \frac{1}{2}\varphi_j(x_0)$ we would then have the contradiction

$$\Sigma_{i=-\mu,\dots,m}\lambda_i\varphi_i(x_0) > \Sigma_{i=-\mu,\dots,m}\lambda_i\alpha_i^* \quad .$$

This concludes the proof of the Convex Multiplier Rule in the case $m > 0$ .

Remark for the case $m = 0$. If $\tilde{M} \neq \Phi$ then $\tilde{L} \neq \Phi$ which contradicts the optimality of $x_0$ . If $\tilde{M} = \Phi$ then $h(M)$ and $K$ are disjoint, and hence $h(M) - K$ and $K$ are disjoint (since $K + K = K$) , and hence $h(M) - K$ and $K$ are separated (since $h(M) - K$ and $K$ are convex), and hence $h(M)$ and $K$ are separated (since $0 \in \bar{K}$ ) .

Proof of the Affine Support Theorem

The function $f = \Sigma_{i=-\mu,\dots,m}\lambda_i h_i$ is concave and continuous on $X$ and we have $f(x) \leq f(x_0)$ for all $x \in N$. According to Proposition 4.1 there exists a continuous affine function $\omega$ on $X$ such that $\omega(x_0) = f(x_0), \omega(x) \leq f(x_0)$ for all $x \in N$ and $f(x) \leq \omega(x)$ for all $x \in X$. Let $f^- = \Sigma_{i=-\mu,\dots,0}\lambda_i h_i$ and let $\omega^- = \omega - \Sigma_{i=1,\dots,m}\lambda_i h_i$. We have then $\omega^-(x_0) = f^-(x_0), \omega^-(x) \leq$ $\leq f^-(x_0)$ for all $x \in N$ and $f^-(x) \leq \omega^-(x)$ for $x \in X$. The function $\lambda_i h_i$ is concave continuous for every $i \in \{-\mu,\dots,0\}$ and the function $\omega^-$ is affine continuous. According to Proposition 4.2 there thus exist continuous affine functions $\omega_{-\mu},\dots,\omega_0$ on $X$ such that $\omega^- = \Sigma_{i=-\mu,\dots,0}\omega_i, \lambda_i h_i(x_0) = \omega_i(x_0)$ whenever $i \in \{-\mu,\dots,0\}$ and $\lambda_i h_i(x) \leq \omega_i(x)$ whenever $i \in \{-\mu,\dots,0\}$ and $x \in X$. For every $i \in \{-\mu,\dots,0\}$ such that $\lambda_i = 0$ we have then $\omega_i \equiv 0$ and we let $\ell_i$ be a continuous affine function such that $\ell_i(x_0) = h_i(x_0)$ and $\ell_i(x) \leq h_i(x)$ for all $x \in X$, (the existence of such function $\ell_i$ is given by Proposition 4.1). For every $i \in \{-\mu,\dots,0\}$ such that $\lambda_i \neq 0$ (and hence $\lambda_i < 0$) we let $\ell_i$ be defined by $\ell_i(x) = \omega_i(x)/\lambda_i$ for every $x \in X$; we have then $\ell_i(x) \leq h_i(x)$ for all $x \in X$. For every $i \in \{1,\dots,m\}$ we let $\ell_i = h_i$. For all $x \in N$ we have

$$\lambda\cdot\ell(x) = \Sigma_{i=-\mu,\dots,0}\omega_i(x) + \Sigma_{i=1,\dots,m}\lambda_i h_i(x) = \omega(x) \leq f(x_0) =$$

$$= \Sigma_{i=-\mu,\dots,m}\lambda_i h_i(x_0) = \lambda\cdot\ell(x_0).$$ This concludes the proof of the Affine Support Theorem.

## §6. Maximum Principle for Systems Described by Nonlinear Differential Equations.

Let $X$ be the linear space of all absolutely continuous functions from $[0,1]$ into $R^n$. We shall endow $X$ with the norm $|x| = \max_{t\in[0,1]}|x(t)|$. Let $F$ be a given set of functions $f$ from $R^n x[0,1]$ into $R^n$ such that for any $x \in X$ the functions $f(x(t),t)$ is measurable in $t$ over $[0,1]$. If $f \in F$ a solution of $f$ will be an element $x \in X$ such that $\dot{x}(t) = f(x(t),t)$ for a.e.$t \in [0,1]$. We shall denote by $L$ the set of all $x \in X$ such that $x$ is a solution of some $f \in F$. We are given a function $\varphi = (\varphi_{-\mu},\ldots,\varphi_{-1},\varphi_0,\varphi_1,\ldots,\varphi_m)$ from $X$ into $R^{\mu+m+1}$ and the optimal control problem is to find an element $x_0 \in L$ such that $\varphi_0(x_0) \leq \varphi_0(x)$ for all $x \in L$ satisfying the constraints $\varphi_i(x) \leq 0$ for $i = -\mu,\ldots,-1$ and $\varphi_i(x) = 0$ for $i = 1,\ldots,m$. An element $x_0 \in L$ satisfying the above conditions is called an optimal solution of the given optimal control problem. We shall assume that such optimal solution $x_0$ exists and we shall denote by $f_0$ an element of $F$ such that $x_0$ is a solution of $f_0$.

Before starting the assumptions on the data of the optimal control problem we given a few definitions. A function $g$ from $[0,1]$ into $R^m$ is piecewise continuous if there exists a finite set $0 = \tau_0 < \tau_1 < \ldots < \tau_k = 1$ and for each $i \in \{1,\ldots,k\}$ a continuous function $g_i$ on $[\tau_{i-1},\tau_i]$ such that $g(t) = g_i(t)$ whenever $t \in (\tau_{i-1},\tau_i)$. A function $g$ from $[0,1]$ into $R^m$ is quasi piecewise continuous if there exists a finite set

$0 = \tau_0 < \tau_1 < \ldots < \tau_k = 1$ such that $g$ is continuous on $(\tau_{i-1}, \tau_i)$ for every $i \in \{1,\ldots,k\}$. A function $g$ from $R^k$ into $R^\ell$ is <u>[strongly] differentiable at $y_0 \in R^k$</u> if there exists a linear mapping $A$ from $R^k$ into $R^\ell$ such that for all $\varepsilon > 0$ there is a $\delta > 0$ with

$$|g(y_0) + A(y-y_0) - g(y)| \leq \varepsilon |y-y_0| \quad \text{whenever} \quad |y-y_0| \leq \delta$$

$$\left[\begin{array}{l}\text{and even} \quad |g(y_1) + A(y_2 - y_1) - g(y_2)| \leq \varepsilon |y_2 - y_1| \\ \text{whenever} \quad |y_1 - y_0| \text{ and } |y_2 - y_0| \leq \delta\end{array}\right].$$

<u>Assumptions on the dynamics (i.e. on $F$ and $f_0$ around $x_0$)</u>.

H1. For every $f \in F$, [the function $f(x_0(t),t)$ is quasi piecewise continuous in $t$ over $[0,1]$ and ] there exists an $\eta > 0$ and a real-valued integrable [quasi piecewise continuous] function $q$ on $[0,1]$ such that

(i) $|f(z,t)| \leq q(t)$ whenever $t \in [0,1]$ and $|z-x_0(t)| \leq \eta$

(ii) $|f(z_2,t)-f(z_1,t)| \leq q(t)|z_2-z_1|$ whenever $t \in [0,1]$ and $|z_1-x_0(t)|$, $|z_2-x_0(t)| \leq \eta$.

H2. The set $F$ is "convex-under-switching" i.e. for all $f_1, f_2 \in F$ and all $\tau \in (0,1)$ the set $F$ also contains the function $f$ defined by

$$\begin{aligned} f(z,t) &= f_1(z,t) \quad \text{if} \quad t \in [0,\tau) \\ &= f_2(z,t) \quad \text{if} \quad t \in [\tau,1] \end{aligned}$$

H3. There exists an integrable function $D$ from $[0,1]$ into $R^{n\times n}$ such that for any $\varepsilon > 0$ there exists a $\delta > 0$ with

$$\int_0^1 |f_0(x_0(t),t) + D(t)(x(t)-x_0(t))-f_0(x(t),t)|dt \leq \varepsilon|x-x_0|$$

whenever $x \in X$ and $|x - x_0| \leq \delta$

[and even

$$\int_0^1 |f_0(x_1(t),t) + D(t)(x_2(t)-x_1(t))-f_0(x_2(t),t)|dt \leq \varepsilon|x_2-x_1|$$

whenever $x_1$ and $x_2 \in X$ with $|x_1-x_0|$ and $|x_2-x_0| \leq \delta$].

<u>Assumption on</u> $\varphi$ .

There is a function $g$ from $R^{n\times n}$ into $R^{\mu+m+1}$ such that $\varphi(x) = g(x(0),x(1))$ for all $x \in X$ . For every $i \in \{-\mu,\ldots,0\}$ we shall assume that $g_i$ is differentiable at $(x_0(0),x_0(1))$ . For every $i \in \{1,\ldots,m\}$ we shall assume that $g_i$ is continuous in a neighborhood of $(x_0(0), x_0(1))$ and [strongly] differentiable at $(x_0(0),x_0(1))$ . For every $i \in \{-\mu,\ldots,m\}$ we shall denote by $a_i$ the gradient of $g_i(u,v)$ with respect to $u$ at the point $(x_0(0),x_0(1))$ and by $b_i$ the gradient of $g_i(u,v)$ with respect to $v$ at the point $(x_0(0), x_0(1))$ .

<u>Maximum Principle.</u> There exist a nonzero vector $\lambda = (\lambda_{-\mu},\ldots,\lambda_m)$ in $R^{\mu+m+1}$ and an absolutely continuous function $p$ from $[0,1]$ into $R^n$ such that

(1) $\int_0^1 p(t)\cdot f_0(x_0(t),t)dt \geq \int_0^1 p(t)\cdot f(x_0(t),t)dt$ for all $f \in F$

(2) $p(0) = -\Sigma_{i=-\mu,\ldots,m}\lambda_i a_i$

(3) $p(1) = \Sigma_{i=-\mu,\ldots,m}\lambda_i b_i$

(4) $\dot{p}(t) = -D^T(t)p(t)$ for a.e. $t \in [0,1]$

(5) $\lambda_i \leq 0$ for $i = -\mu,\ldots,0$

(6) $\lambda_i g_i(x_0(0),\ x_0(1)) = 0$ for $i = -\mu,\ldots,-1$

Proof of the Maximum Principle

Let $N$ be the set of all elements $y \in X$ such that for some $f \in F$ we have $\dot{y}(t) = f(x_0(t),t) + D(t)(y(t) - x_0(t))$ for a.e. $t \in [0,1]$. We see immediately that $x_0 \in L$ and that $x_0 \in N$. To simplify the notation we shall sometimes use the symbol $y_0$ instead of $x_0$. In a later section we shall prove that $\bar{N}$ is convex and that for any positive integer $k$ the set $N$ is a [strong] simplicial approximation of order $k$ to the set $L$ around $x_0$.

From the Multiplier Rule we know that there exists a nonzero vector $\lambda = (\lambda_{-\mu},\ldots,\lambda_m)$ such that

(i) $\Sigma_{i=-\mu,\ldots,m}\lambda_i(a_i\cdot y_0(0) + b_i\cdot y_0(1)) \geq$

$\geq \Sigma_{i=-\mu,\ldots,m}\lambda_i(a_i\cdot y(0) + b_i\cdot y(1))$ for all $y \in N$

(ii) $\lambda_i \leq 0$ for $i = -\mu,\ldots,0$

(iii) $\lambda_i g_i(x_0(0),x_0(1)) = 0$ for $i = -\mu,\ldots,-1$

Let $I$ be the identity matrix in $R^{n\times n}$ and let $G : [0,1] \to R^{n\times n}$ be the unique absolutely continuous solution of the linear matrix differential system

$$G(0) = I$$
$$\dot{G}(t) = D(t)G(t) \quad \text{for a.e.} \quad t \in [0,1]$$

we know that $G^{-1}(t)$ exists for every $t \in [0,1]$ and that $G^{-1}$ is an absolutely continuous solution of the linear matrix differential system

$$G^{-1}(0) = I$$
$$\frac{d}{dt}(G^{-1}(t)) = -G^{-1}(t)D(t) \quad \text{for a.e.} \quad t \in [0,1]$$

Then for all $\xi \in R^n$ and all $f \in F$ the set $N$ contains the element $y_{\xi,f}$ defined by

$$y_{\xi,f}(t) = x_0(t) + G(t)(\xi + \int_0^t G^{-1}(\tau)(f(x_0(\tau),\tau)-f_0(x_0(\tau),\tau))d\tau)$$

for all $t \in [0,1]$ .

For all $\xi \in R^n$ we have then

(a) $y_{\xi,f_0}(0) - y_0(0) = \xi$ and $y_{\xi,f_0}(1) - y_0(1) = G(1)\xi$

And for all $f \in F$ we have

$$\text{(b)} \quad \begin{cases} y_{0,f}(0) - y_0(0) = 0 \quad \text{and} \\ \\ y_{0,f}(1)-y_0(1) = G(1) \int_0^1 G^{-1}(t)(f(x_0(t),t)-f_0(x_0(t),t))dt \end{cases}$$

From (i) and (a) we obtain

$$\Sigma_{i=-\mu,\ldots,m}\lambda_i(a_i.\xi + b_i.G(1)\xi) \leq 0 \quad \text{for all} \quad \xi \in R^n$$

and hence

$$\Sigma_{i=-\mu,\ldots,m}\lambda_i a_i = - \Sigma_{i=-\mu,\ldots,m}\lambda_i G^T(1)b_i$$

From (i) and (b) we obtain

$$\Sigma_{i=-\mu,\ldots,m}\lambda_i(b_i.G(1)\int_0^1 G^{-1}(t)(f(x_0(t),t) - f_0(x_0(t),t))dt) \leq 0$$

$$\text{for all} \quad f \in F .$$

and hence

$$\int_0^1 (G^{-1}(t))^T G^T(1)\Sigma_{i=-\mu,\ldots,m}\lambda_i b_i.(f(x_0(t),t)-f_0(x_0(t),t)dt \leq 0$$

$$\text{for all} \quad f \in F .$$

Let $p$ be a function from $[0,1]$ into $R^n$ defined by

$$p(t) = (G^{-1}(t))^T G^T(1)\ \Sigma_{i=-\mu,\ldots,m}\lambda_i b_i \quad \text{for all} \quad t \in [0,1]$$

The function $p$ is then absolutely continuous and we have

$$\dot{p}(t) = -D^T(t)p(t) \quad \text{for a.e.} \quad t \in [0,1]$$

$$p(0) = G^T(1)\ \Sigma_{i=-\mu,\ldots,m}\lambda_i b_i = - \Sigma_{i=-\mu,\ldots,m}\lambda_i a_i$$

$$p(1) = \Sigma_{i=-\mu,\ldots,m}\lambda_i b_i$$

and

$$\int_0^1 p(t).f_0(x_0(t),t)dt \geq \int_0^1 p(t)f(x_0(t),t)dt \quad \text{for all} \quad f \in F .$$

This concludes the proof of the Maximum Principle.

## §7. Variational Set of Ordinary Differential Equations

The data and assumptions of the present section are the same as in the preceding section. In the present section, however, we shall consider only the dynamics of the control system (i.e. $F$, $f_0$, $x_0$, $L$ and $N$) and not the constraints or objective functions (i.e. $\varphi$, $a_i$, $b_i$) . Let $Z$ be the set of positive integers. We recall that for every $\ell \in Z$ the set $\Lambda_\ell$ is the subset of $R^{\ell+1}$ defined by $\Lambda_\ell = \{\alpha = (\alpha_0,\alpha_1,\ldots,\alpha_\ell) : \alpha_i \geq 0$ for $i = 0,1,\ldots,\ell$ and $\Sigma_{i=0,1,\ldots,\ell}\alpha_i = 1\}$ . The element $(1,0,\ldots,0)$ of $\Lambda_\ell$ is denoted by $\alpha^0$ . On $\Lambda_\ell$ we use the norm $|\alpha'' - \alpha'| = \Sigma_{i=0,1,\ldots,\ell}|\alpha''_i - \alpha'_i|$ and for every $\eta \in [0,2]$ we let $\Lambda_\ell^\eta = \{\alpha : \alpha \in \Lambda_\ell, |\alpha - \alpha^0| = \eta\}$ and $\Lambda_\ell^{\leq\eta} = \{\alpha : \alpha \in \Lambda_\ell, |\alpha - \alpha^0| \leq \eta\}$ .

In this section we shall prove that $\bar{N}$ is convex and that for every $k \in Z$ the set $N$ is a [strong] simplicial approximation of order $k$ to $L$ around $x_0$ , i.e. we shall prove the following result:

Proposition 7.1. The set $\bar{N}$ is convex and for any $\epsilon_1, \epsilon_2 > 0$ any set $\{y_1,\ldots,y_\ell\}$ of elements in general position in $N$ there exist an $\eta \in (0, \epsilon_2]$ and a continuous function $\zeta$ from $\Lambda_\ell^\eta$ into $L$ such that

$$|\zeta(\alpha) - \Sigma_{i=0,1,\ldots,\ell}\alpha_i y_i| \leq \epsilon_1 \eta \quad \text{for all} \quad \alpha \in \Lambda_\ell^\eta$$

[ and

$$|\zeta(\alpha'') - \zeta(\alpha') - \Sigma_{i=0,1,\ldots,\ell}(\alpha_i'' - \alpha_i')y_i| \leq \epsilon_1 |\alpha'' - \alpha'|$$

for all $\alpha', \alpha'' \in \Lambda_\ell^\eta$ . ]

Proof of Proposition 7.1. For every $i \in \{1,\ldots,\ell\}$ let $f_i \in F$ be such that $\dot{y}_i(t) = f_i(x_0(t),t) + D(t)(y_i(t) - x_0(t))$ for a.e. $t \in [0,1]$ . We may assume without loss of generality (to be justified in a remark given later) that [for every $i \in \{0,1,\ldots,\ell\}$ the function $f_i(\dot{x}_0(t),t)$ is piecewise continuous in $t$ over $[0,1]$ and that] there exist $\rho > 0$ and $\sigma < +\infty$ such that for every $i \in \{0,1,\ldots,\ell\}$ we have

(i) $|f_i(z,t)| \leq \sigma$ whenever $t \in [0,1]$ and $|z - x_0(t)| \leq \rho$

(ii) $|f_i(z_2,t) - f_i(z_1,t)| \leq \sigma|z_2 - z_1|$ whenever $t \in [0,1]$ and $|z_1 - x_0(t)|, |z_2 - x_0(t)| \leq \rho$

For all $\alpha \in \Lambda_\ell$ let $\xi_\alpha = \Sigma_{i=0,1,\ldots,\ell}\alpha_i y_i(0)$ and for all $k \in Z$ and all $\alpha \in \Lambda_\ell$ let $f_{\alpha,k}$ be the function from

$R^n \times [0,1]$ into $R^n$ defined by $f_{\alpha,k}(z,t) = f_i(z,t)$

$$\text{if } t \in \cup_{j=1,\dots,k}\left[\frac{j-1}{k} + \frac{1}{k}\Sigma_{r=0,k,\dots,i-1}\alpha_r, \frac{j-1}{k} + + \frac{1}{k}\Sigma_{r=0,1,\dots,i}\alpha_r\right)$$

and

$$f_{\alpha,k}(z,1) = f_\ell(z,t) .$$

Since the set $F$ is convex-under-switching we have $f_{\alpha,k} \in F$ for all $k \in Z$ and all $\alpha \in \Lambda_\ell$ .

In later sections we shall prove Results I, II and III stated below.

<u>Result I</u>. There exists a $\delta^* > 0$ such that for all $\alpha \in \Lambda_\ell^{\leq\delta^*}$ and all $k \in Z$ the differential system

$$z(0) = \xi_\alpha$$

$$\dot{z}(t) = f_{\alpha,k}(z(t),t) \quad \text{for a.e. } t \in [0,1]$$

admits a unique solution in $X$ which is denoted by $x_{\alpha,k}$ . Moreover for all $k \in Z$ we have $|x_{\alpha,k} - x_0| \leq \rho$ whenever $\alpha \in \Lambda_\ell^{\leq\delta^*}$ and there exists a $Q_1 < +\infty$ with $|x_{\alpha'',k} - x_{\alpha'k}| \leq Q_1|\alpha'' - \alpha'|$ whenever $\alpha'$ and $\alpha'' \in \Lambda_\ell^{\leq\delta^*}$ .

<u>Definition</u>. For all $\alpha \in \Lambda_\ell$ and all $k \in Z$ let $y_{\alpha,k}$ be the unique solution in $X$ of the linear inhomogeneous differential system

$$z(0) = \xi_\alpha$$

$$\dot{z}(t) = f_{\alpha,k}(x_0(t),t) + D(t)(z(t)-x_0(t))$$

for a.e. $t \in [0,1]$ .

Result II. For every $\bar{\varepsilon}_1 > 0$ there exists an $\eta \in (0, \delta^*]$ such that for all $k \in Z$ we have

$$|x_{\alpha,k} - y_{\alpha,k}| \le \bar{\varepsilon}_1 \eta \quad \text{whenever} \quad \alpha \in \Lambda_\ell^{\le\eta}$$

[ and

$$|(y_{\alpha'',k} - x_{\alpha'',k}) - (y_{\alpha',k} - x_{\alpha',k})| \le \bar{\varepsilon}_1 |\alpha'' - \alpha'|$$

whenever $\alpha'$ and $\alpha'' \in \Lambda_\ell^{\le\eta}$ ]

Remark III. For every $\bar{\varepsilon}_2 > 0$ there exists a $k \in Z$ such that

$$|y_{\alpha,k} - \Sigma_{i=0,1,\ldots,\ell}\alpha_i y_i| \le \bar{\varepsilon}_2 \quad \text{for all} \quad \alpha \in \Lambda_\ell$$

[ and

$$|y_{\alpha'',k} - y_{\alpha',k} - \Sigma_{i=0,1,\ldots,\ell}(\alpha''_i - \alpha'_i)\, y_i| \le \bar{\varepsilon}_2 |\alpha'' - \alpha'|$$

for all $\alpha', \alpha'' \in \Lambda_\ell$ . ]

We now return to the proof of Proposition 7.1. From Result II we know that there exists an $\eta \in (0, \varepsilon_2]$ such that for all $k \in Z$ we have

$$|x_{\alpha,k} - y_{\alpha,k}| \le \frac{\varepsilon_1 \eta}{2} \quad \text{whenever} \quad \alpha \in \Lambda_\ell^{\le \eta}$$

and

$$\left[ \begin{array}{l} |y_{\alpha'',k} - x_{\alpha'',k}) - (y_{\alpha',k} - x_{\alpha',k})| \le \frac{\varepsilon_1}{2} |\alpha'' - \alpha'| \\ \quad \text{whenever} \quad \alpha' \text{ and } \alpha'' \in \Lambda_\ell^{\le \eta} \quad . \end{array} \right]$$

From Result III we see immediately that $\bar{N}$ is convex and we know that there exists a $k < +\infty$ such that

$$|y_{\alpha,k} - \Sigma_{i=0,1,\ldots,\ell}\, \alpha_i y_i| \le \frac{\varepsilon_1 \eta}{2} \quad \text{whenever} \quad \alpha \in \Lambda_\ell$$

$$\left[ \begin{array}{l} \text{and} \\ |y_{\alpha'',k} - y_{\alpha',k} - \Sigma_{i=0,1,\ldots,\ell}(\alpha_i'' - \alpha_i')y_i| \le \frac{\varepsilon_1}{2} |\alpha'' - \alpha'| \\ \quad \text{whenever} \quad \alpha' \text{ and } \alpha'' \in \Lambda_\ell \quad . \end{array} \right]$$

We have then

$$|x_{\alpha,k} - \Sigma_{i=0,1,\ldots,\ell}\, \alpha_i y_i| \le \varepsilon_1 \eta \quad \text{whenever} \quad \alpha \in \Lambda_\ell^{\le \eta}$$

$$[\text{and} \quad |x_{\alpha'',k} - x_{\alpha',k} - \Sigma_{i=0,1,\ldots,\ell}(\alpha_i'' - \alpha_i')y_i| \le \varepsilon_1 |\alpha'' - \alpha'| \quad \text{for all} \quad \alpha', \alpha'' \in \Lambda_\ell^{\le \eta}] \; .$$

We now define the function $\zeta$ by $\zeta(\alpha) = x_{\alpha,k}$ for all $\alpha \in \Lambda_\ell^{\le \eta}$ .

Since the function $\zeta$ is defined on the set $\Lambda_\ell^{\leq\eta}$ and not only on the set $\Lambda_\ell^{\eta} \subset \Lambda_\ell^{\leq\eta}$ we have obtained, without additional efforts, a result stronger than we actually need. This concludes the proof of Proposition 7.1.

Remark. We justify here the supplementary assumption made at the beginning of the proof of Proposition 7.1. According to our original assumptions we know that for every $i \in \{0,1,\ldots,\ell\}$ there exist an $\eta_i > 0$ and a real valued integrable [quasi piecewise continuous] function $q_i$ on $[0,1]$ such that

(i) $|f_i(z,t)| \leq q_i(t)$ whenever $t \in [0,1]$ and $|z - x_0(t)| \leq \eta_i$

(ii) $|f_i(z_2,t) - f_i(z_1,t)| \leq q_i(t)|z_2 - z_1|$ whenever $t \in [0,1]$ and $|z_1 - x_0(t)|, |z_2 - x_0(t)| \leq \eta_i$

Let $\rho = \min\{\eta_0,\eta_1,\ldots,\eta_\ell\} > 0$ and let $m$ be a real-valued integrable [quasi piecewise continuous] function on $[0,1]$ defined by $m(t) = \max\{1,q_0(t),\ldots,q_\ell(t)\}$ for all $t \in [0,1]$. For all $i \in \{0,1,\ldots,\ell\}$ we have then

(i) $|f_i(z,t)| \leq m(t)$ whenever $t \in [0,1]$ and $|z - x_0(t)| \leq \rho$

(ii) $|f_i(z_2,t) - f_i(z_1,t)| \leq m(t)|z_2 - z_1|$ whenever $t \in [0,1]$ and $|z_1 - x_0(t)|, |z_2 - x_0(t)| \leq \rho$

If we let $\alpha = \int_0^1 m(\theta)\, d\theta$ , if we introduce a new time variable $\tau$ by the relation $\tau = \int_0^1 m(\theta)\, d\theta/\alpha$ , if we rewrite all the differential equations in F with respect to this new time variable $\tau$ and if we change back again the labels from $\tau$ to t , then we are in a position to assume that for all $i \in \{0,1,\ldots,\ell\}$ we have

(i) $|f_i(z,t)| \le \alpha$ whenever $t \in [0,1]$ and $|z - x_0(t)| \le \rho$

(ii) $|f_i(z_2,t) - f_i(z_1,t)| \le \alpha|z_2 - z_1|$ whenever $t \in [0,1]$ and $|z_1 - x_0(t)|$, $|z_2 - x_0(t)| \le \rho$

[For every $i \in \{0,1,\ldots,\ell\}$ the functions $f_i(x_0(t),t)$ are now bounded and quasi piecewise continuous in t over [0,1] . In order to obtain piecewise continuous functions $f_i(x_0(t),t)$ we shall introduce a second change of time variable. This second change of time variable will give us, at no extra cost, the continuity of $f_i(x_0(t),t)$ : a result stronger than the piecewise continuity of $f_i(x_0(t),t)$ that we actually need. Let $\{0 \le \tau_0 < \tau_1 < \ldots < \tau_k \le 1\}$ be the set of all $t \in [0,1]$ at which one at least of the functions $f_i(x_0(t),t)$, $i = 0,1,\ldots,\ell$ is discontinuous in t . Let m* be defined over [0,1] by

$$m^*(t) = 1 + \Sigma_{i=0,1,\ldots,k} \frac{1}{\sqrt{|t-\tau_i|}} \quad \text{if } t \in [0,1] \setminus \{\tau_0,\tau_1,\ldots,\tau_k\}$$

$$= 1 \quad \text{if } t \in \{\tau_0,\tau_1,\ldots,\tau_k\} .$$

The function $m^*$ is integrable and quasi piecewise continuous over $[0,1]$. Let $\sigma^* = \int_0^1 m^*(\theta)\, d\theta$ and let $\tau = \int_0^t m^*(\theta)\, d\theta/\sigma^*$. We shall rewrite all the differential equations in $F$ with respect to this new time variable $\tau$ and change back again the labels from $\tau$ to $t$. We then obtain that for every $i \in \{0,1,\dots,\ell\}$ the function $f_i(x_0(t),t)$ is continuous in $t$ over $[0,1]$. As a matter of fact the only thing we need for our further developments is to know that for every $i \in \{0,1,\dots,\ell\}$ the function $f_i(x_0(t),t)$ is piecewise continuous in $t$ over $[0,1]$.]

We are now in a position to assume that there exists a $\sigma < +\infty$ such that for all $i \in \{0,1,\dots,\ell\}$ we have

(i) $|f_i(z,t)| \leq \sigma$ whenever $t \in [0,1]$ and $|z - x_0(t)| \leq \rho$

(ii) $|f_i(z_2,t) - f_i(z_1,t)| \leq \sigma |z_2 - z_1|$ whenever $t \in [0,1]$ and $|z_1 - x_0(t)|, |z_2 - x_0(t)| \leq \rho$

We remark here that the original assumptions are not affected by {this} [these] change[s] of the time variable. Moreover we can often avoid {this} [these] change[s] of the time variable when the data are a little smoother than the minimum required in our assumptions.

## §8. Dependence on the data of solutions of differential equations.

The data and notations are the same as in the preceding two sections. We are given an element $x_0 \in X$, elements $\xi_0, \xi_1, \ldots, \xi_\ell$ in $R^n$ and elements $f_0, f_1, \ldots, f_\ell$ in $F$. We assume that

H1. $x_0(0) = \xi_0$ and $\dot{x}_0(t) = f_0(x_0(t),t)$ for a.e. $t \in [0,1]$.

H2. There exist real numbers $\rho > 0$ and $\sigma < +\infty$ such that for all $i \in \{0,1,\ldots,\ell\}$ we have

(a) $|f_i(z,t)| \leq \sigma$ whenever $t \in [0,1]$ and $|z - x_0(t)| \leq \rho$

(b) $|f_i(z_2,t) - f_1(z_2,t)| \leq \sigma|z_2 - z_1|$ whenever $t \in [0,1]$ and $|z_1 - x_0(t)|,\ |z_2 - x_0(t)| \leq \rho$

H3. There exists an integrable function $D$ from $[0,1]$ into $R^{n\times n}$ such that for every $\epsilon > 0$ there exists a $\delta \in (0,\rho]$ with

$$\int_0^1 |f_0(x_0(t),t) + D(t)(x(t) - x_0(t)) - f_0(x(t),t)|dt \leq \epsilon|x-x_0|$$

whenever $x \in X$ and $|x - x_0| \leq \delta$

[ and even

$$\int_0^1 |f_0(x_1(t),t) + D(t)(x_2(t) - x_1(t)) - f_0(x_2(t),t)|dt \leq$$

$$\leq \epsilon|x_2 - x_1| \text{ whenever } x_1, x_2 \in X \text{ and}$$

$$|x_1 - x_0|,\ |x_2 - x_0| \leq \delta$$ ]

As in the preceding section we let $\Lambda_\ell = \{\alpha = (\alpha_0, \alpha_1, \ldots, \alpha_\ell) : \alpha_i \geq 0$ for $i = 0,1,\ldots,\ell$ and $\Sigma_{i=0,1,\ldots,\ell}\alpha_i = 1\}$. The element $(1,0,0,\ldots,0)$ of $\Lambda_\ell$ is denoted by $\alpha^0$. On $\Lambda_\ell$ we define the metric $|\alpha'' - \alpha'| = \Sigma_{i=0,1,\ldots,\ell}|\alpha''_i - \alpha'_i|$. For every $\eta \in [0,2]$ we let $\Lambda_\ell^{\leq\eta} = \{\alpha : |\alpha - \alpha^0| \leq \eta\}$. For every $\alpha \in \Lambda_\ell$ let $\xi_\alpha = \Sigma_{i=0,1,\ldots,\ell}\alpha_i\xi_i$. We have then $|\xi_{\alpha''} - \xi_{\alpha'}| \leq |\alpha'' - \alpha'| \max_{i=1,\ldots,\ell}|\xi_i - \xi_0|$. For every $k \in Z$ and every $\alpha \in \Lambda_\ell$ let $f_{\alpha,k}$ be defined by

$$f_{\alpha,k}(z,t) = f_i(z,t)$$

$$\text{if} \quad t \in \bigcup_{j=1,\ldots,k}\Big[\frac{j-1}{k} + \frac{1}{k}\,\Sigma_{r=0,1,\ldots,i-1}\alpha_r, \frac{j-1}{k} + \frac{1}{k}\,\Sigma_{r=0,1,\ldots,i}\alpha_r\Big)$$

and by

$$f_{\alpha,k}(z,1) = f_\ell(z,1) .$$

We have then $\mu\{t : t \in [0,1], f_{\alpha',k}(z,t) \neq f_{\alpha'',k}(z,t)$ for some $z \in R^n\} \leq 2\ell|\alpha'' - \alpha'|$. Let $Q_1 = e^\sigma(4\ell\sigma + \max_{i=1,\ldots,\ell}|\xi_i - \xi_0|)$ and let $\delta^* = \rho/2Q_1$.

<u>Proposition 8.1</u>. If $\alpha'$ and $\alpha'' \in \Lambda_\ell$, $k \in Z$, $\tau > 0$ and if $x'$ and $x''$ are absolutely continuous functions from $[0, \tau]$ into $R^n$ such that

(i) $x'(0) = \xi_{\alpha'}$ and $x''(0) = \xi_{\alpha''}$

(ii) $\dot{x}'(t) = f_{\alpha',k}(x'(t),t)$ and $\dot{x}''(t) = f_{\alpha'',k}(x''(t),t)$ for a.e. $t \in [0, \tau]$

(iii) $|x'(t) - x_0(t)|$ and $|x''(t) - x_0(t)| \leq \rho$ for all $t \in [0, \tau]$

then $|x''(t)-x'(t)| \leq Q_1 |\alpha''-\alpha'|$ for all $t \in [0,\tau]$ .

<u>Proof of Proposition 8.1</u>. For all $t \in [0, \tau]$ we have

$$|x''(t)-x'(t)| \leq |\xi_{\alpha''}-\xi_{\alpha'}| + \int_0^t |f_{\alpha'',k}(x''(\theta),\theta)-f_{\alpha',k}(x'(\theta),\theta)|\, d\theta$$

$$\leq |\xi_{\alpha''} - \xi_{\alpha'}| + \int_0^t |f_{\alpha'',k}(x''(\theta),\theta) - f_{\alpha',k}(x''(\theta), \theta)|d\theta$$

$$+ \int_0^t |f_{\alpha',k}(x''(\theta),\theta) - f_{\alpha',k}(x'(\theta), \theta)|d\theta$$

$$\leq |\alpha''-\alpha'| \max_{i=1,\ldots,\ell} |\xi_i-\xi_0| + 2\ell |\alpha''-\alpha'| 2\sigma + \int_0^t \sigma |x''(\theta)-x'(\theta)|d\theta .$$

From Gronwall's inequality this implies that for all $t \in [0, \tau]$ we have $|x''(t)-x'(t)| \leq e^{\sigma}(4\ell\,\sigma + \max_{i=1,\ldots,\ell} |\xi_i-\xi_0|)|\alpha''-\alpha'| = Q_1 |\alpha'' - \alpha'|$ . This concludes the proof of Proposition 8.1.

<u>Proposition 8.2</u>. For all $\alpha \in \Lambda_{\ell}^{\leq\delta^*}$ and all $k \in Z$ the differential system

$$z(0) = \Sigma_{i=0,1,\ldots,\ell}\, \alpha_i \xi_i$$

$$\dot{z}(t) = f_{\alpha,k}(z(t),t) \quad \text{for a.e. } t \in [0,1]$$

admits a unique solution in $X$ which is denoted by $x_{\alpha,k}$ . Moreover for all $k \in Z$ we have $|x_{\alpha,k} - x_0| \leq \rho$ whenever

$\alpha \in \Lambda_{\ell}^{\leq\delta*}$ and $|x_{\alpha'',k} - x_{\alpha',k}| \leq Q_1 |\alpha'' - \alpha'|$ whenever $\alpha'$ and $\alpha'' \in \Lambda_{\ell}^{\leq\delta*}$.

Proof of Proposition 8.2. Let $k \in Z$ and let $\alpha \in \Lambda_{\ell}^{\leq\delta*}$. Since $|\xi_\alpha - \xi_0| < \rho$ our assumptions guarantee that there exists an $\varepsilon > 0$ such that the differential system

$$z(0) = \xi_\alpha$$

$$\dot{z}(t) = f_{\alpha,k}(z(t),t) \quad \text{for a.e. } t \in [0, \varepsilon]$$

admits a unique absolutely continuous solution on $[0, \varepsilon]$ which will be denoted by $x^{\varepsilon}_{\alpha,k}$ and which satisfies the inequality $|x^{\varepsilon}_{\alpha,k}(t) - x_0(t)| < \rho$ for all $t \in [0, \varepsilon]$. Let $\tau = \sup\{\varepsilon : \varepsilon \in (0,1], x^{\varepsilon}_{\alpha,k}$ exists, $|x^{\varepsilon}_{\alpha,k}(t) - x_0(t)| < \rho$ for all $t \in [0, \varepsilon]\}$. Since $|\dot{x}^{\varepsilon}_{\alpha,k}(t)| \leq \sigma$ for a.e. $t \in [0,\varepsilon]$ it follows that $x^{\tau}_{\alpha,k}$ exists and that $|x^{\tau}_{\alpha,k}(t) - x_0(t)| \leq \rho$ for all $t \in [0, \tau]$. We are now faced with the following alternative: either we have $\tau = 1$ or we have $\tau < 1$ and $|x^{\tau}_{\alpha,k}(\tau) - x_0(\tau)| = \rho$. In the first case $x_{\alpha,k}$ is uniquely defined over $[0,1]$ and $|x_{\alpha,k} - x_0| \leq \rho$. The second case cannot happen since for every $t \in [0, \tau]$ we know, from Proposition 8.1, that $|x^{\tau}_{\alpha,k}(t) - x_0(t)| \leq Q_1 |\alpha - \alpha_0| \leq \rho/2$. The last statement of Proposition 8.2 is then an immediate consequence of Proposition 8.1. This concludes the proof of Proposition 8.2.

For every $\alpha \in \Lambda_\ell$ and every $k \in Z$ the linear inhomogeneous differential system

$$z(0) = \Sigma_{i=0,1,\ldots,\ell}\alpha_i\xi_i$$

$$\dot z(t) = f_{\alpha,k}(x_0(t),t) + D(t)(z(t)-x_0(t)) \quad \text{for}$$

$$\text{a.e.} \quad t \in [0,1]$$

has a unique solution in $X$ which will be denoted by $y_{\alpha,k}$ .

Proposition 8.3. For every $\epsilon > 0$ there exists a $\delta \in (0, \delta^*]$ such that for all $k \in Z$ we have

$$|y_{\alpha,k} - x_{\alpha,k}| \le \epsilon|\alpha - \alpha^0| \quad \text{whenever} \quad \alpha \in \Lambda_\ell^{\le\delta}$$

[and even

$$|(y_{\alpha'',k}-x_{\alpha'',k}) - (y_{\alpha',k}-x_{\alpha',k})| \le \epsilon|\alpha''-\alpha'| \quad \text{whenever}$$

$$\alpha',\ \alpha'' \in \Lambda_\ell^{\le\delta}$$
]

Proof of Proposition 8.3. We give below the proof of the bracketed version of Proposition 8.3. (i.e. the proof to be read by Reader B). For the standard version of Proposition 8.3. (i.e. for Reader A) repeat the same proof with $\alpha$ instead of $\alpha''$ and $\alpha^0$ instead of $\alpha'$ . Let $\bar\epsilon > 0$ be such that

$e^{\int_0^1 |D(t)|dt} \ell_1 \bar{\varepsilon} < \frac{\varepsilon}{2}$ . Let $\bar{\delta} \in (0, \rho]$ be such that

$$\int_0^1 |f_0(x_{\alpha',k}(t),t) + D(t)(x_{\alpha'',k}(t) - x_{\alpha',k}(t)) - f_0(x_{\alpha'',k}(t),t)|dt$$

$$\leq \bar{\varepsilon}|x_{\alpha''k} - x_{\alpha',k}| \quad \text{whenever} \quad |x_{\alpha',k} - x_0|, |x_{\alpha'',k} - x_0| \leq \bar{\delta} .$$

Let $\delta^+ \in (0, \delta^*]$ be such that $\ell_1 \delta^+ \leq \bar{\delta}$ and let $\delta \in (0, \delta^+]$ be such that $8\ell\sigma\delta \leq \bar{\varepsilon}$ which implies

$e^{\int_0^1 |D(t)|dt} \ell_1 \, 8\ell\sigma\delta \leq \frac{\varepsilon}{2}$ .

For all $\alpha', \alpha'' \in \Lambda_\ell^{\leq\delta}$ , all $k \in Z$ and all $\theta \in [0,1]$ we have then $|(y_{\alpha'',k}(\theta) - x_{\alpha'',k}(\theta)) - (y_{\alpha',k}(\theta) - x_{\alpha',k}(\theta))|$

$$\leq \int_0^\theta |D(t)||(y_{\alpha'',k}(t) - x_{\alpha'',k}(t)) - (y_{\alpha',k}(t) - x_{\alpha',k}(t))|dt + B + C + D$$

where

$$B = \int_0^1 |f_0(x_{\alpha',k}(t),t) + D(t)(x_{\alpha'',k}(t)-x_{\alpha',k}(t))-f_0(x_{\alpha'',k}(t),t)|dt$$

$$\leq \bar{\varepsilon}|x_{\alpha'',k} - x_{\alpha',k}| \leq \bar{\varepsilon}\,\ell_1|\alpha'' - \alpha'|$$

since $|\alpha' - \alpha^0|$ and $|\alpha'' - \alpha^0| \leq \delta \leq \delta^+ \leq \delta^*$

and hence $|x_{\alpha',k} - x_0|$ and $|x_{\alpha'',k} - x_0| \leq \ell_1\delta \leq \ell_1\delta^+ \leq \bar{\delta}$

$$C = \int_0^1 |(f_{\alpha'',k}(x_{\alpha',k}(t),t) - f_{\alpha'',k}(x_{\alpha'',k}(t),t)) - (f_0(x_{\alpha',k}(t),t) - f_0(x_{\alpha'',k}(t),t))|dt$$

$$\leq 2\ell|\alpha''-\alpha^0|2\sigma|x_{\alpha'',k}-x_{\alpha',k}| \leq 4\ell\sigma|\alpha'' - \alpha^0|\ell_1|\alpha'' - \alpha'|$$

and

$$D = \int_0^1 |(f_{\alpha'',k}(x_0(t),t) - f_{\alpha'',k}(x_{\alpha',k}(t),t)) - (f_{\alpha',k}(x_0(t),t) -$$

$$- f_{\alpha',k}(x_{\alpha',k}(t),t))|dt$$

$$\leq 2\ell|\alpha'' - \alpha'|2\sigma|x_{\alpha',k} - x_0| \leq 4\ell\sigma|\alpha' - \alpha^0|Q_1|\alpha'' - \alpha'| .$$

From Gronwall's inequality we have then

$$|(y_{\alpha'',k} - x_{\alpha'',k}) - (y_{\alpha',k} - x_{\alpha',k})| \leq \ell^{\int_0^1|D(t)|dt}(B+C+D) \leq \varepsilon|\alpha''-\alpha'| .$$

This concludes the proof of Proposition 8.3.

## §9. Convex-under-switching combination of linear inhomogeneous ordinary differential equations.

We are given points $\xi_0,\xi_1,\ldots,\xi_\ell$ in $R^n$, integrable [piecewise continuous] functions $u_0,u_1,\ldots,u_\ell$ from $[0,1]$ into $R^n$ and an integrable function $D$ from $[0,1]$ into $R^{n\times n}$.

For every $i \in \{0,1,\ldots,\ell\}$ let $y_i$ be the unique absolutely continuous solution on $[0,1]$ of the linear inhomogeneous differential system:

$$y_i(0) = \xi_i$$

$$\dot{y}_i(t) = u_i(t) + D(t)(y_i(t) - y_0(t)) \quad \text{for a.e. } t \in [0,1]$$

Let $\Lambda_\ell = \{\alpha = (\alpha_0,\alpha_1,\ldots,\alpha_\ell) : \alpha_i \geq 0$ for $i = 0,1,\ldots,\ell$ and $\Sigma_{i=0,1,\ldots,\ell}\alpha_i = 1\}$. On $\Lambda_\ell$ we use the norm $|\alpha'' - \alpha'| = \Sigma_{i=0,1,\ldots,\ell}|\alpha''_i - \alpha'_i|$. For every $\alpha \in \Lambda_\ell$ let $\xi_\alpha = \Sigma_{i=0,1,\ldots,\ell}\alpha_i\xi_i$. For every $k \in Z$, every $j \in \{1,\ldots,k\}$ every $\alpha \in \Lambda_\ell$ and every $i \in \{0,1,\ldots,\ell\}$ let

$$A(k,j,\alpha,i) = [\tfrac{j-1}{k} + \tfrac{1}{k}\,\Sigma_{r=0,1,\ldots,i-1}\alpha_r,\ \tfrac{j-1}{k} + \tfrac{1}{k}\,\Sigma_{r=0,1,\ldots,i}\alpha_r)\ .$$

For every $k \in Z$ and every $\alpha \in \Lambda_\ell$ let $u_{\alpha,k}$ be a function from $[0,1]$ into $R^n$ defined by

$$u_{\alpha,k}(t) = u_i(t) \quad \text{if} \quad t \in \cup_{j=1,\ldots,k}A(k,j,\alpha,i)$$

and

$$u_{\alpha,k}(1) = u_\ell(1) \; .$$

For every $\alpha \in \Lambda_\ell$ and every $k \in Z$ let $y_{\alpha,k}$ be the unique absolutely continuous solution on $[0,1]$ of the linear inhomogeneous differential system

$$y_{\alpha,k}(0) = \xi_\alpha$$

$$\dot{y}_{\alpha,k}(t) = u_{\alpha,k}(t) + D(t)(y_{\alpha,k}(t) - y_0(t))$$

$$\text{for a.e. } t \in [0,1] \; .$$

In this section we shall prove the following result.

Proposition 9.1. For every $\varepsilon > 0$ there exists a positive integer $N$ such that for all $k \geq N$ we have

$$|y_{\alpha,k} - \Sigma_{i=0,1,\ldots,\ell}\alpha_i y_i| \leq \varepsilon \quad \text{whenever} \quad \alpha \in \Lambda_\ell$$

[and

$$|y_{\alpha'',k} - y_{\alpha',k} - \Sigma_{i=0,1,\ldots,\ell}(\alpha_i'' - \alpha_i')y_i| \leq \varepsilon|\alpha'' - \alpha'|$$

whenever $\alpha', \alpha'' \in \Lambda_\ell$ .]

Before giving the proof of Proposition 9.1 we shall establish the following two intermediary results:

Lemma 1. If the functions $u_0, u_1, \ldots, u_\ell$ are piecewise constant and if $D(t) \equiv 0$ then for every $\varepsilon > 0$ there exists a positive integer $N$ such that for all $k \in N$ we have

$$|y_{\alpha'',k} - y_{\alpha',k} - \Sigma_{i=0,1,\ldots,\ell}(\alpha_i'' - \alpha_i')y_i| \leq \varepsilon|\alpha'' - \alpha'|$$

whenever $\alpha', \alpha'' \in \Lambda_\ell$

and, in particular,

$$|y_{\alpha,k} - \Sigma_{i=0,1,\ldots,\ell}\alpha_i y_i| \leq \varepsilon \quad \text{whenever} \quad \alpha \in \Lambda_\ell \ .$$

<u>Lemma 2</u>. If the functions $u_0, u_1, \ldots, u_\ell$ are integrable [piecewise continuous] and if $D(t) \equiv 0$ then for every $\varepsilon > 0$ there exists a positive integer $N$ such that for all $k \in N$ we have

$$|y_{\alpha,k} - \Sigma_{i=0,1,\ldots,\ell}\alpha_i y_i| \leq \varepsilon \quad \text{whenever} \quad \alpha \in \Lambda_\ell$$

[ and

$$|y_{\alpha'',k} - y_{\alpha',k} - \Sigma_{i=0,1,\ldots,\ell}(\alpha_i'' - \alpha_i')y_i| \leq \varepsilon|\alpha'' - \alpha'|$$

whenever $\alpha', \alpha'' \in \Lambda_\ell$ . ]

<u>Proof of Lemma 1.</u> Let $M = \sup_{t\in[0,1], i\in\{0,1,\ldots,\ell\}}|u_i(t)|$ and let $Q$ be the number of elements in $[0,1]$ where one at least of the functions $u_i$ is discontinuous. Let $N < +\infty$ be such that $6MQ(\ell + 1)/N \leq \varepsilon$ . Let $k$ be a positive integer with $k \geq N$ . Let $A_k$ be the set of all $j \in \{1,2,\ldots,k\}$ such that $u_i$ is constant on $[\frac{j-1}{k}, \frac{j}{k})$ for all $i \in \{0,1,\ldots,\ell\}$ and let

$B_k = \{1,2,\ldots,k\} \setminus A_k$ . The set $B_k$ contains at most $Q$ elements. For every $j \in \{1,2,\ldots,k\}$ every $i \in \{0,1,\ldots,\ell\}$ and every $\alpha', \alpha'' \in \Lambda_\ell$ let $E(j,i,\alpha',\alpha'') =$

$$\left|\int_{A(k,j,\alpha'',i)} u_i(t)dt - \int_{A(k,j,\alpha',i)} u_i(t)dt - (\alpha_i'' - \alpha_i') \int_{\frac{j-1}{k}}^{\frac{j}{k}} u_i(t)dt\right| .$$

If $j \in A_k$ we have $E(j,i,\alpha',\alpha'') = 0$ and if $j \in B_k$ we have $E(j,i,\alpha',\alpha'') \leq \frac{6M}{k} |\alpha'' - \alpha'|$ . We have then

$|y_{\alpha'',k} - y_{\alpha',k} - \Sigma_{i=0,1,\ldots,\ell}(\alpha_i'' - \alpha_i')y_i| \leq$

$\Sigma_{j\in\{1,\ldots,k\}}\Sigma_{i\in\{0,1,\ldots,\ell\}}E(j,i,\alpha',\alpha'') \leq \frac{6MQ(\ell+1)}{k}|\alpha''-\alpha'| \leq \varepsilon|\alpha''-\alpha'|$ .

This concludes the proof of Lemma 1.

<u>Proof of Lemma 2</u>. Let $u_0^*, u_1^*,\ldots,u_\ell^*$ be piecewise constant functions such that for all $i \in \{0,1,\ldots,\ell\}$ we have $\int_0^1 |u_i^*(t) - u_i(t)|dt \leq \frac{\varepsilon}{3\ell}$ [and even $|u_i^*(t) - u_i(t)| \leq \frac{\varepsilon}{12\ell}$ for all $t \in [0,1]$] . For every $i \in \{0,1,\ldots,\ell\}$, every $\alpha \in \Lambda_\ell$ and every positive integer $k$ let $y_i^*$, $u_{\alpha,k}^*$ and $y_{\alpha,k}^*$ be obtained in the same manner as $y_i$, $u_{\alpha,k}$ and $y_{\alpha,k}$ but with the functions $u_0,u_1,\ldots,u_\ell$ replaced by $u_0^*,u_1^*,\ldots,u_\ell^*$ . For all $k \in Z$ and all $\alpha \in \Lambda_\ell$ we have $\int_0^1 |u_{\alpha,k}^*(t) - u_{\alpha,k}(t)|dt \leq \frac{\varepsilon}{3}$ [and even $|u_{\alpha,k}^*(t) - u_{\alpha,k}(t)| \leq \frac{\varepsilon}{12\ell}$ for all $t \in [0,1]$] . For all $k \in Z$, $\alpha', \alpha'' \in \Lambda_\ell$ we have

$\mu\{t : t \in [0,1],\ (u^*_{\alpha'',k}(t) - u_{\alpha'',k}(t)) \neq (u^*_{\alpha',k}(t) - u_{\alpha',k}(t))\}$

$\leq 2\ell\,|\alpha'' - \alpha'|$ .

For all $k \in Z$ we have then $|y^*_{\alpha,k} - y_{\alpha,k}| \leq \int_0^1 |u^*_{\alpha,k}(t) - u_{\alpha,k}(t)|dt \leq \frac{\varepsilon}{3}$

for all $\alpha \in \Lambda_\ell$ [and even $|(y^*_{\alpha'',k} - y_{\alpha'',k}) - (y^*_{\alpha',k} - y_{\alpha',k})| \leq$

$\int_0^1 |((u^*_{\alpha'',k}(t) - u_{\alpha'',k}(t)) - (u^*_{\alpha',k}(t) - u_{\alpha',k}(t))|dt \leq$

$$2\ell\,|\alpha'' - \alpha'|\frac{2\varepsilon}{12\ell} = \frac{\varepsilon}{3}|\alpha'' - \alpha'|$$

for all $\alpha'$ and $\alpha'' \in \Lambda_\ell$] . From Lemma 1 we know that there exists an $N < +\infty$ such that for all $k \in N$ we have

$$|y^*_{\alpha'',k} - y^*_{\alpha',k} - \Sigma_{i=0,1,\ldots,\ell}(\alpha''_i - \alpha'_i)y^*_i| \leq \frac{\varepsilon}{3}|\alpha'' - \alpha'|$$

whenever $\alpha', \alpha'' \in \Lambda_\ell$

and in particular

$$|y^*_{\alpha,k} - \Sigma_{i=0,1,\ldots\ell}\alpha_i y^*_i| \leq \frac{\varepsilon}{3} \quad \text{whenever} \quad \alpha \in \Lambda_\ell$$

For all $k \geq N$ we have then $|y_{\alpha,k} - \Sigma_{i=0,1,\ldots,\ell}\alpha_i y_i| \leq$

$|y^*_{\alpha,k} - \Sigma_{i=0,1,\ldots\ell}\alpha_i y^*_i| + |y^*_{\alpha,k} - y_{\alpha,k}| +$

$$+ |\Sigma_{i=0,1,\ldots,\ell}\alpha_i(y^*_i - y_i)| \leq \frac{\varepsilon}{3} + \frac{\varepsilon}{3} + \frac{\varepsilon}{3} \leq \varepsilon$$

whenever $\alpha \in \Lambda_\ell$

$$\left[\begin{array}{l} \text{and even} \quad |y_{\alpha'',k} - y_{\alpha',k} - \Sigma_{i=0,1,\ldots,\ell}(\alpha''_i - \alpha'_i)y_i| \le \\ |y^*_{\alpha'',k} - y^*_{\alpha',k} - \Sigma_{i=0,1,\ldots,\ell}(\alpha''_i - \alpha'_i)y^*_i| + \\ \quad + |(y^*_{\alpha'',k} - y_{\alpha'',k}) - (y^*_{\alpha',k} - y_{\alpha',k})| + \\ \quad\quad + |\Sigma_{i=0,1,\ldots,\ell}(\alpha''_i - \alpha'_i)(y_i - y_i)| \\ \le \frac{\varepsilon}{3}|\alpha'' - \alpha'| + \frac{\varepsilon}{3}|\alpha'' - \alpha'| + \frac{\varepsilon}{3}|\alpha'' - \alpha'| = \varepsilon|\alpha'' - \alpha'| \end{array}\right]$$

This concludes the proof of Lemma 2.

Proof of Proposition 9.1. Let $G : [0,1] \to R^{n\times n}$ be the unique absolutely continuous solution of the linear matrix differential system:

$$G(0) = I$$

$$\dot{G}(t) = D(t)\,G(t) \quad \text{for a.e.} \quad t \in [0,1]$$

We know that for every $t \in [0,1]$ the matrix $G(t)$ admits an inverse $G^{-1}(t)$ and that $G^{-1}$ is an absolutely continuous solution of the linear matrix differential system

$$G^{-1}(0) = I$$

$$\frac{d}{dt}\,G^{-1}(t) = -\,G^{-1}(t)D(t) \quad \text{for a.e.} \quad t \in [0,1]\,.$$

For every $i \in \{0,1,\ldots,\ell\}$ let $z_i$ be a function from $[0,1]$

into $R^n$ defined by

$$z_i(t) = G^{-1}(t)(y_i(t) - y_0(t)) + y_0(0)$$

For every $i \in \{0,1,\ldots,\ell\}$ the function $z_i$ is absolutely continuous and we have

$$\dot{z}_i(t) = G^{-1}(t)(u_i(t) - u_0(t)) \quad \text{for a.e.} \quad t \in [0,1] .$$

For every $i \in \{0,1,\ldots,\ell\}$ let $v_i$ be an integrable [piecewise continuous] function from $[0,1]$ into $R^n$ defined by $v_i(t) = G^{-1}(t)(u_i(t) - u_0(t))$ for all $t \in [0,1]$ . For every $\alpha \in \Lambda_\ell$ and every positive integer $k$ let $z_{\alpha,k}$ be obtained in the same manner as $y_{\alpha,k}$ but with the functions $u_0,u_1,\ldots,u_\ell$ replaced by $v_0,v_1,\ldots,v_\ell$ and the function $D$ set to be identically zero. Let $|G| = \max_{t\in[0,1]}|G(t)|$ where $|G(t)| = \max_{|x|\leq 1}|G(t)\,x|$ . According to Lemma 2 there exists a positive integer $N$ such that for all $k \in N$ we have

$$|z_{\alpha,k} - \Sigma_{i=0,1,\ldots,\ell}\alpha_i z_i| \leq \varepsilon/|G| \quad \text{whenever} \quad \alpha \in \Lambda_\ell$$

[ and $|z_{\alpha'',k} - z_{\alpha',k} - \Sigma_{i=0,1,\ldots,\ell}(\alpha''_i-\alpha'_i)z_i| \leq \varepsilon|\alpha''-\alpha'|/|G|$ whenever $\alpha', \alpha'' \in \Lambda_\ell$ ]

For every $k \geq N$ and every $\alpha \in \Lambda_\ell$ and every $t$ $[0,1]$ we have

$$y_{\alpha,k}(t) = y_0(t) + G(t)(z_{\alpha,k}(t) - y_0(0)) .$$

For every $k \geq N$ we have then

$$|y_{\alpha,k} - \Sigma_{i=0,1,\ldots,\ell}\alpha_i y_i| \leq |G||z_{\alpha,k} - \Sigma_{i=0,1,\ldots,\ell}\alpha_i z_i| \leq \varepsilon$$

whenever $\alpha \in \Lambda_\ell$

$$\left[\begin{aligned} \text{and } & |y_{\alpha'',k} - y_{\alpha',k} - \Sigma_{i=0,1,\ldots,\ell}(\alpha_i'' - \alpha_i')y_i| \\ & \leq |G||z_{\alpha'',k} - z_{\alpha',k} - \Sigma_{i=0,1,\ldots,\ell}(\alpha_i'' - \alpha_i')z_i| \\ & \leq \varepsilon|\alpha'' - \alpha'| \text{ whenever } \alpha', \alpha'' \in \Lambda_\ell \end{aligned}\right].$$

This concludes the proof of Proposition 9.1.

## REFERENCES


1. Carathéodory, C., Calculus of Variations and Partial Differential Equations of the First Order, Holden-Day, San Francisco, 1967.

2. Halkin, H., On the necessary condition for optimal control of nonlinear systems, J. An. Math., 12, 1964, pp. 1-82.

3. Halkin, H., Nonlinear nonconvex programming in an infinite dimensional space, in "Mathematical Theory of Control", A. V. Balakrishnan and L. W. Neustadt, eds., Academic Press, 1968.

4. John, F., Extremum Problems with Inequalities as Subsidiary Conditions, in K. O. Friedricks, O. E. Neugebauer, and J. J. Stoker, (eds.) "Studies and Essays: Courant Anniversary Volume", pp. 187-204, Interscience Publishers, New York, 1948.

5. McShane, E. J., On Multipliers for Lagrange Problems, Amer. J. Math., 61, 1939, pp. 809-819.

6. Pontryagin, L. S., and al., The mathematical theory of optimal processes, Wiley, New York, 1962.

# VARIOUS TOPICS IN THE THEORY OF OPTIMAL CONTROL OF DISTRIBUTED SYSTEMS

**J.L. Lions**

**College de France and IRIA**

IRIA - Laboria
Domaine de Voluceau
78 Rocquencourt
France

## CONTENTS

Introduction.

These notes correspond to a set of nine lectures given at the 14th Biennial Seminar of the Canadian Math. Congress, London, Ontario, August 1973; in the oral presentation, some well-known facts were more carefully stated than here, where on the contrary we present in fuller details some new results.

The general goal of these lectures is to present some topics of the theory of optimal control of distributed parameter systems. A number of interesting topics have not been considered here; some of them are listed below, together with some bibliographical indications.

Chapter 1 gives the relationship between the optimal control of distributed systems and free boundary problems. In order to make the exposition not too technical, we restrict ourselves to the case when the state is given by a linear elliptic equation. The methods introduced are general. We also give, at the end of Chapter 1, a partially formal "proof" of a Bang-Bang theorem.

In Chapter 2 and 3 we consider some problems of a "priori feedback"; by this we mean situations where, on physical grounds, we devise a policy which should be "close" to the optimal policy. This leads to problems which generally enter the framework of Variational Inequalities (V.I.) or Quasi Variational Inequalities (Q.V.I.). One has then to show that these problems admit a solution.

The V.I. were introduced in Stampacchia [1], Lions-Stampacchia [1] for the solution of unilateral problems in elasticity.

Variants and new types of V.I. were introduced in Duvaut-Lions [1] for the study of a number of situations arising in mechanics. Properties of the solutions were studied by a number of authors; we refer in particular to H. Brézis [1] (and the bibliography of this paper) for the regularity of the solutions. The V.I. were used for problems in optimal control of distributed systems in Lions [1]. We wish to mention here the work of C. Baiocchi [1] which by a very interesting change of unknown function, reduces a class of free boundary problem arising in hydrodynamics to V.I.; other problems of free boundary type have been recently solved starting from this idea, by Brézis-Stampacchia [1] Duvaut [1].

We have shown in Bensoussan-Lions [2] how the technique of V.I. of evolution plays an essential role in stopping time problems; for stationary cases, this was first noticed by W. Fleming [1].

The Q.V.I. have recently been introduced by Bensoussan and the author (cf. Bensoussan-Lions [1] [3] [4] [5], Bensoussan-Goursat-Lions [1]) for the study of impulse control problems (which are briefly considered in Chapter 8); the Q.V.I. are met here in a different case. A systematic study of Q.V.I. and their applications will be made in Bensoussan-Lions [4].

For V.I. and Q.V.I. we study here particular cases which allow a self contained presentation.

Chapter 4 studies the case where constraints depend on the state of the system. After briefly recalling some well-known methods, we show how duality theory, in the sense of Fenchel and

Rockafellar (cf. Rockafellar [1], Ekeland-Teman [1]) permits one to suppress the constraints on the state; this idea is due to Mossino [1], where examples different from those of the text, are studied.

Chapter 5 briefly presents some cases where the state equation is nonlinear. This is in general the case in most of the applications, and references are very numerous. We give short indications of an example arising in heat theory (J.P. Yvon [1]) and some examples arising in bio-chemistry (J.P. Kernevez [1], Kernevez-Thomas [1], Brauner-Penel [1] [2], (1) ). We also show how the ideas of Chapters 2 and 3 could be useful for nonlinear problems. Without attempting, by any means, a survey of other examples, we wish to mention the nonlinear hyperbolic equations arising in the study of tides, cf. G. Duff [1] and in transport theory, cf. Bamberger-Yvon [1] . Highly nonlinear equations also arise in the control of fusion; cf. Mercier-Temam [1], P.K.C. Wang [1]. Let us also mention problems (arising in applications) where the state is given by the first eigenvalue or the first eigenfunction of an elliptic problem - making the state nonlinear in the control; we refer to F. Mignot [1].

Chapter 6 considers the case where the control is a geometrical variable. This problem can arise in several ways. It can be that we have to choose in an optimal way some points or some curves inside or on the boundary of a given domain (cf. Amouroux [4],

---

(1) See also the report of Banks [1] in these proceedings.

Saguez [1], Van de Wielle [1], also for numerical aspects of these problems); see also the problem of "stabilization by scanning", for which we refer to Bamberger [1], Jaffré [1], Saint Jean Paulin [1]. It can also be that we have to choose the domain itself; these are problems of optimum design; there is a fantastic number of publications on this subject; it seems to be useful to think of these problems in terms of optimal control (cf. O. Pironneau [1] [2]).

Chapter 7 studies the cases where the cost of the control is "small". We show how these problems "reduce" to problems of singular perturbations generally of a non classical type: in particular they are often related to the theory of asymptotic expansions for the solution of singular boundary problems for pseudo differential operators on varieties. (cf. also Lions [4] [6]).

Singular (or regular) perturbations arise naturally in the optimal control of "large" systems which are "weakly coupled". We refer to O'Malley [1], Lions [1] and the bibliographies therein.(1)

Chapter 8 considers impulse control problems, introduced by Bensoussan-Lions [1]. The motivation covers, for instance, problems of optimal dynamic policies in management. A complete study of the relationships between the classical literature on the subject (see the end of Chapter 2) and the topics presented here will

---

(1) See also the report O'Malley [3] in these proceedings.

be presented in the books by Bensoussan-Lions [4]. We give here only brief indications. We show how these problems reduce to Q.V.I.

Chapter 9 gives some indication of the numerical methods by presenting some of the numerical computations made by D. Leroy [1] for a cooling problem.

As we already said, there are a number of topics we do not consider at all. In particular:

1) for general existence theorems, we refer to Bidaut [1], Berkowitz [1], Baranger [1], Murat [1];
2) for the study of feedback, and nonlinear equations of Riccati's type, we refer to Lions [1] [2], and for a direct study of these equations to L. Tartar [1];
3) for hyperbolic systems, only short remarks are presented here; we refer to Lions [1], Russell [1], in particular for the study of observability and controllability;
4) for systems with delays, we refer to Delfour-Mitter [1](1), Oguztorelli [1] [2];
5) for decomposition methods, we refer to the report Bensoussan-Lions-Temam [1], and the bibliography therein;
6) multi-criteria problems are also considered in the paper Bensoussan-Lions-Temam, loc. cit, and in a paper of J.P. Yvon, to appear;
7) for sensitivity analysis, cf. Pritchard [1];

---

(1) See also the report of M. Delfour in these proceedings.

8) problems of identification of distributed systems can be reduced to problems of optimal control with the control in the coefficients; for this approach, cf. G. Chavent [1].

# CHAPTER 1

# OPTIMAL CONTROL AND FREE SURFACES

## 1. Elliptic State Equation

### 1.1 Optimal System

Let $\Omega$ be a bounded open set in $R^n$ , with smooth boundary $\Gamma$ .

In $\Omega$ we consider a linear second order partial differential operator A of elliptic type. In order to make things as simple as possible we shall take

$$(1.1) \qquad A = -\Delta + I, \quad \Delta = \frac{\partial^2}{\partial x_1^2} + \ldots + \frac{\partial^2}{\partial x_n^2}, \quad I = \text{identity}.$$

The state of the system is given by the solution of the Neumann problem

$$(1.2) \qquad Ay = f , \quad \text{in } \Omega ,$$

$$(1.3) \qquad \frac{\partial y}{\partial \nu} = v \quad \text{on } \Gamma .$$

In (1.2) $f$ is given in $L^2(\Omega)$ and in (1.3) $v$ is given in $L^2(\Gamma)$ . ( $\frac{\partial}{\partial \nu}$ denotes the normal derivative of $y$ , taken toward the exterior of $\Omega$ .)

Equations (1.2) (1.3) admit a unique solution in the Sobolev space $H^1(\Omega)$ (1); it can be defined by the "variational formula":

$$a(y,\phi) = (f,\phi) + \int_\Omega v\phi d\Gamma \qquad \forall\ \phi \in H^1(\Omega) \tag{1.4}$$

where

$$a(\phi,\psi) = \sum_{i=1}^{n} \int_\Omega \frac{\partial\phi}{\partial x_i} \frac{\partial\psi}{\partial x_i} dx + \int_\Omega \phi\psi dx ,$$

$$(f,\phi) = \int_\Omega f\phi dx .$$

We shall denote $y$ by $y(v)$ . The mapping

$$v \to y(v) \tag{1.5}$$

is affine continuous from $L^2(\Gamma) \to H^1(\Omega)$ . []

The cost function we consider is defined by

$$J(u) = \int_\Gamma |y(v) - z_d|^2 d\Gamma + N\int_\Gamma v^2 d\Gamma , \tag{1.6}$$

where $z_d$ is given in $L^2(\Gamma)$ and where $N$ is a given positive number.(2) []

---

(1) $H^1(\Omega) = \{\phi | \phi, \frac{\partial\phi}{\partial x_1},\ldots,\frac{\partial\phi}{\partial x_n} \in L^2(\Omega)\}$ . All functions we consider are supposed to be real valued.

(2) We shall consider in the following sections the case "N=0" .

The problem of control is as follows: Let $U_{ad}$ be a non-empty closed convex set of $U = L^2(\Gamma)$ ; we want to minimize $J$ over $U_{ad}$ :

$$(1.7) \qquad \inf J(v), \; v \in U_{ad} . \qquad []$$

Remark 1.1

In Chapter 4 we shall consider cases when $U_{ad}$ is defined by constraints on $v$ and on $y(v)$ ; in this Chapter, $U_{ad}$ is defined by constraints on $v$ only. []

The solution of Problem (1.7) is straightforward.

Theorem 1.1. Problem (1.7) admits a unique solution $u \in U_{ad}$ which is characterized by the optimality condition:

$$(1.8) \qquad \int_\Gamma (y-z_d)(y(v)-y)d\Gamma + N\int_\Gamma u(v-u)d\Gamma \geq 0 \;\forall\; v \in U_{ad} ,$$

where

$$y(u) = y .$$

Proof

1) The function $v \to J(v)$ is continuous from $U \to R$ , strictly convex, and it satisfies

$$J(v) \geq N\|v\|_U^2 ,$$

so that the existence and uniqueness of $u$ is well known.

2) The function $J$ is differentiable and one has

$$(1.9) \quad \frac{1}{2}(J'(u),v-u) = \int_\Gamma (y(u)-z_d)(y(v)-y(u))d\Gamma + N\int_\Gamma u(v-u)d\Gamma \ .$$

The <u>optimal control</u> $u$ is characterized by the <u>variational inequality</u>

$$(1.10) \qquad (J'(u),v-u) \geq 0 \quad \forall \ v \in U_{ad} \ , \quad u \in U_{ad} \ ,$$

and hence (1.8) follows. [ ]

The next step is to transform (1.8) into a more convenient form by introducing the <u>adjoint state</u> $p$ which is defined in the following manner:

$$(1.11) \qquad \begin{aligned} &Ap = 0 \ ^{(1)}, \\ &\frac{\partial p}{\partial \nu} = y - z_d \quad \text{on } \Gamma \ . \end{aligned}$$

If we multiply (1.11) by $y(v) - y(u) = y(v) - y$ and if we apply Green's formula, we find

$$\begin{aligned} 0 &= -\int_\Gamma (y-z_d)(y(v)-y)d\Gamma + \int_\Gamma p\left(\frac{\partial y(v)}{\partial \nu} - \frac{\partial y(u)}{\partial \nu}\right)d\Gamma \\ &= -\int_\Gamma (y-z_d)(y(v)-y)d\Gamma + \int_\Gamma p(v-u)d\Gamma \quad \text{by (1.3)} \end{aligned}$$

(1) In general one would use here the operator $A^*$ adjoint to $A$ ; in the particular case we are considering, $A^* = A$ .

so that (1.8) becomes:

$$(1.12) \qquad \int_\Gamma (p + Nu)(v-u)d\Gamma \geq 0 \quad \forall \quad v \in U_{ad} \;, \; u \in U_{ad} \;.$$

We summarize:

<u>Theorem 1.2. The unique solution</u> u <u>of problem</u> (1.7) <u>is characterized by the following system</u>

$$(1.13) \qquad \begin{cases} Ay = f \;, \; Ap = 0 \quad \text{in} \quad \Omega \;, \\ \dfrac{\partial y}{\partial \nu} = u \;, \; \dfrac{\partial p}{\partial \nu} = y - z_d \quad \text{on} \quad \Gamma \;, \\ \int_\Gamma (p + Nu)(v-u)d\Gamma \geq 0 \quad \forall \quad v \in U_{ad}, \; u \in U_{ad} \;. \quad [] \end{cases}$$

System (1.13) is called the <u>optimality system</u>.

<u>Remark 1.2</u>

Since solving (1.13) <u>is equivalent</u> to solving the problem of Optimal control, (1.13) admits a unique solution. []

In order to study the structure of (1.13) let us consider some particular cases.

## 1.2 <u>The case without constraints</u>.

<u>If</u> $U_{ad} = U$ (the "no constraints" case), the inequality in (1.13) is equivalent to

$$p + Nu = 0 \; . \tag{1.14}$$

The optimality system is equivalent to the linear boundary value problem

$$\begin{aligned} &Ay = f, \quad Ap = 0 \;, \text{ in } \Omega \\ &\frac{\partial y}{\partial \nu} + \frac{1}{N}p = 0 \;, \; \frac{\partial p}{\partial \nu} = y - z_d \quad \text{on } \Gamma \end{aligned} \tag{1.15}$$

and then $u$ is given by (1.14).

1.3 A case with constraints. Commutation lines.

We consider now the case when $U_{ad}$ is given by

$$U_{ad} = \{v | v \in L^2(\Gamma) \;, \; 0 \le v < M\} \; . \tag{1.16}$$

Then one easily checks that the inequality in (1.13) is equivalent to:

$$\begin{cases} 0 < u < M \Rightarrow p + Nu = 0 \\ u = 0 \quad \Rightarrow \quad p \geq 0 \\ u = M \quad \Rightarrow \quad p + NM \leq 0 \; . \end{cases} \tag{1.17}$$

Therefore we have three regions on $\Gamma$ :

$$\begin{aligned} &\Gamma_0 = \{x | u(x) = 0\} \;, \; \Gamma_M = \{x | u(x) = M\} \;, \\ &S = \{x | 0 < u(x) < M \;, \end{aligned} \tag{1.18}$$

which are defined on $\Gamma$ up to a set of measure $0$ on $\Gamma$.

The boundaries of these regions are the commutation lines.

Remark 1.3

One sees here the analogy between this problem and problems of free surfaces, or free boundaries, as they appear in classical physics. We shall return to this remark on several occasions.

## 2. The case $N = 0$ without constraints.

### 2.1 Orientation

In many applications, one meets the case when $N = 0$ in (1.6) and also the case when $N$ is "small" (1).

We consider in this section Problem (1.7) with

$$U_{ad} = U \; , \quad N = 0 \; . \tag{2.1}$$

### 2.2 A "non-existence" result.

We are going to show that

(2.2) "in general" (2) problem (1.7) with (2.1) does not admit a solution.

---

(1) Which corresponds to a "cheap" control. We shall return on this point.

(2) This is made precise below.

Let us check first that

$$\inf J(v) = 0 . \tag{2.3}$$

We consider a sequence $z_{dj}$ of <u>smooth</u> functions on $\Gamma$ such that

$$z_{dj} \to z_d \quad \text{in} \quad L^2(\Gamma) . \tag{2.4}$$

Let $\phi_j$ be the (smooth) solution of

$$A\phi_j = f , \quad \phi_j = z_{dj} \quad \text{on} \quad \Gamma ; \tag{2.5}$$

(actually $\phi_j \in H^2(\Omega)$ , Sobolev space of order 2).

Let us <u>define</u>

$$v_j = \frac{\partial \phi_j}{\partial \nu} , \tag{2.6}$$

so that $v_j \in L^2(\Gamma)$ (actually $v_j \in H^{1/2}(\Gamma)$) .

We have of course

$$J(v_j) = \int_\Gamma |z_{dj} - z_d|^2 d\Gamma \tag{2.7}$$

and $J(v_j) \to 0$ and (2.3) follows. $\square$

Let us admit for a moment that there exists a $u$ such that

(2.8) $$J(u) = 0 .$$

Then $y(u) = y$ satisfies

(2.9) $$Ay = f \text{ in } \Omega , \quad y = z_d \text{ on } \Gamma$$

and then $u$ is _formally_ defined by

(2.10) $$u = \frac{\partial y}{\partial \nu} .$$

But in general if $z_d \in L^2(\Gamma)$ then $u$, as defined by (2.10), _does not belong to_ $L^2(\Gamma)$ ; but one shows that $\frac{\partial y}{\partial \nu} \in H^{-1}(\Gamma)$ (1). (cf. Lions-Magenes [1]) and that $\frac{\partial y}{\partial \nu} \in L^2(\Gamma)$ if $z_d \in H^1(\Gamma)$ .
Summing up we have:

_Theorem 2.1. Problem_ (1.7) _under condition_ (2.1) _satisfies_ (2.3). _If_ $z_d$ _satisfies the supplementary condition:_ $z_d \in H^1(\Gamma)$ , _then there exists an optimal control, given by_ (2.10).
_If_ $z_d \in L^2(\Gamma)$, $z_d \notin H^1(\Gamma)$ , _then there does not exist an optimal control in_ $L^2(\Gamma)$ . _But one can define a "relaxed" optimal control in_ $H^{-1}(\Gamma)$ , _still given by_ (2.10).

_Remark 2.1_

The fact that, in general, $u$ exists in a _larger_ space than

(1) $H^{-1}(\Gamma)$ = dual space of $H^1(\Gamma)$ = distributions of order $-1$ on $\Gamma$ .

$L^2(\Gamma)$ (where one has the optimal control when $N > 0$), is typical of <u>singular perturbations</u>. This will be made precise later.

## 3. The case $N = 0$ with constraints.

### 3.1 A result of existence and uniqueness.

We consider now the case $N = 0$ and

(3.1) $$U_{ad} = \{v \mid 0 \leq v \leq M\} .$$

Since $U_{ad}$ is <u>bounded</u> in $L^2(\Gamma)$, the existence of a $u$ in $U_{ad}$ which minimizes

(3.2) $$J(v) = \int_\Gamma |y(v) - z_d|^2 d\Gamma$$

is immediate.

One checks that the function $v \to J(v)$ is strictly convex so that

(3.3) there exists a unique $u$ in $U_{ad}$ (defined by (3.1)) such that $J(u) = \inf J(v)$ . []

Moreover, the conclusions of Theorem 2.2 are still valid, with $N = 0$ ; therefore $u$ is characterized by the system:

$$(3.4)\qquad \begin{aligned} &Ay = f\ ,\quad Ap = 0\\ &\frac{\partial y}{\partial \nu} = u\ ,\quad \frac{\partial p}{\partial \nu} = y - z_d \quad \text{on } \Gamma\ ,\\ &\int_\Gamma p(v-u)d\Gamma \geq 0 \quad \forall\, v \in U_{ad}\ ,\ u \in U_{ad}\ . \end{aligned}$$

3.2 <u>A "Bang-Bang" result.</u>

Let $\Phi$ be the solution of

$$(3.5)\qquad A\Phi = f\ ,\quad \Phi = z_d \quad \text{on } \Gamma\ .$$

<u>In the particular case</u> when

$$(3.6)\qquad 0 \leq \frac{\partial \Phi}{\partial \nu} \leq M \quad \text{on } \Gamma\ ,$$

the optimal solution is given by

$$(3.7)\qquad u = \frac{\partial \Phi}{\partial \nu}\ .$$

<u>We study now the case when</u> (3.6) <u>does not hold</u>.

The interpretation of the inequality in (3.4) is given by (1.17), where $N = 0$.

We are going to show (in a somewhat formal manner) that

(3.8) $S$ is empty

where $S$ = set of $\Gamma$ when $0 < u < M$ , $p = 0$ .

Therefore $\Gamma = \Gamma_0 \cup \Gamma_M$ , $u = 0$ on $\Gamma_0$ , $u = M$ on $\Gamma_M$ , a result of the Bang-Bang type. []

Let us show first that $S = \Gamma$ is impossible. Instead assume to the contrary that $S = \Gamma$ ; then $p = 0$ on $\Gamma$ and then $p = 0$ in $\Omega$ and $\frac{\partial p}{\partial \nu} = y - z_d = 0$ on $\Gamma$ , i.e. $y = \Phi$ (cf. (3.5)), a case that we have excluded. []

We show now that there does not exist a set, say $S_1$ , with $S_1 \subset S$, measure $(S_1) > 0$ , such that $y = z_d$ on $S_1$ . Indeed, if such an $S_1$ exists, then

$$p = 0 \ \underline{\text{and}} \ \frac{\partial p}{\partial \nu} = 0 \ \text{on} \ S_1 ,$$

hence it follows (by the uniqueness of the Cauchy problem) that $p = 0$ in $\Omega$ and again, $y = \Phi$ , contrary to our assumptions.

Consequently, either $S$ is of measure 0 (and the proof is completed), or there exists $S_+ \subset S$ (or $S_- \subset S$) , of positive measure and such that

$$y > z_d \ \text{on} \ S_+ \ \ (\text{or} \ y < z_d \ \text{on} \ S_- ).$$

We show that the existence of such an $S_+$ (or $S_-$ ) is impossible. Let us prove it for $S_+$ . Indeed, on $S_+$ (we assume that everything is "regular"; this is the formal aspect of this "proof"):

$$p = 0 \quad (\text{since} \quad S_+ \subset S) \;, \quad \frac{\partial p}{\partial \nu} > 0 \quad (\text{since} \quad \frac{\partial p}{\partial \nu} = y - z_d) \;;$$

therefore

$$p < 0 \text{ in a neighborhood } \sigma \text{ of } S_+$$

and $\Delta p = p$ in $\sigma$ implies $\Delta p < 0$ in $\sigma$.
But $\Delta p < 0$ in $\sigma$ and $p = 0$ on $S_+$ imply $\frac{\partial p}{\partial \nu} \leq 0$ on $S_+$, a contradiction. []

# CHAPTER 2

# A PRIORI FEEDBACK AND VARIATIONAL INEQUALITIES[1]

## 1. A Stationary Problem

### 1.1 Formulation of the Problem.

Let us consider again the problem of Section 3, Chapter 1. We want to minimize

$$J(v) = \int_\Gamma |y(v) - z_d|^2 \, d\Gamma \tag{1.1}$$

when $v = \frac{\partial y(v)}{\partial \nu}$ satisfies $0 \leq v \leq M$ .

We can argue, in a formal manner, as follows:

We want to keep $y(v)$ "as close as possible" to $z_d$ ; therefore if $y(v)$ "becomes" $> z_d$ , then we take $v$ minimum, i.e., $v = 0$ and if $y(v)$ "becomes" $< z_d$ , then we take $v$ maximum, i.e., $v = M$ . Therefore the optimal solution should be "close" to a function $\Phi$ - if such a function exists - which satisfies:

$$A\Phi = f \quad \text{in} \quad \Omega \tag{1.2}$$

---

[1] We shall write V.I. for Variational Inequalities.

and

(1.3)
$$\begin{cases} 0 \le \dfrac{\partial\Phi}{\partial\nu} \le M \,, \\ \Phi > z_d \Rightarrow \dfrac{\partial\phi}{\partial\nu} = 0 \,, \\ \Phi < z_d \Rightarrow \dfrac{\partial\phi}{\partial\nu} = M \,. \end{cases}$$

<u>Remark 1.1</u>

The same reasoning applies to <u>other functionals</u>, such as for instance:

(1.4)
$$J_1(v) = \int_\Gamma |y(v) - z_d| d\Gamma \,.$$

<u>Remark 1.2</u>

We show below that (1.2) (1.3) admits a unique solution. Then we <u>define</u> a <u>sub-optimal control</u> by

(1.5)
$$w = \frac{\partial\Phi}{\partial\nu} \,.$$

<u>Open Problem 1.1.</u>

Obtain estimates of $J(w) - J(u)$, where $u$ is the optimal control.

<u>Open Problem 1.2.</u>

Does there exist a cost function $G(v)$ such that

$$G(w) = \inf G(v) \,, \ 0 \le v \le M \ ?$$

Remark 1.3

In the above considerations, we chose a "close to the optimal policy" in terms of the state, which somewhat justifies the terminology "a priori feedback"; this terminology will be more adequate in the evolution problems we consider in the following sections.

1.2 Formulation as a V.I.

Let us define:

$$(1.6) \qquad j(\phi) = M\int_\Gamma (\phi - z_d)^- d\Gamma \ , \ \phi \in H^1(\Omega) \ , \ ^{(1)}.$$

The function $\phi \to j(\phi)$ is continuous, convex from $H^1(\Omega) \to \mathbb{R}$, and it is not differentiable.

We are now going to show that (1.2)(1.3) is equivalent to the following V.I.:

$$(1.7) \qquad a(\Phi,\phi-\Phi) + j(\phi) - j(\Phi) \geq (f,\phi-\Phi) \ \forall \ \phi \in H^1(\Omega) \ ,$$

where $a(y,\phi)$ is defined by (1.4), Chapter 1.

Proof

Let us first take $\phi = \Phi \pm \psi$, $\psi \in D(\Omega)$ (2). Then $j(\phi) = j(\Phi)$ and (1.7) reduces to

(1) $\lambda^+ = \sup(\lambda,0)$, $\lambda^- = \sup(-\lambda,0)$ .

(2) Space of $C^\infty$ functions in $\Omega$, with compact support.

(1.8) $$a(\Phi,\psi) = (f,\psi) \ \forall\ \phi \in D(\Omega) ,$$

(1.9) $$A\Phi = f .$$

We now multiply (1.9) by $\phi - \Phi$ and we apply Green's formula to obtain:

$$- \int_\Gamma \frac{\partial\phi}{\partial\nu} (\phi-\Phi)d\Gamma + a(\Phi,\phi-\Phi) = (f,\phi-\Phi)$$

$$\leq a(\Phi,\phi-\Phi) + j(\phi) - j(\Phi) , \text{ using } (1.7).$$

Hence

(1.10) $$\int_\Gamma \left(\frac{\partial\Phi}{\partial\nu}(\phi-\Phi) + M(\phi-z_d)^- - M(\Phi-z_d)^-\right) d\Gamma \geq 0 \quad \forall\ \phi \in H^1(\Omega) .$$

Let us take[1]

(1.11) $$\phi = z_d + \lambda\psi, \ \psi \geq 0 \ \text{ on } \Gamma , \ \lambda \geq 0 ;$$ (1.10) becomes

(1.12) $$\lambda\int_\Gamma \frac{\partial\Phi}{\partial\nu} \psi \ d\Gamma + X \geq 0 ,$$

where

(1.13) $$X = \int_\Gamma \left(\frac{\partial\Phi}{\partial\nu}(z_d - \Phi) - M(\Phi-z_d)^-\right) d\Gamma .$$

---

[1] One extends (1.10) to functions which are defined on $\Gamma$ only and belong to $L^2(\Gamma)$ . Then (1.11) is valid.

Letting $\lambda \to +\infty$ in (1.12), we see that

$$\int_\Gamma \frac{\partial\Phi}{\partial\nu}\psi \, d\Gamma \geq 0 \quad \forall \quad \psi \geq 0, \quad \text{i.e.,} \tag{1.14}$$

$$\frac{\partial\Phi}{\partial\nu} \geq 0 \quad \text{on} \quad \Gamma \, .$$

Next (1.12) gives

$$X \geq 0 \, . \tag{1.15}$$

We now take in (1.10) $\phi$ given by

$$\phi = z_d - \lambda\psi \, , \; \psi \geq 0 \quad \text{on} \quad \Gamma \, , \; \lambda \geq 0 \, . \tag{1.16}$$

It becomes $\lambda\int_\Gamma (M - \frac{\partial\Phi}{\partial\nu})\psi \, d\Gamma + X \geq 0$ . Hence

$$\frac{\partial\Phi}{\partial\nu} \leq M \quad \text{on} \quad \Gamma \, . \tag{1.17}$$

But

$$\frac{\partial\Phi}{\partial\nu}(z_d - \Phi) - M(\Phi - z_d)^- = -\frac{\partial\Phi}{\partial\nu}(\Phi - z_d)^+ - (M - \frac{\partial\Phi}{\partial\nu})(\Phi - z_d)^- \leq 0$$

(using (1.16), (1.17)), so that $X \leq 0$ and this together with (1.15) shows that

$$\frac{\partial\Phi}{\partial\nu}(\Phi - z_d)^+ + (M - \frac{\partial\Phi}{\partial\nu})(\Phi - z_d)^- = 0 \, . \tag{1.18}$$

One can invert the computations, so that (1.7) is equivalent to

(1.9) (1.16) (1.17) (1.18). But one easily checks that (1.3) is equivalent to (1.16) (1.17) (1.18). []

Remark 1.4.

If $j$ were differentiable, (1.7) would be equivalent to the non-linear boundary value problem:

$$(1.19) \qquad a(\Phi,\phi) + (j'(\Phi),\phi) = (f,\phi) \quad \forall \quad \phi \in H^1(\Omega) \;.$$

In fact $j$ is not differentiable, so that $j'$ is actually a multi-valued operator so that (1.7) can be thought of as a multi-valued equation.

Remark 1.5.

Since $a(y,\phi)$ is symmetric, (1.7) is equivalent to minimizing

$$(1.20) \qquad K(\phi) = \frac{1}{2}\, a(\phi,\phi) + j(\phi) - (f,\phi) \quad \text{over} \quad H^1(\Omega) \;.$$

But (1.7) makes sense in the nonsymmetric case. In any case, in the symmetric case we are considering now it follows from (1.20) that

Theorem 1.1. Problem (1.7) admits a unique solution.

Remark 1.6.

In the non-symmetric case, under the hypothesis

$$(1.21) \qquad a(\phi,\phi) \geq \alpha\|\phi\|^2_{H^1(\Omega)}, \quad \alpha > 0, \ \forall \ \phi \in H^1(\Omega) ,$$

one can show the existence and uniqueness of a solution of (1.7) - cf. J.L. Lions and G. Stampacchia [1].

Remark 1.7.

V.I. of the above type have been introduced, in a different context, in Duvaut-Lions [1], for problems in Mechanics.

## 2. An Evolution Problem

### 2.1 Formulation of the problem.

Let us suppose that the state is given by the solution of

$$(2.1) \qquad \frac{\partial y}{\partial t} - \Delta y = f \quad \text{in} \quad Q = \Omega \times (0,T) ,$$

$$(2.2) \qquad y(x,0) = y_0(x) \quad \text{in} \quad \Omega ,$$

$$(2.3) \qquad \frac{\partial y}{\partial \nu} = v \quad \text{on} \quad \Sigma = \Gamma \times (0,T) ;$$

We assume that $f$ and $y_0$ are given in $L^2(Q)$ and $L^2(\Omega)$ resp. and that $v$ belongs to $U = L^2(\Sigma)$ .

The problem (2.1) (2.2) (2.3) admits a unique solution $y = y(v)$ , which is such that (in particular)

$$y(v) \in L^2(0, T ; H^1(\Gamma)) , \tag{2.4}$$

the mapping $v \to y(v)$ being affine continuous from $L^2(\Sigma)$ into $L^2(0,T;H^1(\Gamma))$ .

The variational formulation of the problem is:

$$\left(\frac{\partial y}{\partial t}, \phi\right) + a(y,\phi) = (f,\phi) + \int_\Gamma v\phi d\Gamma \quad \forall \ \phi \in H^1(\Gamma) \tag{2.5}$$

where

$$a(y,\phi) = \sum_{i=1}^{n} \int_\Omega \frac{\partial y}{\partial x_i} \frac{\partial \phi}{\partial x_i} dx . \tag{2.6}$$

Cost function:

$$J(v) = \int_\Sigma |y(v) - z_d|^2 d\Sigma , \tag{2.7}$$

where $z_d$ is given in $L^2(\Sigma)$ .

Constraints: We suppose that $v \in U_{ad}$ , where

$$U_{ad} = \{v | v \in L^2(\Sigma) , 0 \leq v \leq M\} . \tag{2.8}$$

[]

The problem is:

(2.9) $$\inf J(v) , \quad v \in U_{ad} .$$

<u>This problem admits a unique solution</u> $u$ .

One can derive an optimality system along lines similar to those of Chapter 1. We intend to give now a "<u>sub-optimal</u>" <u>solution</u>, which is based on "<u>a priori feedback</u>".

2.2 <u>A priori feedback</u>.

We follow the same arguments as in Section 1.1.

If $y(v)$ "becomes" $> z_d$ , we use the minimum of $v$ , i.e. $u = 0$ and if it "becomes" $< z_d$ , we take $u = M$ .

Therefore we are led to find - <u>if possible!</u> - a function $\Phi$ which satisfies

(2.10) $$\frac{\partial \Phi}{\partial t} - \Delta\Phi = f \quad \text{in} \quad Q ,$$

(2.11) $$\Phi(x,0) = y_0(x) \quad \text{in} \quad \Omega ,$$

and on $\Sigma$ the conditions (similar to (1.3)):

(2.12) $$\begin{cases} 0 \leq \frac{\partial \Phi}{\partial \nu} \leq M , \\ \Phi > z_d \Rightarrow \frac{\partial \Phi}{\partial \nu} = 0 , \\ \Phi < z_d \Rightarrow \frac{\partial \Phi}{\partial \nu} = M . \end{cases}$$ [ ]

We have Remarks entirely similar to Remarks 1.1 and 1.2, and open problems similar to 1.1, 1.2 Section 1.

## 2.3 V.I. of evolution.

We define $j$ by (1.6).

One shows, as in Section 1.2, that the problem (2.10), (2.11), (2.12) is equivalent to

$$(2.13) \qquad \left(\frac{\partial\Phi}{\partial t}, \phi-\Phi\right) + a(\Phi,\phi-\Phi) + j(\phi) - j(\Phi) \geq (f,\phi-\Phi) \;\forall\; \phi \in H^1(\Omega)$$

with condition (2.11).

This is a V.I. inequality of evolution (cf. Lions-Stampacchia [1], Duvaut-Lions [1]).

One can show the existence and uniqueness of a (weak) solution of (2.13).

We refer to the Bibliography for proofs of this result. The most "elementary" method of proof is by using finite differences. Let us introduce a mesh $\Delta t$ and let us define $\Phi^n$ as the "approximation" of $\Phi(n\Delta t)$. We start with

$$(2.14) \qquad \Phi^0 = y_0$$

and we inductively define $\Phi^n$, $n \geq 1$, by:

$$(2.15) \qquad \left(\frac{\Phi^n - \Phi^{n-1}}{\Delta t}, \phi-\Phi^n\right) + a(\Phi^n, \phi-\Phi^n) + j(\phi) - j(\Phi^n)$$

$$\geq (f^n, \phi-\Phi^n)$$

where, for instance, $f^n = \frac{1}{\Delta t}\int_{(n-1)\Delta t}^{n\Delta t} f(\sigma)d\sigma$ .

Since $a$ is symmetric, solving (2.15) is equivalent to minimizing

$$(2.16)\quad \frac{1}{2}\, a(\phi,\phi) + \frac{1}{2\Delta t}\,(\phi,\phi) + j(\phi) - (f,\phi) + \frac{1}{\Delta t}(\Phi^{n-1},\phi)$$

over $H^1(\Omega)$ , and therefore, uniquely defines $\Phi^n$ .

One can show next (cf. Glowinski, Lions, Trémolieres [1]) that the step function $\Phi_{\Delta t}$ , which equals $\Phi^n$ in $[n\Delta t,(n+1)\Delta t]$ , converges (in $L^2(0,T;\, H^1(\Omega))$) to $\Phi$ , as $\Delta t \to 0$ .

## 3. Another type of V.I. of evolution.

### 3.1 A forcing problem. (cf. Duvaut-Lions [1]).

Suppose we consider a system whose state is given by (2.1), subject to (2.2), and that we want $\frac{\partial y}{\partial t}$ to be $\geq 0$ on $\Sigma$ , with "minimal" expenditure of $\frac{\partial y}{\partial \nu}$ . Then if $\frac{\partial y}{\partial t} > 0$ , we shall take $\frac{\partial y}{\partial \nu} = 0$ and if $\frac{\partial y}{\partial t} = 0$ , then $\frac{\partial y}{\partial \nu} > 0$ .

Summing up:

$$(3.1)\quad \frac{\partial y}{\partial t} \geq 0\ ,\quad \frac{\partial y}{\partial \nu} \geq 0\ ,\quad \frac{\partial y}{\partial t}\,\frac{\partial y}{\partial \nu} = 0 \quad \text{on } \Sigma\ .$$

We want to find a function $y$ which satisfies (2.1) (2.2) and (3.1).

3.2 V.I. of evolution.

One can check that the above problem is equivalent to finding $y$ such that

$$(3.2)\qquad \left(\frac{\partial y}{\partial t}, \phi - \frac{\partial y}{\partial t}\right) + a\left(y, \phi - \frac{\partial y}{\partial t}\right) \geq \left(f, \phi - \frac{\partial y}{\partial t}\right) \quad \forall \ \phi \in K$$

$$(3.3)\qquad \frac{\partial y}{\partial t} \in K ,$$

$$(3.4)\qquad y(x,0) = y_0(x) ,$$

where $K$ is the (non empty) closed convex subset of $H^1(\Omega)$ defined by

$$(3.5)\qquad K = \{\phi \,|\, \phi \in H^1(\Omega) , \quad \phi \geq 0 \text{ on } \Gamma\} .$$

One can show (cf. Duvaut-Lions [1]) that there exists a unique solution of (3.2) (3.3) (3.4).

Remark 3.1.

The problem (3.2), (3.3), (3.4) is a V.I. of evolution of a different type from the one considered in Section 2.

# CHAPTER 3

## A PRIORI FEEDBACK AND QUASI VARIATIONAL INEQUALITIES[1]

## 1. A stationary problem.

### 1.1 Formulation of the problem.

Let us again consider the state to be given by

$$Ay(v) = f, \quad A = -\Delta + I , \tag{1.1}$$

$$\frac{\partial}{\partial \nu} y(v) = v \quad \text{on} \quad \Gamma . \tag{1.2}$$

We set, for $\phi \in H^1(\Omega)$

$$\mathfrak{m}(\phi) = \text{mean value of } \phi \text{ on } \Gamma = \frac{1}{\text{meas.}\Gamma} \int_\Gamma \phi d\Gamma \tag{1.3}$$

and we consider the cost function

$$J(v) = \int_\Gamma |y(v) - \mathfrak{m}y(v)|^2 d\Gamma \tag{1.4}$$

the problem being to minimize $J$ over $U_{ad}$ :

$$U_{ad} = \{v | 0 \leq v \leq M\} . \qquad \square \tag{1.5}$$

(1) We shall write Q.V.I. for Quasi Variational Inequalities.

It is very simple to check that there exists $u \in U_{ad}$ such that

(1.6) $$J(u) = \inf J(v) , \quad u \in U_{ad} . \qquad []$$

The uniqueness of $u$ is not true in general. Indeed, let us consider all solutions of

(1.7) $$\begin{cases} A\phi = f \\ \phi = \text{constant on } \Gamma . \end{cases}$$

If we define $\Phi$ and $\psi$ by

(1.8) $$\begin{cases} A\Phi = f \text{ in } \Omega , \quad \Phi = 0 \text{ on } \Gamma , \\ A\psi = 0 \text{ in } \Omega , \quad \psi = 1 \text{ on } \Gamma \end{cases}$$

then

(1.9) $$\phi = \Phi + \lambda\psi , \quad \lambda \in R$$

satisfies (1.7) and if we define

(1.10) $$v = \frac{\partial\Phi}{\partial\nu} + \lambda \frac{\partial\psi}{\partial\nu}$$

then $J(v) = 0$ , and $v$ is an optimal control iff it belongs to $U_{ad}$ .

Summing up: if there exist constants $\lambda$ such that

$$(1.11) \qquad 0 \leq \frac{\partial \Phi}{\partial \nu} + \lambda \frac{\partial \psi}{\partial \nu} \leq M \quad \text{on} \quad \Gamma$$

then all the optimal controls are given by (1.10). []

One can write the optimality system as follows; let $u$ be an optimal control and let us set $y(u) = y$. We introduce the adjoint state $p$ by

$$(1.12) \qquad \begin{aligned} & Ap = 0 \quad \text{in} \quad \Omega \\ & \frac{\partial p}{\partial \nu} = y - \mathfrak{m}(y) \quad \text{on} \quad \Gamma \end{aligned}$$

and the optimality condition is

$$(1.13) \qquad \int_\Gamma p(v-u)d\Gamma \geq 0 \quad \forall \quad v \in U_{ad} .$$

Indeed the condition $\frac{1}{2}(J'(u), v-u) \geq 0 \quad \forall \quad v \in U_{ad}$ becomes

$$\int_\Gamma (y-\mathfrak{m}(y))(y(v)-y-\mathfrak{m}(y(v)-y))d\Gamma \geq 0$$

i.e.

$$\int_\Gamma (y-\mathfrak{m}(y))(y(v)-y)d\Gamma \geq 0 ;$$

multiplying (1.12) by $y(v)-y$ and applying Green's formula gives (1.13). []

Let us give now a solution based on direct considerations similar to those in Chapter 2, Section 1.

1.2 A priori feedback.

If $y$ "becomes" less than its mean value $M(y)$, we apply the maximum $v$, i.e. $\frac{\partial y}{\partial \nu} = M$, and if $y$ "becomes" greater than $M(y)$, we apply the minimum of $v$, i.e. $\frac{\partial y}{\partial \nu} = 0$.

We are thus lead to solve - if possible - the problem:

$$(1.14) \qquad A\Phi = f \text{ in } \Omega ,$$

$$(1.15) \qquad \begin{cases} 0 \leq \frac{\partial \Phi}{\partial \nu} \leq M , \\ \Phi < \mathfrak{m}(\Phi) \Rightarrow \frac{\partial \Phi}{\partial \nu} = M \\ \Phi > \mathfrak{m}(\Phi) \Rightarrow \frac{\partial \Phi}{\partial \nu} = 0 . \end{cases}$$

Remark 1.1

If we find a solution $\Phi$ of (1.14), (1.15), then we can take as sub-optimal control

$$(1.16) \qquad w = \frac{\partial \Phi}{\partial \nu} .$$

Remark 1.2 (Similar to Remark 1.1, Section 1, Chapter 2).

The same reasoning would apply to other functionals, such as, for instance

(1.17) $$J_1(v) = \int_\Gamma |y - \mathfrak{m}(y)| d\Gamma .$$

<u>Open problems</u>: Similar to Problems 1.1 and 1.2, section 1.1, Chapter 2.

We are now going to show the existence of solution(s) of (1.16), (1.15), under the hypothesis

(1.18) $$f \geq 0 .$$

1.3 <u>Formulation as a Q.V.I.</u>

For $\phi,\psi$ in $H^1(\Omega)$ we define

(1.19) $$j(\phi,\psi) = M\int_\Gamma (\phi - \mathfrak{m}(\psi))^- d\Gamma .$$

We are going to check <u>that</u> (1.14)(1.15) <u>is equivalent to</u>

(1.20) $$a(\Phi,\phi-\Phi) + j(\phi,\Phi) - j(\Phi,\Phi) \geq (f,\phi-\Phi) \;\forall\; \phi \in H^1(\Omega) .$$

<u>Remark 1.3</u>

The inequality (1.20) is a Q.V.I. The structure of (1.20) is clearly an <u>extension</u> of the structure of V.I.

<u>Remark 1.4</u>

Q.V.I. were introduced by Bensoussan and the author (cf. Bensoussan-Lions [1] [3], Bensoussan-Goursat-Lions [1]) for <u>impulse control problems</u> (see Chapter 8) - Q.V.I. of type (1.20)

are slightly different from those previously introduced. A complete report will be included in Bensoussan-Lions [4]. []

Proof of the equivalence between (1.14)(1.15) and (1.20).

Taking firstly $\phi = \Phi \pm \psi$, $\psi \in D(\Omega)$, in (1.20), shows that $\Phi$ satisfies (1.16). Multiplying (1.16) by $\phi - \Phi$ and applying Green's formula gives

$$(1.21) \qquad \int_\Gamma \frac{\partial \Phi}{\partial \nu} (\phi - \Phi) d\Gamma + j(\phi, \Phi) - j(\Phi, \Phi) \geq 0 \ \forall \ \phi .$$

We take next in (1.21)

$$(1.22) \qquad \phi = m(\Phi) \pm \lambda \psi , \quad \lambda \in R_+ , \ \psi \geq 0 \ \text{on} \ \Gamma .$$

We obtain the two conditions

$$(1.23) \qquad \begin{cases} \lambda \int_\Gamma \frac{\partial \Phi}{\partial \nu} \psi \, d\Gamma + X \geq 0 , \\ \lambda \int_\Gamma \psi (M - \frac{\partial \Phi}{\partial \nu}) d\Gamma + X \geq 0 , \end{cases}$$

where

$$(1.24) \qquad X = \int_\Gamma [\frac{\partial \Phi}{\partial \nu} (m(\Phi) - \Phi) - M(\Phi - m(\Phi))^- ] d\Gamma .$$

It follows that we have (1.15) (as in Chapter 2, Section 1.2).

Remark 1.5

It does not seem possible to formulate the problem (1.14) (1.15) as a V.I.

1.4 Existence of solutions of the Q.V.I. (I)

We now show the existence of a maximal solution of (1.20) by an iterative procedure.

We start with $\Phi^o$ , a solution of the Neumann problem

(1.25) $$A\Phi^o = f \text{ in } \Omega , \frac{\partial\Phi^o}{\partial\nu} = M .$$

We then define $\Phi^1$ as the solution of the V.I.:

(1.26) $$a(\Phi^1,\phi-\Phi^1) + j(\phi,\Phi^o) - j(\Phi^1,\Phi^o) \geq (f,\phi-\Phi^1), \forall \phi .$$

[We remark that solving (1.26) is equivalent to minimizing

$$\frac{1}{2}a(\phi,\phi) + j(\phi,\Phi^o) - (f,\phi) \text{ over } H^1(\Omega)] .$$

We then define inductively

(1.27) $$a(\Phi^n,\phi-\Phi^n) + j(\phi,\Phi^{n-1}) - j(\Phi^n,\Phi^{n-1}) \geq (f,\phi-\Phi^n) \forall \phi .$$

Let us show that

(1.28) $$\Phi^o \geq \Phi^1 \geq \ldots \geq \Phi^{n-1} \geq \Phi^n \geq \ldots \geq 0 . \qquad []$$

Let us remark that (1.25) is equivalent to

$$a(\Phi^o,\phi) = (f,\phi) + \int_\Gamma M\phi d\Gamma \; . \tag{1.29}$$

We take in (1.26)

$$\phi = \Phi^1 = -(\Phi^o - \Phi^1)^- \quad \text{i.e.} \quad \phi = \inf(\phi^o, \phi^1) \tag{1.30}$$

and $\phi = (\Phi^o - \Phi^1)^-$ in (1.29). We obtain

$$a(\Phi^o - \Phi^1 \; , \; (\Phi^o - \Phi^1)^-) - Y \geq 0 \; , \tag{1.31}$$

where

$$Y = M\int_\Gamma (\Phi^o - \Phi^1)^- \, d\Gamma + j(\Phi^1,\Phi^o) - j(\inf(\Phi^o,\Phi^1),\Phi^o) \; . \tag{1.32}$$

It is a simple exercise to check that

$$Y \geq 0 \tag{1.33}$$

and since $a(\psi,\psi^-) = -a(\psi^-,\psi^-)$ , (1.31) gives

$$a((\Phi^o - \Phi^1)^- \; , \; (\Phi^o - \Phi^1)^-) + Y \leq 0 \; . \tag{1.34}$$

But (1.33) (1.34) imply that $(\Phi^o - \Phi^1)^- = 0$ , i.e. $\Phi^o \geq \Phi^1$ .

Let us admit recursively that $\Phi^1 \geq \Phi^2 \geq \ldots \geq \Phi^{n-1}$ and let us show that

$$\Phi^{n-1} \geq \Phi^n . \tag{1.35}$$

We set $\mathfrak{m}(\Phi^j) = \mathfrak{m}^j$ and we notice that $\Phi^{n-2} \geq \Phi^{n-1}$ implies

$$\mathfrak{m}^{n-2} \geq \mathfrak{m}^{n-1} . \tag{1.36}$$

We choose in (1.27) $\phi$ by $\phi - \Phi^n = -(\Phi^{n-1} - \Phi^n)^-$ , i.e. $\phi = \inf(\Phi^{n-1}, \Phi^n) = \inf$, and in the analogous V.I. for $\Phi^{n-1}$ we choose $\phi - \Phi^{n-1} = (\Phi^{n-1} - \Phi^n)^-$ i.e.

$$\phi = \sup(\Phi^{n-1}, \Phi^n) = \sup .$$

Adding up the results, we obtain

$$a((\Phi^{n-1} - \Phi^n)^- , (\Phi^{n-1} - \Phi^n)^-) + Z \leq 0 \tag{1.37}$$

where

$$Z = j(\Phi^{n-1},\Phi^{n-2}) + j(\Phi^n,\Phi^{n-1}) - j(\inf,\Phi^{n-1}) - j(\sup,\Phi^{n-2}) .$$

Therefore $Z = M\int_\Gamma z d\Gamma$ , where

$$z = (\Phi^{n-1}-\mathfrak{m}^{n-2})^- + (\Phi^n-\mathfrak{m}^{n-1})^- - (\inf - \mathfrak{m}^{n-1})^- - (\sup - \mathfrak{m}^{n-2})^- .$$

We look at the various possible cases.

If $\Phi^{n-1} \geq \mathfrak{m}^{n-2}$ and $\Phi^n \geq \mathfrak{m}^{n-1}$ then $z = 0$.

If $\Phi^{n-1} < \mathfrak{m}^{n-2}$ and $\Phi^n < \mathfrak{m}^{n-1}$ then again $z = 0$.

If $\Phi^{n-1} \geq \mathfrak{m}^{n-2}$ and $\Phi^n < \mathfrak{m}^{n-1}$ then

$$z = \mathfrak{m}^{n-1} - \Phi^n - \mathfrak{m}^{n-1} + \inf(\Phi^n, \Phi^{n-1}) = 0 \text{, since}$$

$\mathfrak{m}^{n-1} \leq \mathfrak{m}^{n-2}$ so that $\Phi^n < \Phi^{n-1}$.

Let us consider the last case $\Phi^{n-1} < \mathfrak{m}^{n-2}$, $\Phi^n \geq \mathfrak{m}^{n-1}$. Since $\Phi^n \geq \mathfrak{m}^{n-1}$ we have $\inf \geq \mathfrak{m}^{n-1}$ and

$$z = \mathfrak{m}^{n-2} - \Phi^{n-1} - (\sup - \mathfrak{m}^{n-2})^- .$$

If $\sup \geq \mathfrak{m}^{n-2}$ then $z \geq 0$.

If $\sup < \mathfrak{m}^{n-2}$ then $z = \sup - \Phi^{n-1} \geq 0$.

Therefore $z \geq 0$ in all possible cases, so that $Z \geq 0$ and (1.37) imply that $(\Phi^n - \Phi^{n-1})^- = 0$, hence (1.35) follows. Let us choose now $\phi$ in (1.27) by

$$\phi - \Phi^n = (\Phi^n)^- , \quad \text{i.e. } \phi = (\Phi^n)^+ .$$

We obtain

$$(1.38) \qquad a((\Phi^n)^-, (\Phi^n)^-) + j(\Phi^n,\Phi^{n-1}) - j((\Phi^n)^+,\Phi^{n-1}) + \\ + (f,(\Phi^n)^+) \leq 0 .$$

But $j(\Phi^n,\Phi^{n-1}) - j((\Phi^n)^+, \Phi^{n-1}) \geq 0$ and since we assumed that $f \geq 0$, $(f,(\Phi^n)^+) \geq 0$, so that (1.38) implies that $(\Phi^n)^- = 0$, i.e. $\Phi^n \geq 0$, and the proof of (1.28) is completed. []

Let us show now

<u>Theorem 1.1</u> <u>As</u> $n \to \infty$ , <u>the sequence</u> $\Phi^n$ <u>converges in</u> $H^1(\Omega)$ <u>weakly and in</u> $L^p(\Omega)$ <u>strongly for all finite</u> $p$ , <u>to a solution</u> $\Phi$ <u>of the</u> Q.V.I. (1.20). <u>The solution</u> $\Phi$ <u>is maximal among all possible solutions</u> $\psi$ <u>of</u> (1.20) <u>which satisfy</u> $0 \le \psi \le \Phi^o$ .

<u>Proof</u>

Taking $\phi = 0$ in (1.27) gives (since $j(0, \Phi^{n-1}) = 0$)

$$a(\Phi^n,\Phi^n) + j(\Phi^n,\Phi^{n-1}) \le (f,\Phi^n) \ ;$$

hence it follows that

$$\|\Phi^n\|_{H^1(\Omega)} \le c \ . \tag{1.39}$$

From (1.28) and (1.39) it follows that $\Phi^n \to \Phi$ in the sense given in the Theorem; if $\Phi^n \to \Phi$ in $H^1(\Omega)$ weakly, then $\Phi^n \to \Phi$ in $L^2(\Gamma)$ <u>strongly</u> (cf. Lions-Magenes [1]). Therefore

$$j(\phi,\Phi^{n-1}) \to j(\phi,\Phi)$$

$$j(\Phi^n,\Phi^{n-1}) \to j(\Phi,\Phi)$$

and (1.27) gives

$$a(\Phi,\phi) + j(\phi,\Phi) - j(\Phi,\Phi) - (f,\phi-\Phi) \ge \underline{\lim}\ a(\Phi^n,\Phi^n) \ge a(\Phi,\Phi) \ .$$

Now let $\psi$ be a solution of (1.20), such that $\Phi^o \geq \psi$. Then one checks inductively that $\Phi^n \geq \psi \ \forall n$, so that $\Phi \geq \psi$ and the proof is completed.

1.5 <u>Existence of solutions of the Q.V.I.</u> (II)

We started the induction by the "upper solution" (1.25). We can start with a "lower solution" $\psi^o$ defined by

$$(1.40) \qquad A\psi^o = f \text{ in } \Omega , \quad \frac{\partial \psi^o}{\partial \nu} = 0 \text{ on } \Gamma .$$

We then define $\psi^1, \psi^2, \ldots$ by

$$(1.41) \qquad a(\psi^n, \phi - \psi^n) + j(\phi, \psi^{n-1}) - j(\psi^n, \psi^{n-1}) \geq (f, \phi - \psi^n) \ \forall \phi$$

and we verify that

$$(1.42) \qquad 0 \leq \psi^o \leq \psi^1 \leq \ldots \leq \psi^{n-1} \leq \psi^n \leq \ldots \leq \Phi^o .$$

Then $\psi^n$ <u>converges</u> (in $H^1(\Omega)$ weakly and in $L^p(\Omega)$ strongly) <u>to a minimal solution of</u> (1.20).

<u>Remark 1.6</u>

The considerations of Section 1.1 show that in general there is no uniqueness of the solution of the Q.V.I. We conjecture that, if there does not exist a constant $\lambda$ satisfying (1.11), the Q.V.I. (1.20) admits a unique solution.

Remark 1.7

On $\Gamma$ there are three regions

$$\Gamma_o = \{x \mid \Phi > \mathfrak{m}(\Phi)\} , \quad \Gamma_M = \{x \mid \Phi < \mathfrak{m}(\Phi)\} ,$$

$$S = \{x \mid \Phi = \mathfrak{m}(\Phi)\} .$$

We see again that this problem is of the type of free boundary problems.

## 2. An evolution problem.

### 2.1 Formulation of the problem.

We suppose that the state of the system is given by

$$\frac{\partial y}{\partial t} - \Delta y = f \quad \text{in} \quad Q = \Omega \times (0,T) , \tag{2.1}$$

$$y(x,0) = y_o(x) \quad \text{in} \quad \Omega , \tag{2.2}$$

$$\frac{\partial y}{\partial \nu} = v \quad \text{on} \quad \Sigma = \Gamma \times (0,T) . \tag{2.3}$$

We define

$$\mathfrak{m}y(t) = \mathfrak{m}y = \frac{1}{\text{meas.}\Gamma} \int_\Gamma y(\cdot,t)d\Gamma \tag{2.4}$$

and we consider the cost function

$$J(v) = \int_{\Sigma} |y(v) - \mathfrak{M}y(v)|^2 d\Gamma \tag{2.5}$$

that we want to minimize over $U_{ad}$ defined by

$$U_{ad} = \{v \mid 0 \leq v \leq M\} . \tag{2.6}$$

An "a priori feedback" solution is given by the solution - if possible! - of

$$\frac{\partial \Phi}{\partial t} - \Delta \Phi = f \quad \text{in} \quad Q , \tag{2.7}$$

$$\Phi(x,0) = y_o(x) \quad \text{in} \quad \Omega , \tag{2.8}$$

with the boundary conditions

$$\begin{cases} 0 \leq \frac{\partial \Phi}{\partial \nu} \leq M , \\ \Phi < \mathfrak{M}(\Phi) \Rightarrow \frac{\partial \Phi}{\partial \nu} = M , \\ \Phi > \mathfrak{M}(\Phi) \Rightarrow \frac{\partial \Phi}{\partial \nu} = 0 . \end{cases} \tag{2.9}$$

2.2 <u>Formulation as a Q.V.I. of evolution.</u>

One checks, by similar arguments to those of Section 1, that problem (2.7) (2.8) (2.9) is equivalent to

$$(2.10)\qquad \left(\frac{\partial\Phi}{\partial t}, \phi-\Phi\right) + a(\Phi,\phi-\Phi) + j(\phi,\Phi) - j(\Phi,\Phi) \geq (f,\phi-\Phi) \quad \forall \ \phi \in H^1(\Omega)$$

with (2.8), where $j(\phi, \psi)$ is defined as in (1.19). The inequality (2.10) is called a Q.V.I. of evolution.

One can show (we refer to Bensoussan-Lions [4] for proofs) that there exist maximal and minimal solutions of (2.10), assuming that $f \geq 0$ , $y_o \geq 0$ .

# CHAPTER 4

# CONSTRAINTS ON THE STATE

## 1. Direct Study of an Example.

### 1.1 Formulation of the Problem.

Let $\Omega$ be a bounded open set in $R^n$ with smooth boundary $\Gamma$ and let $\mathcal{L}$ be an $(n-2)$ dimensional smooth variety contained in $\Gamma$. In physical situations, $n = 3$, and $\mathcal{L}$ will be a curve on $\Gamma$ (cf. fig. 1).

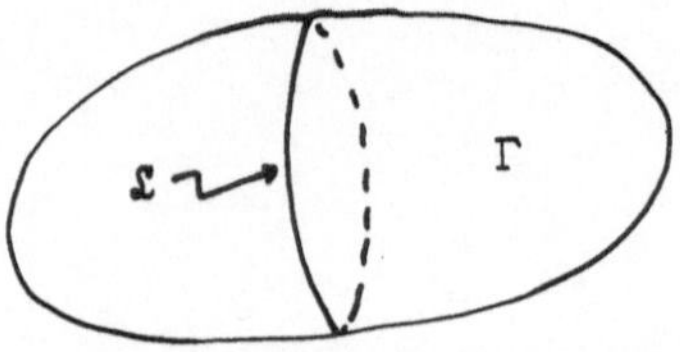


Fig. 1

The state $y(v)$ is given by

(1.1) $$Ay(v) = f \quad \text{in } \Omega \text{ (1)}$$

(1.2) $$\frac{\partial y(v)}{\partial \nu} = v \quad \text{on } \Gamma .$$

---

(1) To simplify the exposition, we always assume that $A = -\Delta + I$.

Let $U_1$ be a closed convex subset of $U = L^2(\Gamma)$ and let $g$ be a given function on $\mathfrak{L}$. We define

$$(1.3) \qquad U_{ad} = \{v \mid v \in U_1 \,,\, y(v) = g \text{ on } \mathfrak{L}\} .$$

<u>We suppose that</u> $U_{ad}$ <u>is non-empty</u> (cf. an example below). We check that definition (1.3) <u>makes sense</u>; indeed, for $f \in L^2(\Omega)$, $v \in L^2(\Gamma)$, one has $y(v) \in H^1(\Gamma)$ so that $y(v)|_{\mathfrak{L}}$ is meaningful. One also gets <u>a necessary condition</u> for $U_{ad}$ to be non-empty:

$$(1.4) \qquad g \in H^{1/2}(\mathfrak{L}) .$$

This follows from general trace theorems (cf. Lions-Magenes [1]).

<u>Example 1.1.</u> If $U_1 = L^2(\Gamma)$, condition (1.4) is sufficient for $U_{ad}$ to be non-empty. Indeed, we define $G \in H^1(\Gamma)$ such that $G|_{\mathfrak{L}} = g$ and we solve

$$(1.5) \qquad A\Phi = f \text{ in } \Omega \,,\quad \Phi|_{\Gamma} = G .$$

Then $\frac{\partial\Phi}{\partial\nu} \in L^2(\Gamma)$ and if we choose $v = \frac{\partial\Phi}{\partial\nu}$, we see that $v \in U_{ad}$. []

The set $U_{ad}$ is <u>closed</u> and <u>convex</u>.

Let us consider the cost function:

(1.6) $$J(v) = \int_\Gamma |y(v) - z_d|^2 d\Gamma + N\int_\Gamma v^2 d\Gamma \ .$$

Exactly as in Chapter 1, we see that <u>there exists a unique</u> $u$ <u>in</u> $U_{ad}$ <u>such that</u>

(1.7) $$J(u) = \inf J(v) \ , \ v \in U_{ad} \ .$$

If we set $y(u) = y$, $u$ is characterized by

$$\frac{1}{2}(J'(u), v-u) \geq 0 \ \ \forall \ v \in U_{ad} \ , \ \text{ i.e.,}$$

(1.8) $$\int_\Gamma (y-z_d)(y(v)-y)d\Gamma + N\int_\Gamma u(v-u)d\Gamma \geq 0 \ \ \forall \ v \in U_{ad} \ .$$

1.2 <u>Optimality System</u>

Introducing (as in Chapter 1) the adjoint state $p$ by

(1.9) $$Ap = 0, \ \frac{\partial p}{\partial \nu} = y - z_d \ \text{ on } \Gamma \ ,$$

one sees that (1.8) is equivalent to

(1.10) $$\int_\Gamma (p + Nu)(v-u)d\Gamma \geq 0 \ \ \forall \ v \in U_{ad} \ .$$

The difficulty lies in the interpretation of (1.10), when $U_{ad}$ is defined by constraints expressed not only in terms of $v$ but <u>also</u> in terms of $y(v)$ .

We give below some possible methods to take care of this difficulty.

## 2. Reformulation of the problem in terms of pseudo-differential operations on $\Gamma$ .

### 2.1 The Operator $\mathcal{A}$

Let $\Phi$ be the solution of

(2.1) $\quad A\Phi = 0,\ \Phi = \phi$ on $\Gamma$ , where $\phi$ is given in $H^1(\Gamma)$, say.

Then we define

(2.2) $$\mathcal{A}\phi = \frac{\partial\Phi}{\partial\nu}\ .$$

We see that

(2.3) $$\mathcal{A} \in \mathcal{L}(H^1(\Gamma)\ ;\ L^2(\Gamma))$$

and $\mathcal{A}$ is an isomorphism from $H^1(\Gamma)$ onto $L^2(\Gamma)$ (1). The operator $\mathcal{A}$ is a pseudo differential operator on $\Gamma$ .

### 2.2 Reformulation of the Problem.

We assume now that

(1) Actually $\mathcal{A}$ is an isomorphism from $H^s(\Gamma)$ onto $H^{s-1}(\Gamma)$ $\forall$ $s \in R$ .

(2.4) $$U_{ad} = \{v | v \in L^2(\Gamma),\ y(v)|_{\mathcal{L}} = g,\ g \in H^{1/2}(\mathcal{L})\} .$$

We do not restrict the generality by taking $f = 0$. Then if we set

(2.5) $y(u)|_\Gamma = \phi_0$ we see that (1.8) is equivalent to

(2.6) $$(\phi_0 - z_d, \phi - \phi_0)_\Gamma + N(\mathcal{A}\,\phi_0, \mathcal{A}(\phi - \phi_0))_\Gamma \geq 0$$

$$\forall\ \phi \in H^1(\Gamma),\ \phi|_{\mathcal{L}} = g\ ,\ \text{where}\ (\phi, \psi)_\Gamma = \int_\Gamma \phi\psi\, d\Gamma\ .$$

But (2.6) is equivalent to

(2.7) $$N\mathcal{A}^*\mathcal{A}\phi_0 + \phi_0 = z_d \quad \text{on} \quad \Gamma - \mathcal{L}\ ,$$

$$\phi_0|_{\mathcal{L}} = g$$

where $\mathcal{A}^*$ is the adjoint of $\mathcal{A}$ ; in our particular case $\mathcal{A}^* = \mathcal{A}$ .

We remark that (2.7) is a <u>non-homogeneous boundary value problem on the variety</u> $\Gamma$ . []

Remark 2.1

Suppose that $U_{ad}$ is defined by

(2.8) $$U_{ad} = \{v | g_0 \leq y(v) \leq g_1 \quad \text{on} \quad \mathcal{L}\ ,$$

where $g_0$ and $g_1$ are given in $H^{1/2}(\mathcal{L}),\ g_0 \leq g_1\}$ .

Then the optimal control $u$ is defined by solving first the V.I.

$$(2.9)\qquad (\phi_0 - z_d, \phi-\phi_0)_\Gamma + N(\mathcal{A}\phi_0, \mathcal{A}(\phi-\phi_0))_\Gamma \geq 0$$

$$\forall \phi \in H^1(\Gamma) \text{ such that } g_0 \leq \phi \leq g_1 \text{ on } \mathcal{L}$$

$$\text{and } g_0 \leq \phi_0 \leq g_1 \text{ on } \mathcal{L} .$$

Next we solve $\qquad Ay = 0, \quad y = \phi_0$ on $\Gamma$

and finally $\qquad u = \frac{\partial y}{\partial \nu}$ .

## 3. Approximation by Penalty

### 3.1 The Penalty Method

Let $\varepsilon > 0$ . We consider the cost function

$$(3.1)\qquad J_\varepsilon(v) = \int_\Gamma |y(v)-z_d|^2 d\Gamma + N\int_\Gamma v^2 d\Gamma + \frac{1}{\varepsilon}\int_{\mathcal{L}} (y(v)-g)^2 d\mathcal{L} .$$

The term $\frac{1}{\varepsilon}\int_{\mathcal{L}} (y(v)-g)^2 d\mathcal{L}$ is a penalty term.

We now consider the "approximate" problem without constraints on the state:

$$(3.2)\qquad \inf J_\varepsilon(v), \quad v \in U_1 .$$

Problem (3.2) admits a unique solution $u_\varepsilon \in U_1$ . []

For Problem (3.2) we can write the optimality system as follows; we set $y(u_\varepsilon) = y_\varepsilon$ . Then

$$(3.3) \qquad Ay_\varepsilon = f, \quad A^*p_\varepsilon = 0$$

$$(3.4) \qquad \frac{\partial y_\varepsilon}{\partial \nu} = u_\varepsilon, \quad \frac{\partial p_\varepsilon}{\partial \nu} = y_\varepsilon - z_d + \frac{1}{\varepsilon}(y_\varepsilon - g)\mu_{\mathfrak{L}}$$

where $\mu_{\mathfrak{L}}$ is the distribution on $\Gamma$ given by

$$\langle\mu_{\mathfrak{L}}, \psi\rangle_\Gamma = \int_{\mathfrak{L}} \psi \, d\Gamma \ ;$$

the optimality condition is now

$$(3.5) \qquad \int_\Gamma (p_\varepsilon + Nu_\varepsilon)(v - u_\varepsilon) d\Gamma \geq 0 \quad \forall \ v \in U_1 \ , \ u_\varepsilon \in U_1 \ . \qquad []$$

Example 3.1

Let us suppose that

$$(3.6) \qquad U_1 = \{v \,|\, 0 \leq v \leq M\} \ .$$

Then (3.5) is equivalent to

$$(3.7) \qquad \begin{cases} 0 < u_\varepsilon < M \Rightarrow p_\varepsilon + Nu_\varepsilon = 0 \ , \\ u_\varepsilon = 0 \Rightarrow p_\varepsilon \geq 0 \\ u_\varepsilon = M \Rightarrow p_\varepsilon + NM \leq 0 \ . \end{cases}$$

## 3.2 A Convergence Result

If we assum that $U_{ad} \neq \emptyset$, then

$$u_\varepsilon \to u \quad \text{in} \quad L^2(\Gamma) \quad \text{strongly as} \quad \varepsilon \to 0 . \tag{3.8}$$

Proof

Let $v_0$ be chosen in $U_{ad}$. Then

$$J_\varepsilon(u_\varepsilon) \leq J_\varepsilon(v_0) = J(v_0) \tag{3.9}$$

so that:

$$\|u_\varepsilon\|_{L^2(\Gamma)} \leq C , \tag{3.10}$$

$$\|y(u_\varepsilon) - g\|_{L^2(\mathcal{L})} \leq C\varepsilon^{1/2} . \tag{3.11}$$

It follows from (3.10) that we can extract a subsequence, still denoted by $u_\varepsilon$, such that

$$u_\varepsilon \to w \quad \text{in} \quad L^2(\Gamma) \quad \text{weakly,} \quad w \in U_1 . \tag{3.12}$$

It follows that $y(u_\varepsilon) = y_\varepsilon \to y(w)$ in $H^1(\Omega)$ weakly and $y(u_\varepsilon)|_\Gamma \to y(w)|_\Gamma$ in $H^1(\Gamma)$ weakly, so that $y_\varepsilon|_{\mathcal{L}} \to y(w)|_{\mathcal{L}}$ in $H^{1/2}(\mathcal{L})$ weakly; by virtue of (3.11) it follows that

$$y(w)|_{\mathcal{L}} = g , \quad \text{i.e.,} \quad w \in U_{ad} . \tag{3.13}$$

We observe now that

$$J(w) \leq \underline{\lim}\ J(u_\varepsilon) \leq \underline{\lim}\ J_\varepsilon(u_\varepsilon) \leq J(v_0) \quad \forall \quad v_0 \in U_{ad} \,.$$

Hence $w = u$ and (3.8) follows.

## 4. Duality

### 4.1 A Formal Computation

Let us consider on a real Hilbert space $V$ a function of the form:

(4.1) $v \to \mathcal{G}(v) = F(v) + G(\Lambda v)$ from $V \to R$ ; in (4.1) we have

(4.2) $\Lambda \in \mathcal{L}(V;Y)$, $Y$ = real Hilbert space,

(4.3) $F$ and $G$ are lower semi-continuous convex functions on $V$ and $Y$ resp. such that $-\infty < F(v) \leq +\infty$, $-\infty < G(y) \leq +\infty$ and $F$ and $G$ are not identically $+\infty$ [(1)].

We consider the problem

(4.4) $\qquad P: \ \inf\ \mathcal{G}(v), \quad v \in V$ .

---

(1) One says that $F$ (and $G$ ) is a proper convex function.

Remark 4.1

It is quite clear that the problems considered up to now can enter the preceding framework, and in a number of different ways.[]

We want to introduce--in a formal manner--a dual problem $P^*$ of $P$.

Let us introduce

$$(4.5) \qquad \Phi(v,q) = F(v) + G(q) \quad \text{over} \quad V \times Y ,$$

and let us remark that

$$(4.6) \qquad \inf_v G(v) = \inf_v \inf_q \sup_{q^*} [\Phi(v,\Lambda v-q) - \langle q^*,q\rangle]$$

where $q^* \in Y^*$ = dual of $Y$ (it can be identified with $Y$ or not) and where $\langle q^*,q\rangle$ denotes the scalar product in the duality $Y^*,Y$; formula (4.6) is immediate since the $\sup_{q^*}$ is $\infty$ unless $q = 0$.

But now if we commute--in a formal manner--the inf and the sup in (4.6), we obtain:

$$(4.7) \qquad \inf G(v) = \sup_{q^*} \inf_{v,q} [\Phi(v,\Lambda v-q) - \langle q^*,q\rangle]$$

$$= \sup_{q^*} \{-\sup_{v,q} [\langle q^*,q\rangle - \Phi(v,\Lambda v-q)\} .$$

If we set

(4.8) $\Lambda v - q = \rho$ (4.7) can be written

(4.9) $$\inf G(v) = \sup_{q^*} \{-\sup_{v,q} [\langle q^*, \Lambda v - \rho\rangle - \Phi(v,\rho)]\} .$$

But using (4.5) one has

$$\sup_{v,\rho} [\langle q^*, \Lambda v - \rho\rangle - \Phi(v,\rho)] = \sup_{v,\rho} [\langle \Lambda^* q^*, v\rangle - F(v) + \langle -q^*, \rho\rangle - G(\rho)].$$

If we recall the general definition of <u>the conjugate function</u> $\psi^*$ of $\psi$ :

(4.10) $$\psi^*(q^*) = \sup_{q} [\langle q^*, q\rangle - \psi(q)]$$ we obtain

(4.11) $$\sup_{v,\rho} [\langle q^*, \Lambda v - \rho\rangle - \Phi(v,\rho)] = F^*(\Lambda^* q^*) + G^*(-q^*).$$

Hence (4.9) gives

(4.12) $$\inf_{v} G(v) = \sup_{q^*} [-F^*(\Lambda^* q^*) - G^*(-q^*)] .$$

<u>We define the dual problem</u> $P^*$ <u>of</u> $P$ <u>as</u>:

(4.13) $$P^*: \sup_{q^*} [-F^*(\Lambda^* q^*) - G^*(-q^*)] .$$

Formula (4.12) is (formally) equivalent to

(4.14) $$\text{Inf } P = \sup P^* .$$

### 4.2 Sufficient condition for (4.14) to be true.

We give now a sufficient condition for the formal computation of the preceding Section to be justified. For proofs and more general results, we refer to T.R. Rockafellar [1]. Cf. also I. Ekeland and R. Temam [1]. We consider

$$h(q) = \inf_v \Phi(v, \Lambda v - q) = \inf_v [F(v) + G(\Lambda v - q)] \tag{4.15}$$

and we suppose that

(4.16) $h(0) < \infty$,

$h$ is lower semi-continuous at the origin.

Then (4.14) is true.

Remark 4.2

The dual problem $P^*$ does not necessarily admit a solution (cf. Example I below).

### 4.3 Example I

We consider again problem (1.7).

We define, with the notations of Section 4.1:

$$V = L^2(\Gamma), \quad Y = L^2(\Gamma) \times L^2(\mathcal{L}) ,$$

$$\Lambda v = \{y(v)|_\Gamma , y(v)|_{\mathcal{L}}\} \quad \text{(supposing } f = 0 \text{)}, \tag{4.17}$$

(4.18) $$F(v) = \frac{1}{2} N\int_\Gamma v^2 d\Gamma \ ,$$

(4.19) $$\begin{cases} G(q) = G_1(q_1) + G_2(q_2), \ q = \{q_1,q_2\} \in Y \ , \\ G_1(q_1) = \frac{1}{2}\int_\Gamma |q_1 - z_d|^2 d\Gamma \ , \\ G_2(q_2) = \begin{cases} 0 & \text{if} \quad q_2 = g \\ +\infty & \text{if} \quad q_2 \neq g \ . \end{cases} \end{cases}$$

Then problem (1.7) is equivalent to

$$P: \inf_v \ [F(v) + G(\Lambda v)] \ .$$

It is simple to verify that (4.16) holds true. <u>We can therefore apply</u> (4.14). []

We compute the dual functions $F^*, G^*$ and the operator $\Lambda^*$ ; we identify $V$ and $Y$ with their dual.

For $q = \{q_1, q_2\} \in L^2(\Gamma) \times L^2(\mathcal{L})$ , let us <u>define</u> $\Phi = \Phi(q)$ as the solution of

(4.20) $$\begin{cases} A^*\Phi = 0 \quad \text{in} \quad \Omega \ , \\ \dfrac{\partial \Phi}{\partial \nu} = q_1 + \mu_{\mathcal{L}} q_2 \quad \text{on} \quad \Gamma \end{cases}$$

where $\mu_{\mathcal{L}}$ is defined by

$$\langle \mu_{\mathcal{L}}, \psi \rangle = \int_{\mathcal{L}} \psi \, d\mathcal{L} \ .$$

Problem (4.20) is solved in a weak sense, as in Lions-Magenes [1]. It follows that (4.20) admits a unique solution which admits a trace $\Phi/\Gamma$ in $L^2(\Gamma)$ . One has:

$$(4.21) \qquad \Lambda^* q = \Phi(q)\big|_\Gamma \; .$$

Indeed $0 = (A^*\Phi,\ y(v)) = -\int_\Gamma \frac{\partial\Phi}{\partial\nu}\, y(v)\,d\Gamma + \int_\Gamma \Phi v\ d\Gamma$ .

Hence $\langle\Phi, v\rangle_\Gamma = \int_\Gamma q_1 y(v)\,d\Gamma + \int_{\mathcal{L}} q_2\ y(v)\ d\mathcal{L} = \langle q, \Lambda v\rangle_\Gamma$ . []

One has

$$(4.22) \qquad F^*(v) = \frac{1}{2N}\int_\Gamma v^2\,d\Gamma \; ,$$

$$(4.23) \qquad G_1^*(q_1) = \sup_{q_1'} \left[\int_\Gamma q_1 q_1'\ d\Gamma - \frac{1}{2}\int_\Gamma |q_1' - z_d|^2\ d\Gamma\right]$$

$$= \frac{1}{2}\int_\Gamma q_1^2\,d\Gamma + \int_\Gamma q_1 z_d\ d\Gamma$$

$$(4.24) \qquad G_2^*(q_2) = \sup_{q_2'} \left[\int_{\mathcal{L}} q_2 q_2'\ d\mathcal{L} - G_2(q_2')\right] = \int_{\mathcal{L}} q_2 g\ d\mathcal{L} \; .$$

Therefore problem $P^*$ becomes:

$$(4.25) \qquad \sup_q \left[- \frac{1}{2N}\int_\Gamma \Phi(q)^2\ d\Gamma - \frac{1}{2}\int_\Gamma q_1^2\ d\Gamma + \int_\Gamma q_1 z_d\ d\Gamma + \int_{\mathcal{L}} q_2 g\,d\mathcal{L}\right].$$

Consequently (with the notations of Section 1)

$$(4.26) \qquad \operatorname{Inf} \tfrac{1}{2}J(v) = -\inf_q \left[\frac{1}{2N}\int_\Gamma \Phi(q)^2 d\Gamma + \frac{1}{2}\int_\Gamma q_1^2 d\Gamma - \int_\Gamma q_1 z_d \, d\Gamma - \int_{\mathcal{L}} q_2 g \, d\mathcal{L}\right] .$$

<u>In the dual problem,</u> $q \in L^2(\Gamma) \times L^2(\mathcal{L})$ <u>without constraints;</u> <u>the "state"</u> $\Phi(q)$ <u>is defined by</u> (4.20).

We remark that the dual problem is <u>not coercive</u>(1) (if $g \neq 0$) so that it does not necessarily admit a solution.

<u>Remark 4.3</u>

As was said in the Introduction, the idea of "<u>suppressing the constraints on the state</u>" by duality is due to J. Mossino [1](2). The disadvantage of this method is that the dual problem does not necessarily admit a solution. But (4.26) is nevertheless useful, in particular for numerical methods (cf. Mossino [1]).

<u>Remark 4.4</u>

Let us briefly show why $\Phi(q)_\Gamma \in L^2(\Gamma)$ . For $h \in L^2(\Gamma)$ , we define $\psi$ by

$$(4.27) \qquad A\psi = 0 \text{ in } \Omega \, , \; \frac{\partial\psi}{\partial\nu} = h \text{ on } \Gamma \, .$$

We know that $\psi \in H^2(\Omega)$ so that

---

(1) i.e. The functional to be minimized in the dual problem is not necessarily infinite at infinity.

(2) Which applies this idea to different interesting examples.

(4.28) $$\psi|_\Gamma \in H^{3/2}(\Gamma),\ \psi|_\mathcal{L} \in H^1(\mathcal{L}) .$$

If $\Phi$ is a solution of (4.20), Green's formula (cf. Lions-Magenes [1] for the justification) gives:

(4.29) $$\int_\Gamma \Phi h \, d\Gamma = \int_\Gamma q_1 \psi \, d\Gamma + \int_\mathcal{L} q_2 \psi \, d\mathcal{L} .$$

The mapping $h \to \{\psi|_\Gamma, \psi|_\mathcal{L}\}$ is continuous from $L^2(\Gamma) \to H^{3/2}(\Gamma) \times H^1(\mathcal{L})$, so that (4.29) shows that $\phi|_\Gamma \in L^2(\Gamma)$ even if $q_1 \in H^{-3/2}(\Gamma)$ and $q_2 \in H^{-1}(\mathcal{L})$ :

(4.30) $$\|\Phi\|_{\Gamma^2(\Gamma)} \leq c\,[\|q_1\|_{H^{-3/2}(\Gamma)} + \|q_2\|_{H^{-1}(\mathcal{L})}]$$

and the desired result follows. []

An an exercise we give another example of "suppressing the state-constraints by duality".

## 4.4 Example II

Let the state be given by

(4.31) $$\begin{cases} A\, y(v) = 0 & \text{in } \Omega , \\ y(v) = v & \text{on } \Gamma \text{ where } v \in L^2(\Gamma) ; \end{cases}$$

$y(v)$ is a weak solution of the non-homogenous Dirichlet problem.

One can then show (cf. Lions-Magenes [1]) the existence and uniqueness of a solution $y(v) \in L^2(\Omega)$ , such that

(4.32) $$\frac{\partial y(v)}{\partial \nu} \in H^{-1}(\Gamma) .$$

Therefore if $g$ <u>is given in</u> $H^1(\Gamma)$ , one can define $\langle \frac{\partial y(v)}{\partial \nu}, g \rangle$ (scalar product between $H^{-1}(\Gamma)$ and $H^1(\Gamma)$) .

The <u>constraint-set</u> $U_{ad}$ is defined by

(4.33) $$U_{ad} = \{v \mid \langle \frac{\partial y(v)}{\partial \nu}, g \rangle_\Gamma = \alpha\}, \ \alpha \text{ given in } R .$$

This defines a <u>non-empty</u> closed convex subset of $L^2(\Gamma)$ , since if we take $h \in H^{-1}(\Gamma)$ such that $\langle h,g \rangle_\Gamma = \alpha$ and if we solve $A\Phi = 0$ in $\Omega$ , $\frac{\partial \Phi}{\partial \nu} = h$ on $\Gamma$ , then $\Phi|_\Gamma \in U_{ad}$ . []

The <u>cost function</u> we consider is

(4.34) $$J(v) = \int_\Omega |y(v) - z_d|^2 dx + N\int_\Gamma v^2 d\Gamma . \qquad []$$

We use the notations of Section 4.1:

$$V = L^2(\Gamma), \ Y = L^2(\Omega) \times R ,$$

(4.35) $$\Lambda v = \{y(v)|_\Omega , \langle \frac{\partial y(v)}{\partial \nu}, g \rangle\}$$

(4.36) $$F(v) = \frac{1}{2} N\int_\Gamma v^2 d\Gamma ,$$

$$(4.37)\qquad \begin{cases} G(q) = G_1(q_1) + G_2(q_2) \ , \\ G_1(q_1) = \frac{1}{2}\int_\Omega |q_1 - z_d|^2 dx \ , \\ G_2(q_2) = \begin{cases} 0 & \text{if } q_2 = \alpha \ , \\ +\infty & \text{if } q_2 \neq \alpha \ . \end{cases} \end{cases}$$

Problem P is

$$(4.38)\qquad \inf_v \, [F(v) + G(\Lambda v)] \ . \qquad \square$$

We identify V and Y with their duals.

For $q = \{q_1, q_2\} \in Y$ , we define $\Phi(q)$ by

$$(4.39)\qquad \begin{cases} A^*\Phi(q) = q_1 \quad \text{in } \Omega \ , \\ \Phi(q)|_\Gamma = -q_2 g \ . \end{cases}$$

If we define $z(q_1)$ by

$$(4.40)\qquad \begin{cases} A^*z(q_1) = q_1 \quad \text{in } \Omega \ , \\ z(q_1)|_\Gamma = 0 \end{cases}$$

and $\zeta$ by

$$(4.41)\qquad A^*\zeta = 0 \ \text{ in } \Omega \ , \quad \zeta = -g \ \text{ on } \Gamma$$

then

(4.42) $$\Phi(q) = z(q_1) + q_2\zeta .$$

We check, by applying Green's formula, that

(4.43) $$\Lambda^* q = -\frac{\partial\Phi}{\partial\nu}(q) .$$ []

We have:

(4.44) $$F^*(v) = \frac{1}{2N}\int_\Gamma v^2 d\Gamma ,$$

(4.45) $$\begin{cases} G_1^*(q_1) = \frac{1}{2}\int_\Omega q_1^2\, dx + \int_\Omega q_1\, z_d\, dx \\ G_2^*(q_2) = q_2\alpha . \end{cases}$$

Therefore problem P* can be written:

(4.46) $$\sup_q \left[-\frac{1}{2N}\int_\Gamma \left(\frac{\partial\Phi}{\partial\nu}(q)\right)^2 d\Gamma - \frac{1}{2}\int_\Omega q_1^2 dx + \int_\Omega q_1 z_d + q_2\alpha\right] .$$

We replace in (4.46) $\Phi$ by (4.42). Let us set:

(4.47) $$b = \int_\Gamma \left(\frac{\partial\zeta}{\partial\nu}\right)^2 d\Gamma ,$$

$$c(q_1) = \int_\Gamma \frac{\partial z(q_1)}{\partial\nu}\frac{\partial\zeta}{\partial\nu}\, d\Gamma,\quad d(q_1) = \int_\Gamma \left(\frac{\partial z(q_1)}{\partial\nu}\right)^2 d\Gamma .$$

Then (4.46) is equivalent to

(4.48) $$-\inf_q G(q_1,q_2) ,$$ with

(4.49) $$G(q_1,q_2) = \frac{b}{2N} q_2^2 + (\frac{c(q_1)}{N} - \alpha)q_2 + \frac{1}{2N}d(q_1) + + \frac{1}{2}\int_\Omega q_1^2 dx - \int_\Omega q_1 z_d \, dz .$$

If we minimize in $q_2$ first we find

(4.50) $$\inf_{q_2} G(q_1,q_2) = H(q_1) ,$$ where

(4.51) $$H(q_1) = \frac{1}{2N}\left[d(q_1) - \frac{c^2(q_1)}{b}\right] + \frac{1}{2}\int_\Omega q_1^2 \, dx - -\int_\Omega q_1 z_d \, dx + \frac{\alpha}{b}c(q_1) - \frac{N\alpha^2}{2b} .$$

Therefore the problem reduces to the minimization without constraints of the coercive[1] cost function $H(q_1)$ , the state being given by (4.40).

One can easily write the optimality system for this problem.

---

(1) Observe that $d(q_1) - \frac{c^2(q_1)}{b} \geq 0$ .

# CHAPTER 5

# NON-LINEAR STATE EQUATIONS

## 1. Differentiable Problem

### 1.1 Control Problems in Biochemistry.

A large number of problems of optimal control appears in biochemistry. We refer to J.O. Kernevez [1], J.O. Kernevez and D. Thomas [1]. We briefly present here an example treated in C.M. Brauner and P. Penel [1]. See the references to these authors in the Bibliography.

The state $\{i,y,e,z\}$ (1) is given by the non-linear parabolic system

$$(1.1)\quad \begin{cases} \dfrac{\partial i}{\partial t} - \dfrac{\partial^2 i}{\partial x^2} + \sigma(ey + e + z-1) = 0, \quad x \in (0,1),\ t > 0\ , \\[2ex] \dfrac{\partial y}{\partial t} - \dfrac{\partial^2 y}{\partial x^2} + \sigma(ei-z) = 0\ , \\[2ex] \dfrac{\partial e}{\partial t} + \mu(ey + ei + e-1) + \lambda(e + z-1) = 0\ , \\[2ex] \dfrac{\partial z}{\partial t} + \mu(z - ei) = 0 \end{cases}$$

---

(1) i = concentration of "inhibiteur"
y = concentration of "substrat"
e = enzyme concentration
z = biochemical complex.

where $\sigma,\lambda,\mu > 0$ , with the boundary conditions

$$(1.2) \qquad \begin{cases} i(0,t) = i(1,t) = v(t) & (= \text{control}) \\ y(0,t) = \alpha,\ y(1,t) = \mu\ , & \alpha,\mu > 0 \ \text{ given} , \end{cases}$$

and the initial conditions

$$(1.3) \qquad i(x,0) = 0, y(x,0) = 0,\ e(x,0) = 1,\ z(x,0) = 0\ . \quad []$$

The first point to prove is that this problem is "well set". Indeed, one can show (cf. Brauner-Penel, loc. cit.) that there exists a unique solution $i(v)$, $y(v)$, $e(v)$, $z(v)$ of the system (1.1), (1.2), (1.3), in the space

$$(L^2(0,T;H_0^1(\Omega)))^2 \times L^2(Q) \times L^\infty(Q),\ \Omega = (0,1),\ Q = \Omega \times (0,T).\ []$$

The set of admissible controls is given by

$$(1.4) \qquad U_{ad} = \{v \,|\, 0 < \varepsilon \leq v(t) \leq M\}$$

and the cost function is given by

$$(1.5) \qquad J(v) = \int_0^T |y(\tfrac{1}{2},t;\ v) - z_d(t)|^2 dt + N\int_0^T v^2(t)dt\ . \qquad []$$

Of course the function $v \to J(v)$ has no reason to be convex. One can show the existence of (at least) an optimal control $u$ :

$$(1.6) \qquad J(u) = \inf J(v),\ v \in U_{ad},\ u \in U_{ad} .$$

The problem of uniqueness in $U_{ad}$ is open. []

The non-linearities in (1.1) being of <u>polynomial</u> type, it is natural to expect--and one indeed proves--that <u>the state is differentiable in</u> $v$, for instance in $(L^2(Q))^4$. One can therefore write a <u>necessary condition</u> to be satisfied by $u$:

$$(1.7) \qquad \int_0^T (y(\tfrac{1}{2},t;u)-z_d)(\hat{y}(u),v-u)dt + N\int_0^T u(v-u)dt \geq 0 \ \forall\ v \in U_{ad} ,$$

where

$$(1.8) \qquad (\hat{y}(u),v-u) = \frac{d}{d\lambda}\, y(u+\lambda(v-u))\big|_{\lambda=0} .$$

For explicit computations, we refer to Brauner-Penel, loc.cit.

## 1.2 <u>Control problems in heat transfer.</u>

We consider here a simplified problem related to the study of J.P. Yvon [1] (for the optimal heating of a furnace).

The <u>state</u> $y = y(v)$ is given by the solution of

$$(1.9) \qquad \frac{\partial y}{\partial t} - \Delta y = f \quad \text{in} \quad Q = \Omega\times(0,T)$$

$$(1.10) \qquad y(x,0) = y_0(x) \quad \text{in} \quad \Omega ,$$

(1.11) $y(x,t)$ does not depend on $x$ on $\Gamma$ ,

$$\int_\Gamma \frac{\partial y}{\partial \nu}(x,t)d\Gamma + c\, y^4 = v(t) \quad \text{on } \Gamma , \quad c > 0 ,$$

where $v$ is the control variable, subject to

(1.12) $$v \in U_{ad} \iff 0 \leq v(t) \leq M .$$

The functions $f$ and $y_0$ are given in, say, $L^2(Q)$ and $L^2(\Omega)$ , and are $\geq 0$ . If we admit for a while that (1.9) (1.10) (1.11) uniquely define $y(v)$ , <u>the cost function</u> is given by

(1.13) $$J(v) = \int_\phi |y(x,t;v) - z_d|^2 dxdt + N\int_0^T v^2 dt . \qquad []$$

<u>Proof of existence and uniqueness of</u> $y(v)$ .

Let us introduce

(1.14) $$V = \{\phi | \phi \in H^1(\Omega),\ \phi = \text{constant}^{(1)} \text{on } \Gamma\} .$$

We verify that (1.9) (1.10) (1.11) is equivalent to[2]

(1.15) $$\left(\frac{\partial y}{\partial t}, \phi\right) + a(y,\phi) + c\int_\Gamma y^4 \phi \, d\Gamma = (f,\phi) + v\int_\Gamma \phi \, d\Gamma \;\forall\, \phi \in V,$$

---

(1) Which depends on $\Phi$ .

(2) $\int_\Gamma y^4 \phi \, d\Gamma = (y|_\Gamma)^4 \, (\phi|_\Gamma)$ meas. $\Gamma$ .

where $(f,\phi) = \int_\Omega f\,\phi\,dx,\ a(y,\phi) = \sum_{i=1}^{n} \int_\Omega \frac{\partial y}{\partial x_i} \frac{\partial \phi}{\partial x_i}\,dx$ ,

(1.16) $y$ takes its values in $V$ , and $y$ satisfies (1.10).

We solve (1.15) in two steps. We consider first the modified problem:

$$
(1.17) \quad \begin{cases} (\frac{\partial z}{\partial t},\phi)+a(z,\phi)+c\int_\Gamma |z|^3 z\,\phi\,d\Gamma = (f,\phi)+v\int_\Gamma \phi\,d\Gamma \ \forall\, \phi \in V , \\ z(t) \in V , \\ z|_{t=0} = y_0 . \end{cases}
$$

The operator from $V$ into $V'$ defined by

$z \to a(z,\phi) + c\int_\Gamma |z|^3 z\,\phi\,d\Gamma = (\mathcal{A}(z),\phi)$ is monotone (in the sense that $(\mathcal{A}(z_1) - \mathcal{A}(z_2), z_1-z_2) \geq 0 \ \forall\ z_1,z_2)$ and coercive. Therefore it follows from Minty [1], Browder [1], (cf. exposition in Lions [5]) that (1.17) admits a unique solution. In the second step we show that $z \geq 0$ so that $y(v) = z$ . Indeed, if we replace in (1.17) $\Phi$ by $z^-$ , we obtain:

$$
(1.18) \quad -(\frac{\partial z}{\partial t},z^-) + a(z^-,z^-) + c\int_\Gamma |z|^3 (z^-)^2 d\Gamma + (f,z^-)+v\int_\Gamma z^- d\Gamma = 0
$$

and $\int_0^T -(\frac{\partial z}{\partial t}, z^-)dt = \int_0^T \frac{1}{2}\frac{d}{dt}|z^-|^2 dt \geq 0$ since $z^-|_{t=0} = 0$ .

All terms in (1.18) are $\geq 0$ , so that $z^- = 0$ . []

One can then prove (cf. J.P. Yvon, loc.cit.) the existence of $u \in U_{ad}$ such that $J(u) = \inf J(v),\ v \in U_{ad}$. One can also show that $y$ is differentiable (in $L^2(Q)$) in $v$, so that one can again write necessary conditions. We refer to Yvon, loc.cit., where one will also find numerical computations based on these necessary conditions.

### 1.3 Remarks

One also meets in applications problems where the state equation is non-linear hyperbolic or a coupled system partially hyperbolic and partially parabolic. We refer to G. Duff [1] for the control of problems in hydrodynamics and to Bamberger and Yvon [1] for problems in the transportation of gas.

## 2. Non-Differentiable Problems

### 2.1 Example

Let us consider the state $y(v)$ given by the solution of

$$-\Delta y + y = 0 \quad \text{in } \Omega \tag{2.1}$$

$$y \geq 0,\ \frac{\partial y}{\partial \nu} - v \geq 0,\ y\left(\frac{\partial y}{\partial \nu} - v\right) = 0 \quad \text{on } \Gamma,\ v \in L^2(\Gamma)\ . \tag{2.2}$$

This problem admits a unique solution. Indeed, one easily checks that it is equivalent to the V.I.

(2.3) $$a(y,\phi-y) \geq \int_\Gamma v(\phi-y)d\Gamma \quad \forall \phi \in K ,$$

$$y \in K$$

where

(2.4) $$K = \{\phi | \phi \in H^1(\Omega), \ \phi \geq 0 \ \text{ on } \ \Gamma\} ,$$

and where

(2.5) $$a(y,\phi) = \sum_{i=1}^{n} \int_\Omega \frac{\partial y}{\partial x_i} \frac{\partial \phi}{\partial x_i} dx + \int_\Omega y \, \phi \, dx .$$

Since $a(y,\phi)$ is symmetric, (2.3) is equivalent to minimizing

$$\frac{1}{2} a(\phi,\phi) - \int_\Gamma v \, \phi \, d\Gamma \quad \text{over } K .$$ []

The solution $y(v)$ of (2.3) is not differentiable in $v$. Let us consider a very simple particular case: $\Omega = (0, +\infty)$, $v \in R$. The solution of (2.1) (2.2) is

(2.6) $$y(v) = v^+ e^{-x} \quad \text{and}$$

$v \to v^+ e^{-x}$ is not differentiable from $R \to L^2(\Omega)$ .

Remark 2.1

The problem (2.1) (2.2) is a simplification of problems arising in "unilateral mechanics"; cf. G. Duvaut-J.L. Lions [1].

Remark 2.2

The non differentiability will be the general situation for systems governed by V.I.

Remark 2.3

Problems of optimal control for systems governed by V.I. arise in biochemistry. Cf. works of J.O. Yvon and J. Kernevez. []

We consider the cost function.

$$J(v) = \int_\Gamma |y(v) - z_d|^2 d\Gamma + N\int_\Gamma v^2 d\Gamma \ . \tag{2.7}$$

It is simple to verify that there exists a unique optimal control $u \in U_{ad}$ such that

$$J(u) = \inf \ J(v), \ v \in U_{ad} \ . \tag{2.8}$$

The difficulty lies in the computation of necessary conditions.

2.2 Regularization.

Let $\lambda \to r(\lambda)$ be a $C^1$ function from $R \to R$ with the following properties:

$$\begin{cases} r(\lambda) > 0 \quad \text{for} \quad \lambda < 0 \ , \ r(\lambda) = 0 \quad \text{for} \quad \lambda \geq 0 \\ r \ \text{ is monotone}, \quad r(\lambda) \sim -\lambda \quad \text{as} \quad \lambda \to -\infty \ . \end{cases} \tag{2.9}$$

For $\varepsilon > 0$ let $y_\varepsilon = y_\varepsilon(v)$ be <u>the</u> solution of

$$a(y_\varepsilon,\phi) - \frac{1}{\varepsilon}(r(y_\varepsilon),\phi)_\Gamma = (v,\phi)_\Gamma \tag{2.10}$$

(where $(g,\phi)_\Gamma = \int_\Gamma g\,\phi\,d\Gamma$) ; (2.10) admits a unique solution, by virtue of the theory of <u>monotone operators</u>.

We have (cf. Lions [5] for general results along these lines):

$$y_\varepsilon(v) \to y(v) \quad \text{in} \quad H^1(\Omega) \quad \text{as} \quad \varepsilon \to 0 , \tag{2.11}$$

where $y(v)$ is the solution of (2.3).

We therefore define an "approximate" problem by considering

$$J_\varepsilon(v) = \int_\Gamma |y_\varepsilon(v) - z_d|^2\, d\Gamma + N\int_\Gamma v^2\, d\Gamma . \tag{2.12}$$

One can easily check that there exists $u_\varepsilon \in U_{ad}$ such that

$$J_\varepsilon(u_\varepsilon) = \inf J_\varepsilon(v),\ v \in U_{ad} ; \tag{2.13}$$

we can find optimal controls $u$ for (2.12) such that

$$u_\varepsilon \to u \quad \text{in} \quad L^2(\Gamma) \quad \text{weakly.} \qquad [] \tag{2.14}$$

If we set

$$\frac{d}{d\lambda}\, y_\varepsilon(u_\varepsilon + \lambda(v-u_\varepsilon))|_{\lambda=0} = \hat{y} \tag{2.15}$$

then one shows that

$$(2.16) \qquad a(\hat{y},\phi) - \frac{1}{\varepsilon}(r'(y_\varepsilon(u_\varepsilon))\hat{y},\phi)_\Gamma = (v-u_\varepsilon,\phi)_\Gamma \quad \forall\ \phi \in H^1(\Omega)\ .$$

A necessary condition for $u_\varepsilon$ to be optimal is

$$(2.17) \qquad \int_\Gamma (y_\varepsilon - z_d)\hat{y}\ d\Gamma + N\int_\Gamma u_\varepsilon(v-u_\varepsilon)d\Gamma \geq 0,\ y_\varepsilon(u_\varepsilon) = y_\varepsilon\ ,$$

$$\forall\ v \in U_{ad}\ .$$

If we introduce the adjoint state $p_\varepsilon$ by

$$(2.18) \qquad a(p_\varepsilon,\phi) - \frac{1}{\varepsilon}(r'(y_\varepsilon)p_\varepsilon,\phi)_\Gamma = (y_\varepsilon - z_d,\phi)_\Gamma \quad \forall\ \phi \in \dot{H}^1(\Omega)\ ,$$

we have by taking $\phi = \hat{y}$ in (2.18)

$$(y_\varepsilon - z_d,\hat{y})_\Gamma = a(\hat{y},p_\varepsilon) - \frac{1}{\varepsilon}(r'(y_\varepsilon)\hat{y},p_\varepsilon)_\Gamma = (p_\varepsilon,\ v-u_\varepsilon)_\Gamma$$

by virtue of (2.16), so that (2.17) is equivalent to

$$(2.19) \qquad \int_\Gamma (p_\varepsilon + Nu_\varepsilon)(v-u_\varepsilon)d\Gamma \geq 0 \quad \forall\ v \in U_{ad}\ .$$

<u>Summing up</u>: by using a <u>differentiable</u> penalty term in (2.10), one approximates the initial problem by differentiable problems--for which one can write <u>necessary conditions</u>. []

Remark 2.4

These necessary conditions are convenient in numerical computations, as it is shown in J.P. Yvon [2].

## 3. A Priori Feedback

### 3.1 Reduction to V.I.

The ideas on a priori feedback introduced in Chapters 2 and 3 extend to situations where the state equation is non-linear. Let us give first an example where the problem reduces to a V.I. for a non-linear operator.

Let the state $y = y(v)$ be given by the non-linear elliptic problem:

$$-\Delta y + y^3 = f \quad \text{in } \Omega , \quad f \in L^2(\Omega) , \tag{3.1}$$

$$\frac{\partial y}{\partial \nu} = v \quad \text{on } \Gamma \tag{3.2}$$

where

$$0 \leq v \leq M . \tag{3.3}$$

This problem admits a unique solution

$$y \in H^1(\Omega) \cap L^4(\Omega) \ ^{(1)} . \tag{3.4}$$

(1) $L^4(\Omega) \subset H^1(\Omega)$ if $n \leq 4$ .

Let the cost function be given by

$$(3.5) \qquad J(v) = \int_\Gamma |y(v) - z_d|^2 d\Gamma \ .$$

By the same arguments as in Chapter 2, Section 1.1, one is led to finding $\Phi$ such that

$$(3.6) \qquad -\Delta\Phi + \Phi^3 = f \ ,$$

$$(3.7) \qquad \begin{cases} \Phi < z_d \Rightarrow \dfrac{\partial\Phi}{\partial\nu} = M \ , \\ \Phi > z_d \Rightarrow \dfrac{\partial\Phi}{\partial\nu} = 0 \ , \\ 0 \leq \dfrac{\partial\Phi}{\partial\nu} \leq M \ . \end{cases} \qquad []$$

One can formulate the problem (3.6) (3.7) by a V.I. If we define

$$a(y,\phi) = \sum_{i=1}^{n} \int_\Omega \frac{\partial y}{\partial x_i} \frac{\partial \phi}{\partial x_i} dx + \int_\Omega y^3 \phi \, dx$$

and

$$j(\phi) = M\int_\Gamma (\phi - z_d)^- \, d\Gamma$$

then the problem is equivalent to

$$(3.8) \qquad a(\Phi,\phi-\Phi) + j(\phi) - j(\Phi) \geq (f,\phi-\Phi) \ \forall \ \phi \in H^1(\Omega) \cap L^4(\Omega) \ .$$

Remark 3.1

The formulation (3.8) is the same as (1.7), Chapter 2, but here $z(y,\phi)$ corresponds to a non-linear operator; in Chapter 2, $a(y,\phi)$ was a bilinear form. []

According to the general theory of V.I. (H. Brezis [1], J.L. Lions [5]) there exists a unique solution of (3.8). []

Remark 3.2

As in the linear case, we would obtain (3.6) (3.7) for cost functions other than (3.7)--such as, for instance, $\int_\Gamma |y(v)-z_d| d\Gamma$ .

Remark 3.3

Once (3.8) is solved, one takes as control

$$w = \frac{\partial \Phi}{\partial \nu} . \tag{3.9}$$

This is a sub-optimal control. It is an open problem to obtain estimates on $\inf_v J(v) - J(w)$ .

3.2 Reduction to Q.V.I.

Let us suppose that the state is still given by (3.1) (3.2). If we define

$$\mathfrak{m}(\phi) = \frac{1}{\text{meas}.\Gamma} \int_\Gamma \phi \, d\Gamma$$

let us suppose that the cost function is given by

$$(3.10) \qquad J(v) = \int_\Gamma |y(v) - m(y(v))|^2 d\Gamma \ .$$

Then, by the same arguments as in Chapter 3, Section 1.2, one is led to finding a function $\Phi$ such that

$$(3.11) \qquad -\Delta\Phi + \Phi^3 = f \quad \text{in} \quad \Omega \ ,$$

$$(3.12) \qquad \begin{cases} 0 \leq \dfrac{\partial\Phi}{\partial\nu} \leq M \ , \\ \Phi < m(\Phi) \Rightarrow \dfrac{\partial\Phi}{\partial\nu} = M \ , \\ \Phi > m(\Phi) \Rightarrow \dfrac{\partial\Phi}{\partial\nu} = 0 \ . \end{cases}$$

Using the notation (1.19), Chapter 3, and the notations of 3.1, one sees that (3.11),(3.12) is equivalent to

$$(3.13) \qquad a(\Phi,\phi-\Phi) + j(\phi,\Phi) - j(\Phi,\Phi) \geq (f,\phi-\Phi) \ \forall\ \phi \in H^1(\Omega) \cap L^4(\Omega) \ .$$

By arguments similar to those of Chapter 3, <u>one proves</u>[(1)] <u>the existence of a maximal</u> (resp. <u>minimal</u>) <u>solution</u> of (3.13).

<u>Remark 3.4</u>

The above considerations extend to non-linear parabolic equations.

---

(1) Assuming that $f \geq 0$ .

# CHAPTER 6

# PROBLEMS WHERE THE CONTROL VARIABLE CONTAINS GEOMETRICAL ARGUMENTS

## 1. Constraints on variable curves.

### 1.1 Statement of the problem.

On the boundary $\Gamma$ of $\Omega \subset \mathbb{R}^n$, let $\mathcal{L}_\lambda$ be a family of "curves" (if $n = 3$ ; $n - 2$ dimensional varieties in general) which depend on $\lambda \in [0,1]$ .

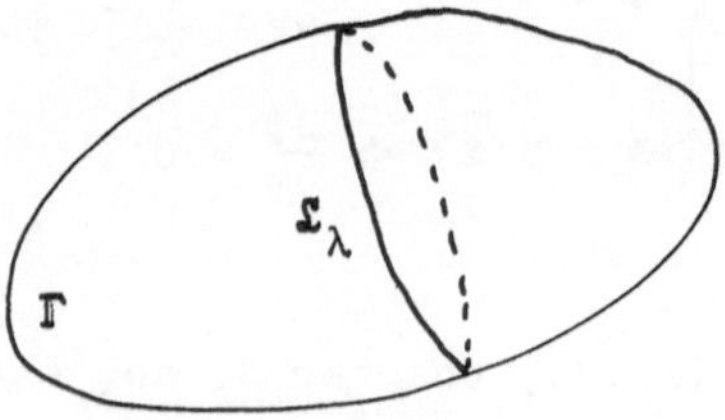


Figure 2

We suppose that the mapping $\lambda \to \mathcal{L}_\lambda$ is "smooth and regular", in a sense made precise below.

Let y be the state, given by

$$(1.1) \qquad -\Delta y + y = f \quad \text{in} \quad \Omega \quad , \qquad f \in L^2(\Omega)$$

$$(1.2) \qquad \frac{\partial y}{\partial \nu} = v \quad \text{on} \quad \Gamma \, , \qquad v \in U = L^2(\Gamma)$$

Let $G$ be a given function in $H^1(\Gamma)$. We denote by $g_\lambda$ the restriction of $G$ to $\mathcal{L}_\lambda$ $(g_\lambda \in H^{1/2}(\mathcal{L}_\lambda))$, and we define

$$(1.3) \qquad U_{ad}(\lambda) = \{v \mid v \in L^2(\Gamma) \ , \ y(v) = g_\lambda \quad \text{on} \quad \mathcal{L}_\lambda\} \ .$$

As we saw in Chapter 4, Section 1 (when $\mathcal{L}_\lambda = \mathcal{L}$ does not depend on $\lambda$ ), $U_{ad}(\lambda)$ is a <u>non empty</u> closed convex subset of $L^2(\Gamma)$. []

The problem we consider is as follows: let the cost function be given by

$$(1.4) \qquad J(v) = \int_\Gamma |y(v) - z_d|^2 \, d\Gamma + N\int_\Gamma v^2 d\Gamma \ ;$$

<u>Then we want to minimize in</u> $\lambda$ $\inf J(v),\ v \in U_{ad}(\lambda)$. Let $u_\lambda$ be the optimal control <u>for fixed</u> $\lambda$ :

$$(1.5) \qquad J(u_\lambda) = \inf J(v),\ v \in U_{ad}(\lambda),\ u_\lambda \in U_{ad}(\lambda) \ ;$$

we want to minimize $\lambda \to J(u_\lambda)$ on $[0,1]$ :

$$(1.6) \qquad \inf_\lambda J(u_\lambda) \quad .$$

<u>Remark 1.1</u>

One will find in Lions [2] cases where the <u>state</u> depends on variable geometrical arguments. For instance, a case which appears in a number of applications is of the type:

$$(1.7) \qquad -\Delta y + y = f + \sum_{i=1}^{q} v_i \, \delta(x-b_i) \, ,$$

$$(1.8) \qquad \frac{\partial y}{\partial \nu} = 0 \, ,$$

where $v_i \in \mathbb{R}$ and $b_i \in \Omega$ ($\delta(x-b_i)$ = mass + 1 at point $b_i$).

Then the state $y$ depends on $v_i$ and $b_i$ and one has to find "optimal positions" of the $b_i$'s.

For physical applications and numerical computations, cf. Saguez [1].

Remark 1.2

For problems of evolution (of hyperbolic type) where one has to find the best position in order to stabilize the system, we refer to Saint-Jean-Paulin [1].

Remark 1.3

Another problem (of parabolic type) is to find the best position (if it exists) of the $b_i$'s in order to have "the largest" possible range at some finite time $T$. This problem is considered in Amouroux [1].

## 1.2 Existence of an optimal position of $\mathcal{L}_\lambda$ .

Let us define $\phi$ by

(1.9) $$-\Delta\phi + \phi = f \quad \text{in} \quad \Omega \; , \; \phi = G \quad \text{on} \quad \Gamma$$

and let us set

(1.10) $$w = \frac{\partial\phi}{\partial\nu} \; .$$

Then $$w \in U_{ad}(\lambda) \;\; \forall\lambda \quad .$$

Consequently

(1.11) $$J(u_\lambda) \leq J(w)$$

so that

(1.12) $$\|u_\lambda\|_{L^2(\Gamma)} \leq c \; .$$

Let $\lambda_n$ be a minimizing sequence:

(1.13) $$J(u_{\lambda_n}) \to \inf_\lambda J(u_\lambda) \; .$$

By virtue of (1.12), we can extract a subsequence, still denoted by $u_{\lambda_n}$ , such that

(1.14) $$u_{\lambda_n} \to u \text{ in } L^2(\Gamma) \text{ weakly, } \lambda_n \to \lambda_o \text{ in } [0,1] .$$

If we set

(1.15) $$y(u_{\lambda_n}) = y_n , \quad y(u) = y ,$$

then

(1.16) $$y_n \to y \text{ in } H^2(\Omega) \text{ weakly}$$

so that $\underline{\lim}\, J(u_{\lambda_n}) \geq J(u)$ and therefore we shall have $J(u) = \inf_{\lambda} J(u_\lambda)$ if we show that $u \in U_{ad}(\lambda_o)$ . But by virtue of (1.16),

(1.17) $$y_n|_\Gamma \to y|_\Gamma \text{ in } H^{3/2}(\Gamma) .$$

By hypothesis $y_n - G = 0$ on $\mathcal{L}_{\lambda_n}$ and if $\mathcal{L}_\lambda$ depends smoothly on $\lambda$ , it follows that $y - G = 0$ on $\mathcal{L}_{\lambda_o}$ , and the result follows. This will be satisfied if, for instance, we suppose that we can find local mappings which transform $\mathcal{L}_\lambda$ into a family of parallel lines. Thus, under this hypothesis, we have shown the existence of an optimal position $\mathcal{L}_{\lambda_o}$ of the $\mathcal{L}_\lambda$ .

## 2. Control in the coefficients.

### 2.1 Statement of the problem.

In many problems(1) the control variable is the domain itself. A possibility (other methods are considered in the next section) is to transform - by variable mappings - the variable domain, $\Omega_\lambda$ say, into a fixed domain $\Omega$ (2). One is then reduced to a problem of optimal control when the control appears in the coefficients of the operator. []

Let us consider a specific example. Let $a_{ij}(x) \in L^\infty(\Omega)$ real valued,

$$(2.1) \qquad \sum_{i,j=1}^{n} a_{ij}(x)\, \xi_i \xi_j \geq \alpha \sum_{i=1}^{n} \xi_i^2 \quad \text{a.e. in } \Omega,\ \alpha > 0,\ \xi_i \in \mathbb{R}$$

and let $v$ = control function subject to

$$(2.2) \qquad 0 < m \leq v(x) \leq M < \infty \quad \text{a.e. in } \Omega \quad (U_{ad}).$$

The state $y = y(v)$ of the system is given by

$$(2.3) \qquad -\sum_{i,j=1}^{n} \frac{\partial}{\partial x_i}\left(a_{ij}\, v\, \frac{\partial y}{\partial x_j}\right) = f, \quad f \in L^2(\Omega),$$

---

(1) All the problems of "optimum design" - where one has to find the "best shape" of a geometrical domain. These problems are not to be confused with the problems of "optimal design" of experiments in Statistics.

(2) Of course this method assumes rather strong a priori hypotheses on the variations of $\Omega_\lambda$.

(2.4) $$y = 0 \quad \text{on} \quad \Gamma .$$

If the cost function is given by

(2.5) $$J(v) = \int_{\Omega} |y(v) - z_d|^2 dx$$

the problem is

(2.6) $$\text{Inf } J(v) \; , \; v \in U_{ad} .$$

Remark 2.1

Problems of the preceding type also arise directly, i.e. without reference to problems of optimum design. Cf. Luré [1].

2.2. Various remarks

It is an open question to know if there exists a $u \in U_{ad}$ which solves (2.6).

It is shown in Murat [1] that, if in (2.3) one adds to the operator the term $vy$, then the problem does not admit in general a solution.

On the other hand it is shown in Baranger [1] that by perturbing (2.5), one gets existence "in general" of an optimal control. Cf. also Lions [3] for a report on these questions.

The difficulty in trying to show the existence of u solving (2.6) is as follows: let $v_n$ be a minimizing sequence, and let us set $y_n = y(v_n)$ . Then, by virtue of (2.2), we can extract a subsequence, still denoted by $v_n$ , such that

(2.7) $$v_n \to v_o \quad \text{in} \quad L^\infty(\Omega) \quad \text{weak star}$$

(2.8) $$y_n \to y_o \quad \text{in} \quad H^1(\Omega) \quad \text{weakly}$$

but this is not enough to conclude that $v_n \frac{\partial y_n}{\partial x_i} \to v_o \frac{\partial y_o}{\partial x_i}$ in any sense, so that it is not proven that $y_o = y(v_o)$ .

This question is related to the problem of G-convergence, introduced and studied in S. Spagnolo [1] [2], A Marino and S. Spagnolo [1] and E. de Giorgi and Spagnolo [1]. In the next section we give some introductory remarks to this interesting theory.

## 2.3 The G-convergence of elliptic[(1)] operators.

Let us consider the family of operators given by

(2.9) $$A_k\varphi = - \sum_{i,j=1}^{n} \frac{\partial}{\partial x_i} \left(a_{ijk}(x) \frac{\partial\varphi}{\partial x_j}\right)$$

where

---

(1) One will find in the work of Spagnolo et al., loc. cit., a similar study - with many more results! - for parabolic operators.

$$(2.10)\quad \begin{cases} m \sum_{i=1}^{n} \xi_i^2 \leq \sum_{i,j=1}^{n} a_{ijk}(x)\, \xi_i \xi_j \leq M \sum_{i=1}^{n} \xi_i^2 \text{ , a.e. in } \Omega \text{ ,} \\ 0 < m \leq M < \infty \text{ ,} \\ a_{ijk}(x) = a_{jik}(x) \text{ .} \end{cases}$$

Let $A$ be given by

$$(2.11) \qquad A\varphi = - \sum_{i,j=1}^{n} \frac{\partial}{\partial x_i} \left( a_{ij} \frac{\partial \varphi}{\partial x_j} \right) \text{ ,}$$

where the $a_{ij}$'s satisfy conditions analogous to (2.10).

<u>One says that</u> $A_k \rightharpoonup A$ <u>in the G-sense</u> ($A_k \overset{G}{\rightharpoonup} A$ , in short) iff.

$$(2.12) \qquad A_k^{-1} f \rightharpoonup A^{-1} f \quad \text{in } L^2(\Omega) \text{ weakly, } \forall \; f \in L^2(\Omega)$$

(where $A_k^{-1}$, $A^{-1}$ are taken with respect to the Dirichlet problem).

<u>The question is to see what are the implications of</u> (2.12) <u>on the coefficients</u> $a_{ijk}$, $a_{ij}$ . []

<u>Solution in a particular case</u>.

Let us assume that $A_k$ is given by

$$(2.13) \qquad A_k \varphi = - \sum_{j=1}^{n} \frac{\partial}{\partial x_j} \left( \beta_k(x) \frac{\partial \varphi}{\partial x_j} \right)$$

where

$$(2.14) \qquad \beta_k(x) = \beta_{k1}(x_1)\beta_{k2}(x_2) \ldots \beta_{kn}(x_n) \, ,$$

$$m \leq \beta_k(x) \leq M$$

and that $A$ is given by

$$(2.15) \qquad A\varphi = - \Sigma \frac{\partial}{\partial x_j} (\lambda_j(x) \frac{\partial \varphi}{\partial x_j}) \, , \quad m \leq \lambda_j(x) \leq M \, .$$

We are going to show

Theorem 2.1. One has $A_k \overset{G}{\rightarrow} A$ iff

$$(2.16) \qquad \frac{1}{\beta_{kj}} - \frac{\beta_k}{\beta_{kj}} \frac{1}{\lambda_j} \rightarrow 0 \text{ in } L^\infty(\Omega) \text{ weak star, } \forall \; j$$

Proof of the sufficiency of (2.16).

We assume that (2.16) holds true. We have to show that for $\forall \; f,g \in L^2(\Omega)$, $(A_k^{-1}f,g) \rightarrow (A^{-1}f,g)$. But since $A_k$, $A$ are symmetric, it is sufficient to prove that

$$(2.17) \qquad (A_k^{-1}f,f) \rightarrow (A^{-1}f,f) \quad \forall \quad f \in L^2(\Omega) \, .$$

If we set

$$(2.18) \qquad A_k^{-1}f = u_k \, , \quad A^{-1}f = u \, ,$$

(2.17) is equivalent to

$$(2.19) \qquad \sum_j \int_\Omega \beta_k \left(\frac{\partial u_k}{\partial x_j}\right)^2 dx \to \sum_j \int_\Omega \lambda_j \left(\frac{\partial u}{\partial x_j}\right)^2 dx .$$

Let us admit for a moment the

<u>Lemma 2.1</u>. <u>Let</u> $\Omega_o$ <u>be an open set such that</u> $\bar{\Omega}_o \subset \Omega$ . <u>Then</u> $\beta_{kj} \frac{\partial u_k}{\partial x_j} \in H^1(\Omega_o)$ <u>and</u>

$$(2.20) \qquad \left\|\beta_{kj} \frac{\partial u_k}{\partial x_j}\right\|_{H^1(\Omega_o)} \le C(\Omega_o) \|f_k\|_{L^2(\Omega)}$$

By virtue of (2.10) $u_k$ is bounded in $H^1(\Omega)$ so that $\beta_{kj} \frac{\partial u_k}{\partial x_j}$ is <u>bounded in</u> $L^2(\Omega)$ and by virtue of (2.20)(1) one can extract a subsequence - still denoted by $u_k$ - such that $\beta_{kj} \frac{\partial u_k}{\partial x_j}$ <u>converges</u> a.e. <u>in</u> $\Omega$ .

It follows that

$$(2.21) \qquad \beta_{kj} \frac{\partial u_k}{\partial x_j} \to \chi_j \quad \text{in} \quad L^2(\Omega) \quad \underline{\text{strongly}}.$$

In order to show (2.19), we have to evaluate

$$(2.22) \qquad X_k = (A_k u_k, u_k) - (Au, u) = (f, u_k) - (f, u) = (Au, u_k) - (A_k u_k, u)$$
$$= \sum_j \int_\Omega \left(\lambda_j \frac{\partial u}{\partial x_j} \frac{\partial u_k}{\partial x_j} - \beta_k \frac{\partial u_k}{\partial x_j} \frac{\partial u}{\partial x_j}\right) dx$$

---

(1) And of the compactness of the identity mapping from $H^1(\Omega_o)$ into $L^2(\Omega_o)$ .

$$= \sum_j \int_\Omega \left(\frac{1}{\beta_{kj}} - \frac{\beta_k}{\beta_{kj}} \frac{1}{\lambda_j}\right)\left(\beta_{kj} \frac{\partial u_k}{\partial x_j}\right)\left(\lambda_j \frac{\partial u}{\partial x_j}\right) dx .$$

By virtue of (2.16) and (2.21) it follows from (2.22) that $X_k \to 0$ ; hence the sufficiency of (2.16) follows provided we prove Lemma 2.1.

Let us suppress the index "k" in Lemma 2.1 and set $B = A_k$ to avoid confusion. We introduce the operator $L_i$ defined by

$$L_i \varphi = - \sum_{j=1}^{n} \frac{\partial}{\partial x_j} \left(\frac{\beta}{\beta_i (x_i)^2} \frac{\partial \varphi}{\partial x_j}\right) . \tag{2.23}$$

A little computing shows that

$$L_i \left(\beta_i \frac{\partial u}{\partial x_i}\right) = \frac{\partial}{\partial x_i} \left(\frac{1}{\beta_i} f\right) \tag{2.24}$$

and therefore classical estimates show that

$$\left\|\beta_i \frac{\partial u}{\partial x_i}\right\|_{H^1(\Omega_o)} \leq C_1(\Omega_o)\left[\left\|\beta_i \frac{\partial u}{\partial x_i}\right\|_{L^2(\Omega)} + \left\|\frac{\partial}{\partial x_i}\left(\frac{1}{\beta_i} f\right)\right\|_{H^{-1}(\Omega)}\right]$$

$$\leq C_2(\Omega_o) \left[\|u\|_{H^1(\Omega)} + \|f\|_{L^2(\Omega)}\right]$$

$$\leq C(\Omega_o) \|f\|_{L^2(\Omega)} . \qquad []$$

<u>Proof of the necessity of</u> (2.16).

If (2.16) is not true then one can extract a subsequence, still denoted by $\beta_k$ , such that

$$(2.25)\qquad \frac{\beta_k - \tilde{\lambda}_j}{\beta_{kj}} \to 0 \quad \text{in } L^\infty(\Omega) \text{ weak star}$$

with $\tilde{\lambda}_j \neq \lambda_j$ for some $j$.

But then, by the first part of the proof,

$$(2.26)\qquad A_k \xrightarrow{G} \tilde{A}, \quad \tilde{A}\varphi = -\sum_j \frac{\partial}{\partial x_j}\left(\tilde{\lambda}_j \frac{\partial \varphi}{\partial x_j}\right)$$

so that $\tilde{A}^{-1} f = A^{-1} f \ \forall \ f \in L^2(\Omega)$, which contradicts the hypothesis that $\tilde{\lambda}_j \neq \lambda_j$ for some $j$. []

Using in particular Theorem 2.1 it is shown in Marino-Spagnolo [1] that given a "general" operator $A$ of the form (2.11) one can find a sequence of operators $B_k$ of the form (2.15) (with different constraints $m, M$) such that $B_k \xrightarrow{G} A$.

## 3. Optimum design.

### 3.1 An example.

Let us consider the classical problem of the shape of a body whose drag is minimum when moved at constant speed in a viscous fluid.

Let $\mathcal{O}$ be an open set in $\mathbb{R}^3$, with boundary $\Gamma \cup S$ (cf. Fig. 3); $\Gamma$ is fixed and $S$ is unknown; actually $S$ is the boundary of the body.

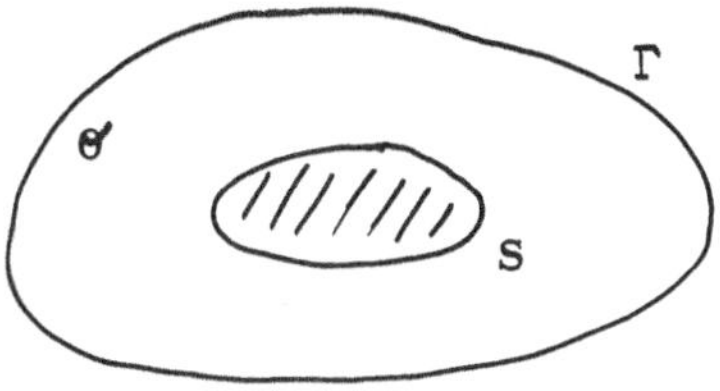


Figure 3

The state of the (linearized) system is $y = y(\mathcal{O})$ given by

(3.1)
$$\begin{cases} \nu\Delta y = -\operatorname{grad} p \ , \\ \operatorname{div} y = 0 \ \text{ in } \ \mathcal{O} \end{cases}$$

with the boundary conditions

(3.2)
$$\begin{cases} y = g \ \text{ on } \ \Gamma \ , \ g \ \text{ given,} \\ y = 0 \ \text{ on } \ S \ . \end{cases}$$

The cost function is given by

(3.3)
$$J(\mathcal{O}) = \frac{2}{\nu} \int_{\mathcal{O}} \sum_{i,j} \sigma_{ij}(y(\mathcal{O}))^2 \, dx \ ,$$

where

(3.4)
$$\sigma_{ij}(y) = \frac{\nu}{2} \left(\frac{\partial y_i}{\partial x_j} + \frac{\partial y_j}{\partial x_i}\right)$$

The problem is to minimize $J(\mathcal{O})$ , when $\mathcal{O}$ (i.e. $S$ ) varies, subject to - <u>for instance</u> -

(3.5) $$\text{meas. } \mathcal{O} = \text{given.}$$ []

It is not known (at least to the author) whether this problem admits a solution without additional assumptions on the family of surfaces described by $S$ .

With additional assumptions of this type the existence of an optimal $S$ is true - and the problem is to obtain <u>first order necessary conditions</u>.

This problem has been studied by O. Pironneau [1] [2].

First order necessary conditions are obtained by writing formulas for the first order variation of $J(\mathcal{O})$ . This type of problem was first considered by J. Hadamard [1]. (Cf. also a paper to appear by Murat and Simon [1]). We refer to Pironneau, loc. cit.

<u>Remark 3.1</u>

There are many problems arising in mechanics of the same type as the preceding example.

<u>Remark 3.2</u>

For a numerical approach to problems of this type, one can consult Pironneau, loc. cit. and Céa, Gioan and Michel [1].

Remark 3.3

All the examples considered up to now assume some smoothness conditions and regularity of behaviour of the boundary of the unknown domain. We give below a problem where no such restriction is needed.

3.2 Relationship to V.I.

The remark below is taken from Bensoussan-Lions [5].

Let $\mathcal{O}$ be an open set in $\mathbb{R}^n$ and let $f$ be a given function in $\mathbb{R}^n$ such that, for instance,

$$\frac{1}{1+|x|} f \in L^2(\mathbb{R}^n) \ . \tag{3.6}$$

The state of the system is given by $y = y(\mathcal{O})$ , a solution of

$$\begin{cases} -\Delta y + \alpha y = f \quad \text{in} \quad \mathcal{O} \ , \quad (\alpha > 0) \\ y = 0 \quad \text{on the boundary} \quad \partial\mathcal{O} \quad \text{of} \quad \mathcal{O} \ . \end{cases} \tag{3.7}$$

We define

$$z_{\mathcal{O}}(x) = \{y(x) \quad \text{in} \quad \mathcal{O} \ , \quad 0 \quad \text{outside} \quad \mathcal{O}\} \tag{3.8}$$

and we set

$$u(x) = \inf_{\mathcal{O}} z_{\mathcal{O}}(x) \ . \tag{3.9}$$

We shall briefly indicate the proof of the following result: u is characterized by

$$(3.10)\qquad \begin{cases} -\Delta u + \alpha u - f \leq 0 , \\ u \leq 0 , \\ (-\Delta u + \alpha u - f)u = 0 \quad \text{in } \mathbb{R}^n \end{cases}$$

with

$$(3.11)\qquad e^{\gamma_1 x} u \in L^2 (\mathbb{R}^n) , \quad \text{for a convenient } \gamma_1 > 0.$$

If we admit for a moment that (3.10) (3.11) admit a unique solution, we define

$$(3.12)\qquad C_o = \{x | u(x) = 0\} , \quad \mathcal{O}_o = \complement C_o .$$

If $f$ is continuous, then $u$ is continuous and (3.12) defines an <u>open</u> set $\mathcal{O}_o$, one has:

$$(3.13)\qquad u(x) = z_{\mathcal{O}_o}(x) ,$$

so that $\mathcal{O}_o$ <u>is the optimal</u> $\mathcal{O}$, <u>whose existence is proven without restrictions on the variations of</u> $\mathcal{O}$.

Remark 3.4

Problems of the preceding type are related to problems of stopping time for which we refer to Bensoussan-Lions [2], where it is shown how V.I. of evolution yield solutions; this was first observed, for stationnary problems, by W. Fleming [1]. Extensions of Bensoussan-Lions to games has been made by A. Friedman [1], who also gave a regularity theorem in [2].

Remark 3.5

In order to simplify the exposition, we replace (3.6) in the following by

$$f \in L^2(\mathbb{R}^n) \tag{3.14}$$

and (3.11) by

$$u \in L^2(\mathbb{R}^n) \; . \qquad \square \tag{3.15}$$

Let us define

$$a(u,v) = \sum_{i=1}^{n} \int_{\mathbb{R}^n} \frac{\partial u}{\partial x_i} \frac{\partial v}{\partial x_i} \, dx + \alpha \int_{\mathbb{R}^n} uv \, dx \quad \text{on} \quad H^1(\mathbb{R}^n) \times H^1(\mathbb{R}^n) \; . \tag{3.16}$$

Then (3.10) is equivalent to

$$\begin{aligned} & a(u,v-u) \geq (f,v-u) \quad \forall \quad v \in K \\ & u \in K \end{aligned} \tag{3.17}$$

where

(3.18) $K = \{v | v \leq 0 \text{ a.e. in } \mathbb{R}^n, \ v \in H^1(\mathbb{R}^n)\}$

This problem admits a unique solution.

It remains to show that (3.9) holds true, with u a solution of (3.17).

If f is continuous, then (Brézis [2]) u is continuous, so that one has (3.13), so that $u \geq \inf z_{\mathcal{O}}$ - and therefore it remains only to show that :

(3.19) if $z_{\mathcal{O}}$ and u are defined by (3.8) and (3.17) resp., then $u \leq z_{\mathcal{O}}$.

Let us denote by $u_{\mathcal{O}}$ the restriction of u to $\mathcal{O}$ and let us define v by

$$v-u = \begin{cases} -(u_{\mathcal{O}} - y_{\mathcal{O}})^+ & \text{in } \mathcal{O} \\ 0 & \text{outside } \mathcal{O}, \end{cases}$$

i.e.

(3.20)
$$v = \begin{cases} \inf(u_{\mathcal{O}}, y_{\mathcal{O}}) & \text{in } \mathcal{O} \\ u & \text{outside } \mathcal{O}. \end{cases}$$

Therefore $v \leq 0$ in $\mathbb{R}^n$ and we can take v in (3.17). If we define $a_{\mathcal{O}}(u,v)$ to be the extremum similar to (3.16) but where integrals are extended over $\mathcal{O}$, we obtain

$$(3.21) \qquad -a_{\mathcal{O}}(u_{\mathcal{O}}, (u_{\mathcal{O}} - y_{\mathcal{O}})^+) \geq -(f,(u_{\mathcal{O}} - y_{\mathcal{O}})^+)_{\mathcal{O}}$$

(where $(f,g)_{\mathcal{O}} = \int_{\mathcal{O}} f(x)g(x)\,dx)$ .

But on the other hand

$$(3.22) \qquad a_{\mathcal{O}}(y_{\mathcal{O}},\varphi) = (f,\varphi)_{\mathcal{O}} \quad \forall\, \varphi \in H_0^1(\mathcal{O})$$

and we observe that $(u_{\mathcal{O}} - y_{\mathcal{O}})^+ = 0$ on $\partial\mathcal{O}$ (since $u \leq 0$) ; we can take $\varphi = (u_{\mathcal{O}} - y_{\mathcal{O}})^+$ in (3.22); we obtain, by adding to (3.21):

$$-a_{\mathcal{O}}(u_{\mathcal{O}} - y_{\mathcal{O}}, (u_{\mathcal{O}} - y_{\mathcal{O}})^+) \geq 0$$

i.e.

$$a_{\mathcal{O}}((u_{\mathcal{O}} - y_{\mathcal{O}})^+, (u_{\mathcal{O}} - y_{\mathcal{O}})^+) \leq 0 \;;$$

hence $(u_{\mathcal{O}} - y_{\mathcal{O}})^+ = 0$ and (3.19) follows.

# CHAPTER 7

# CHEAP CONTROL

## 1. Cheap control and singular perturbations.

### 1.1 Example

We consider again the problem of Chapter 6, Section 1.1: the state $y(v)$ is given by

$$(1.1)\qquad \begin{cases} Ay(v) = 0 & \text{in } \Omega \\ \frac{\partial y}{\partial \nu}(v) = v & \text{on } \Gamma \text{ ;} \end{cases}$$

$v \in L^2(\Gamma)$ is subject to the constraint

$$(1.2)\qquad y(v)|_{\mathcal{L}} = g, \ g \in H^{1/2}(\mathcal{L}) \ .$$

We want to minimize

$$(1.3)\qquad J(v) = \int_\Gamma |y-z_d|^2 \, d\Gamma + \varepsilon\int_\Gamma v^2 d\Gamma$$

where $\varepsilon > 0$ and where we want to see the behaviour of the optimal control $u_\varepsilon$ when $\varepsilon \to 0$ .

Remark 1.1

The term $\varepsilon\int_\Gamma v^2 d\Gamma$ in the cost function is related to the "cost of the control"; the smaller $\varepsilon$ , the cheaper the control, which explains the terminology. []

We shall set

$$y_\varepsilon = y(u_\varepsilon) \quad (1.4)$$

As was seen in Chapter 1, Section 2 (for a different problem, but the same remarks apply), one should expect a singular behaviour of $u_\varepsilon$ as $\varepsilon \to 0$ .

We are going to show this, making precise the manner in which $u_\varepsilon$ behaves in a "singular" manner.

## 1.2 Singular Perturbations.

We use the technique of Chapter 4, Section 2.

If we set:

$$y_\varepsilon|_\Gamma = \phi_\varepsilon \quad (1.5)$$

then $\phi_\varepsilon$ is characterized by

$$\varepsilon(\mathcal{A}\phi_\varepsilon, \mathcal{A}(\phi-\phi_\varepsilon))_\Gamma + (\phi_\varepsilon - z_d, \phi-\phi_\varepsilon)_\Gamma \geq 0 \quad (1.6)$$

$$\forall\ \phi \in H^1(\Gamma) \text{ such that } \phi|_{\mathcal{L}} = g ,$$

where $\mathcal{A}$ is defined by (2.1) (2.2), Chapter 4.

It follows from a general result of D. Huet [1] that

(1.7) $$\phi_\varepsilon \to \phi_o = z_d \quad \text{in} \quad L^2(\Gamma) \quad \text{as} \quad \varepsilon \to 0 .$$

Therefore, $y_\varepsilon \to y_o$ in $L^2(\Omega)$, where $y_0$ is the solution of

(1.8) $$Ay_0 = 0, \quad y_0 = z_d \quad \text{on } \Gamma$$

and then, according to Lions-Magenes [1],

(1.9) $$u_\varepsilon = \frac{\partial y_\varepsilon}{\partial \nu} \to \frac{\partial y_o}{\partial \nu} \quad \text{in} \quad H^{-1}(\Gamma) .$$

The conclusion (1.9) shows the singular behaviour of $u_\varepsilon$ which belongs to $L^2(\Gamma)$ <u>but converges in a larger space</u> $H^{-1}(\Gamma)$ .

[]

Remark 1.2

Problem (1.6) is actually equivalent to the non homogenous boundary value problem

(1.10) $$\begin{cases} \varepsilon \mathcal{A}^* \mathcal{A} \phi_\varepsilon + \phi_\varepsilon = z_d & \text{in } \Gamma - \mathcal{L} , \\ \phi_\varepsilon = g & \text{on } \mathcal{L} . \end{cases}$$

This is a <u>problem of singular perturbations</u> on a variety, for a pseudo-differential operator.

Therefore one has here the boundary layer phenomenon: if we suppose that in (1.10) $z_d \in H^1(\Gamma)$, then

$$\phi_\varepsilon - (z_d + \mathcal{O}_\varepsilon) \to 0 \quad \text{in} \quad H^1(\Gamma) \tag{1.11}$$

where $\mathcal{O}_\varepsilon$ (a corrector term of order 0 ) is of the boundary layer type, that is

$$\mathcal{O}_\varepsilon = (g - z_d)|_{\mathfrak{L}} \quad \exp\left(- \frac{d(x,\mathfrak{L})}{\sqrt{\varepsilon}}\right) , \tag{1.12}$$

where $d(x,\mathfrak{L})$ = geodesic distance on $\Gamma$ from $x$ to $\mathfrak{L}$ . One can write down higher order asymptotic expansions. For the case of ordinary elliptic operators in domains $\subset R^n$ , cf. Visik-Liousternik [1]. For the case of pseudo differential operators on varieties, results are announced in A.S. Demidov [1], L.L. Pokrovski [1].

Remark 1.3

Suppose the constraints are, instead of (1.2),

$$g_o \leq y(v) \leq g_1 \; , \; g_o, g_1 \in H^{1/2}(\mathfrak{L}) \; . \tag{1.13}$$

Then $\phi_\varepsilon = y_\varepsilon|_\Gamma$ is characterized by the V.I. (1.6) where

$$g_o \leq \phi \leq g_1 \; , \quad g_o \leq \phi_\varepsilon \leq g_1 \; .$$

One has the same results (1.7) (1.9).

For expansions in singular perturbations of V.I. we refer to Lions [5].

Remark 1.4

Similar problems often arise in practice, not only for stationary problems but also for evolution problems. Cf. Lions [5] [6]. For numerical computations in problems of this type, cf. Leroy-Yvon [1].

## 2. Cheap Control and Duality

### 2.1 Dual of 1.1

We now apply the technique of Chapter 4, Section 4, to the problem of Section 1.

The dual problem of (1.3) (in the sense of Chapter 4, Section 4), is as follows: let $\phi(q)$ be defined by

$$(2.1) \qquad A^*\phi = 0 \, , \quad \frac{\partial\phi}{\partial\nu} = q_1 + \mu_{\mathcal{L}} q_2 \quad \text{on } \Gamma \, ;$$

then we want to find

$$(2.2) \quad -\inf_{q \in L^2(\Gamma)\times L^2(\mathcal{L})} \left[\frac{1}{2\epsilon}\int_\Gamma \phi(q)^2 d\Gamma + \frac{1}{2}\int_\Gamma q_1^2 d\Gamma - \int_\Gamma q_1 z_d \, d\Gamma - - \int_{\mathcal{L}} q_2 g d\mathcal{L}\right]$$

## 2.2 Formal Limit

The considerations which follow are entirely formal.

Let us admit the existence of a solution $q_\varepsilon$ to (2.2).

Then it is rather natural to think that - due to the penalty term $\frac{1}{2\varepsilon}\int_\Gamma \phi(q)^2 d\Gamma$ - one has

(2.3) $$\phi(q_\varepsilon) \to 0 \qquad (\text{like } \sqrt{\varepsilon}\ ) \text{ in } L^2(\Gamma)$$

and since $A^*\phi(q_\varepsilon) = 0$ , then

(2.4) $$\phi(q_\varepsilon) \to 0 \text{ in } L^2(\Omega)$$

and

(2.5) $$\frac{\partial}{\partial\nu}\phi(q_\varepsilon) \to 0 \text{ in } H^{-1}(\Gamma) .$$

But one can use the "duality relations" between $u_\varepsilon$ and $q_\varepsilon = \{q_{\varepsilon 1}, q_{\varepsilon 2}\}$ .

With the general notations of Chapter 4, we have

(2.6) $$F(u_\varepsilon) + F^*(\Lambda^* q_\varepsilon) = < u_\varepsilon, \Lambda^* q_\varepsilon > ,$$

(2.7) $$G(\Lambda u_\varepsilon) + G^*(-q_\varepsilon) = <\Lambda u_\varepsilon, -q_\varepsilon> .$$

In our particular case, (2.6) becomes

$$\frac{\varepsilon}{2}\int_\Gamma u_\varepsilon^2\, d\Gamma + \frac{1}{2\varepsilon}\int_\Gamma \phi(q_\varepsilon)^2 d\Gamma = \int_\Gamma u_\varepsilon \phi(q_\varepsilon) d\Gamma$$

i.e.

(2.8) $$\varepsilon u_\varepsilon = \phi(q_\varepsilon)|_\Gamma .$$

Relation (2.7) is equivalent to

(2.9) $$G_1(y_\varepsilon) + G_1^*(-q_{1\varepsilon}) = \int_\Gamma y_\varepsilon(-q_{1\varepsilon}) d\Gamma$$

and to

(2.10) $$G_2(y_\varepsilon) + G_2^*(-q_{2\varepsilon}) = \int_\Sigma y_\varepsilon(-q_{2\varepsilon}) d\Sigma$$

which is always satisfied. Relation (2.9) is equivalent to

(2.11) $$y_\varepsilon - z_d + q_{1\varepsilon} = 0 .$$

But (2.5) implies

$$q_{1\varepsilon} + \mu_\Sigma q_{2\varepsilon} \to 0 \quad \text{in} \quad H^{-1}(\Gamma)$$

so that $q_{1\varepsilon} \to 0$ in $H^{-1}(\Gamma)$ and (2.11) gives $y_\varepsilon \to z_d$ in $H^{-1}(\Gamma)$. Actually the method of Section 1 gives more (convergence in $L^2(\Gamma)$ ) and in a non-formal manner, but the preceding remarks show the relationships between <u>cheap control</u> and <u>penalty methods</u>.

## 3. Another example.

### 3.1 Formulation of the problem.

Let the state be given as in Chapter 4, Section 4.4:

$$(3.1)\qquad \begin{cases} Ay = 0 & \text{in } \Omega\,, \\ y = v & \text{on } \Gamma\,, \end{cases}$$

which admits a unique solution in $L^2(\Omega)$ if $v \in L^2(\Gamma)$ .

We suppose that $v$ is subject to

$$(3.2)\qquad < \frac{\partial y}{\partial \nu}(v)\,, g>_\Gamma = \alpha\,,\ g \text{ given in } H^{1/2}(\Gamma),\ \alpha \in \mathbb{R}\,,$$ (1)

and we want to minimize over this set the cost function

$$(3.3)\qquad J(v) = \int |y(v) - z_d|^2 dx + \varepsilon \int v^2 d\Gamma\,. \qquad []$$

The optimal control $u_\varepsilon$ is characterized by

$$(3.4)\qquad \int_\Omega (y_\varepsilon - z_d)(y(v) - y_\varepsilon)dx + \varepsilon \int_\Gamma u_\varepsilon (v - u_\varepsilon) d\Gamma \geq 0$$

$$\forall\ v \in U_{ad}\,, \text{ where } y(u_\varepsilon) = y_\varepsilon\,.$$

---

(1) We denote by $U_{ad}$ the non empty closed convex subset of $L^2(\Gamma)$ defined by (3.2). (It is actually an affine space).

We are going to set

$$(3.5) \qquad \frac{\partial y(v)}{\partial \nu} = \phi \ , \ \frac{\partial y_\varepsilon}{\partial \nu} = \phi_\varepsilon .$$

For $\phi$ given in $H^{-1}(\Gamma)$ , we denote by $\Phi$ <u>the</u> solution of

$$(3.6) \qquad A\Phi = 0 \ , \ \frac{\partial \Phi}{\partial \nu} = \phi \ \text{ on } \Gamma$$

and we set

$$(3.7) \qquad \Phi|_\Gamma = \mathcal{B}\phi \ , \ \Phi = \mathcal{C}\phi \ .$$

We have:

(3.8) $\mathcal{B}$ is an isomorphism from $H^{-1}(\Gamma)$ onto $L^2(\Gamma)$

(3.9) $\mathcal{C} \in \mathcal{L}(H^{-1}(\Gamma) \ ; \ H^{1/2}(\Omega))$.

The optimality condition (3.6) can be written, using these notations:

$$(3.10) \qquad \varepsilon(\mathcal{B}\phi_\varepsilon, \mathcal{B}(\phi-\phi_\varepsilon))_{L^2(\Gamma)} + (\mathcal{C}\phi_\varepsilon - z_d, \mathcal{C}(\phi-\phi_\varepsilon))_{L^2(\Omega} \geq 0$$

$$\forall \ \phi \in H^{-1}(\Gamma) \ \text{ such that } \ \langle \phi, g \rangle = \alpha \ .$$

## 3.2 Singular Behaviour

The form $\varepsilon\,(\mathcal{B}\phi,\mathcal{B}\psi)_{L^2(\Gamma)} + (\mathcal{C}\phi,\mathcal{C}\psi)_{L^2(\Omega)}$ is coercive on $H^{-1}(\Gamma)$. For $\varepsilon = 0$, it equips $H^{-1}(\Gamma)$ with the <u>norm</u> $\|\mathcal{C}\phi\|_{L^2(\Omega)}$ and one shows, using Lions-Magenes [1], that this norm is <u>equivalent</u> to

$$\|\phi\|_{H^{-3/2}(\Gamma)} .$$

Therefore, using D. Huet [1], we obtain:

$$\phi_\varepsilon \to \phi_0 \quad \text{in} \quad H^{-3/2}(\Gamma) \quad \text{as} \quad \varepsilon \to 0 , \tag{3.11}$$

where

$$(\mathcal{C}\phi_0 - z_d, \mathcal{C}\phi)_{L^2(\Omega)} = 0 , \quad \forall \quad \phi \in H^{-3/2}(\Gamma) . \tag{3.12}$$

But (3.12) means that

$$(\phi_0 - z_d, \phi)_{L^2(\Omega)} = 0 , \quad \forall \quad \phi \quad \text{such that} \quad A\phi = 0 . \tag{3.13}$$

But

$$L^2(\Omega) = A^*H_0^2(\Omega) \oplus X ,$$

$$X = \{\phi \mid \phi \in L^2(\Omega),\ A\phi = 0\}$$

so that (3.13) is equivalent to

$$(3.14) \qquad \phi_0 - z_d = A^*\chi\ , \quad \chi \in H_0^2(\Omega)\ .$$

Applying $A$ to (3.14) we obtain

$$(3.15) \qquad AA^*\chi = -Az_d, \quad \chi \in H_0^2(\Omega)$$

which defines $\chi$ . Then

$$(3.16) \qquad \mathcal{B}\phi_0 = (z_d + A^*\chi)_\Gamma$$

and

$$(3.17) \qquad u_\varepsilon = \mathcal{B}\phi_\varepsilon \to u_0 = \mathcal{B}\phi_0 = (z_d + A^*\chi)_\Gamma \quad \text{in} \quad H^{-1/2}(\Gamma)\ .$$

This result shows again a <u>singular behaviour</u>, since $u_\varepsilon \in L^2(\Gamma)$ <u>and converges</u>, as $\varepsilon \to 0$ , in a larger space than $L^2(\Gamma)$ .

<u>Remark 3.1</u>

We leave the application of duality (as in Chapter 4) as an exercise.

# CHAPTER 8

# IMPULSE CONTROL

## 1. The Physical Problem

We follow Bensoussan-Lions [1], Note 1).

### 1.1 A problem in management.

Let $x = \{x_1,\ldots,x_n\} \in R^n$ be the amount of goods at our disposal at time $t$ (the inventory).

We denote by $D(s,t)$ the cumulative demand of these goods, on $[t,s]$ and we suppose that $D(s,t)$ is given by

$$D(s,t) = \int_t^s \mu(\lambda)d\lambda + \sigma(w(s) - w(t)) \tag{1.1}$$

where

$\mu = \{\mu_1,\ldots,\mu_n\}$ is a given continuous function from $R \to R^n$,

$\sigma = (n,n)$ diagonal matrix

$w$ = Wiener process in $R^n$ .

To fix ideas (many other possibilities fit into the same framework) we suppose that we have <u>three types of cost</u>:

each time we place an order, (whatever be the amount), we have to pay a fixed cost, called the <u>set-up cost</u>;

(1.2) We shall denote this set-up cost by $k$ $(k > 0)$ ;

(1.3) a holding cost;

(1.4) a shortage cost.

Remark 1.1

The (natural) hypothesis (1.2) is fundamental for what follows: it excludes a continuous policy of ordering, which would imply an infinite cost.

Notations:

We denote by $v_{xt}$ any policy, of the following type:

$$v_{xt} = \{\theta^1_{xt}, \xi^1_{xt} ; \theta^2_{xt}, \xi^2_{xt}; \ldots ; \theta^{N_{xt}}_{xt}, \xi^{N_{st}}_{xt}\} ; \tag{1.5}$$

in (1.5) $\theta^j_{xt}$ are random in $[t,T]$ (1), $\xi^j_{st}$ are random in $R^n$; $\theta^1_{xt}$ $(\geq t)$ is the $1^{st}$ time we place an order, and we order $\xi^1_{xt}$ ; if $\xi$ denotes any order, the (natural) constraints are:

$$\xi = \{\xi_1, \ldots, \xi_n\}, \quad \xi_j \geq 0 \quad (\text{in short: } \xi \geq 0) ; \tag{1.6}$$

---

(1) $T$ is the horizon, that we suppose to be finite.

in (1.5) $\theta^2_{xt}$ is the $2^d$ time we order, and we order $\xi^2_{xt}$, etc.; $N_{xt}$ is the number of orders which are placed in $[t,T]$.

The <u>state</u> $y_{xt}(s)$ of the inventory at time $s$ is given as follows:

(1.7) $$y_{xt}(s) = x - D(s,t) \quad \text{for} \quad t \leq s < \theta^1_{xt} ,$$

$$y_{xt}(\theta^1_{xt}) = y_{xt}(\theta^1_{xt} - 0) + \xi^1_{xt}$$

etc.

<u>The cost function</u> related to the policy $v_{xt}$ is given by

(1.8) $$J_{xt}(v_{xt}) = E\{kN_{xt} + \int_t^T F(y_{xt}(s),s)ds\}$$

where $F(x,t)$ is a non-negative function which corresponds to the costs (1.3) and (1.4); the term $kN_{xt}$ corresponds to the cost (1.2).

<u>The problem</u>. We set

(1.9) $$u(x,t) = \inf_{v_{xt}} J_{xt}(v_{xt})$$

and we want to characterize $u$ and the optimal policy (if it exists) associated with $u$.

1.2 <u>Policy related to a couple</u> $\{u,C\}$ .

Let $u$ be a continuous function in $Q = R^n \times (0,T)$ , such that the problem

$$(1.10) \qquad \min_{\xi \geq 0} u(x + \xi, t)$$

admits a unique solution $\xi(x,t)$ for all $x$ and $t$ .

Let $C$ be an open subset of $Q$ and $S$ the complementary set of $C$ in $Q$ ; we suppose that the boundary of $C$ is smooth, and that

$$(1.11) \qquad x,t \in S \Rightarrow (x + \xi(x,t),t) \in C .$$

Given $u$ , we <u>define</u> $C$ by

$$(1.12) \qquad C = \{x,t \,|\, u(x,t) < k + \inf_{\xi \geq 0} u(x + \xi,t)\} .$$

Given $\{u,C\}$ , we can define $v_{xt}$ as follows:

$$(1.13) \qquad \theta^1_{xt} = 1^{st} \text{ time } s \geq t \text{ where } x-D(s,t) \text{ gets in } S$$

$$\xi^1_{xt} = \xi(x-D(\theta^1_{xt},t), \theta^1_{xt})$$

and so on.

Remark 1.2

The set C is called the continuation set; the set S defined by

$$(1.14)\qquad S = \{x,t \mid u(x,t) = k + \inf u(x + \xi,t)\}$$

is called the saturation set. []

In what follows we are going to characterize $u(x,t)$, defined by (1.9), in terms of partial differential inequalities.

## 1.3 Partial differential inequalities characterizing u .

One can show (cf. Bensoussan-Lions [1] 1), [3], [4]) that $u(x,t)$ is characterized by the following inequalities and equalities:

$$(1.15)\quad \begin{cases} -\dfrac{\partial u}{\partial t} + A(t)\, u - f \leq 0 , \\ u - M(u) \leq 0 , \\ (-\dfrac{\partial u}{\partial t} + A(t)u-f)\,(u-M(u)) = 0 \quad \text{in} \quad Q = R^n \times (0,T) \end{cases}$$

$$(1.16)\qquad u(x,T) = 0 ;$$

In (1.15),

$$(1.17)\qquad A(t)\phi = -\sum_{i=1}^{n} \frac{1}{2}\, \sigma_i^2 \frac{\partial^2\phi}{\partial x_i^2} + \sum_{i=1}^{n} \mu_i(t) \frac{\partial\phi}{\partial x_i}$$

(1.18) $$M(\phi)(x,t) = \inf_{\xi \geq 0} (k + \phi(x + \xi, t)) .$$

Remark 1.3.

Problem (1.15) (1.16) is (again) of the "free boundary" type. Indeed, by virtue of the third relation in (1.15), there are two regions in $Q$ :

first, the continuation set $C$ where $u < M(u)$ and where

$$- \frac{\partial u}{\partial t} + A(t)u - f = 0 ;$$

and secondly, the saturation set $S$ where $u = M(u)$ .
The interface between $C$ and $S$ is not known - and actually its determination is one of the fundamental questions.

Remark 1.4

In the deterministic case $\sigma = 0$ , and the operator $- \frac{\partial}{\partial t} + A(t)$ becomes hyperbolic.

1.4 Stationary case.

The stationary problem(1) associated with (1.15) (1.16) is: find $u$ such that

(1.19) $$Au - f \leq 0,\ u - M(u) \leq 0,\ (Au-f)(u-M(u)) = 0 \text{ in } R^n \text{ (2)}.$$

(1) The only one we shall consider from now on. We refer to Bensoussan-Lions, loc.cit., for evolution problems.

(2) $A = - \frac{1}{2}\Sigma\sigma_i \frac{\partial^2}{\partial x_i^2} + \Sigma \mu_i \frac{\partial}{\partial x_i} + \alpha,\ \alpha > 0$ (actualization term).

Remark 1.5

One should add to (1.19) conditions at $\infty$ ; if we suppose that $e^{-\gamma_0|x|} f \in L^2(R^n)$ for some $\gamma_0$ small enough (actually, in applications, $\frac{f}{1+|x|} \in L^2(R^n)$) , then we shall look for a solution u of (1.19) satisfying

$$(1.20) \qquad e^{-\gamma|x|} u, \ e^{-\gamma|x|} \frac{\partial u}{\partial x_i} \in L^2(R^n) \quad \text{for } \gamma \text{ small enough.}$$

Remark 1.6.

To simplify the exposition, we shall consider a problem similar to (1.19) but in a bounded open set $\sigma$ of $R^n$ ; then one has to add boundary conditions; the problem is as follows:

$$(1.21) \qquad Au - f \leq 0, \ u - M(u) \leq 0, \ (Au-f)(u-M(u)) = 0 \quad \text{in } \sigma ,$$

where

$$(1.22) \qquad M(u)(x) = \inf \, [k + u(x + \xi)], \ \xi \geq 0, \ x + \xi \in \sigma$$

with the boundary conditions

$$(1.23) \qquad \frac{\partial u}{\partial \nu} \leq 0, \ u - M(u) \leq 0, \ \frac{\partial u}{\partial \nu} (u-M(u)) = 0 \quad \text{on } \Gamma = \partial\sigma .$$

We now give some indications of the solution to this last problem.

## 2. Reduction to Q.V.I.

### 2.1 Formulation in terms of a Q.V.I.

Notations: on $H^1(\sigma)$ we define

$$a(u,v) = \sum_i \frac{\sigma_i^2}{2} \frac{\partial u}{\partial x_i} \frac{\partial v}{\partial x_i} dx + \sum_i \int_\sigma \mu_i \frac{\partial u}{\partial x_i} v dx + \alpha \int_\sigma uv dx , \tag{2.1}$$

$$(f,v) = \int_\sigma fv dx . \tag{2.2}$$

We are going to show the equivalence of the problem (1.21)-(1.23) with

$$\begin{aligned} &a(u,v-u) \geq (f,v-u) \quad \forall \quad v \leq M(u) , \\ &u \leq M(u) . \end{aligned} \tag{2.3}$$

Remark 2.1

One sees the analogy between (2.3) and (1.20), Chapter 3. They both enter the theory of Q.V.I.

A V.I. associated with (2.3) is the following:

$$\begin{aligned} &a(\bar{u},v-\bar{u}) \geq (f,v-\bar{u}), \quad \forall \quad v \leq M(v), \text{ [1]} \\ &\bar{u} \leq M(\bar{u}) . \end{aligned} \tag{2.4}$$

---

(1) And not $v \leq M(\bar{u})$ .

Sketch of the proof of equivalence.

In the region C where $u < M(u)$, we can take $v = u \pm \varepsilon\phi$ where $\phi$ is smooth with "small" support in C, for $\varepsilon$ small enough, and this implies that $Au = f$.

In the region S where $u = M(u)$, we can take $v = u - \varepsilon\phi$, where $\phi$ is smooth with "small" support in $S, \phi \geq 0$, $\varepsilon > 0$ small enough, and this implies that $Au - f \leq 0$.

The boundary inequalities follow from Green's formula.

## 2.2 Idea of the solution of the Q.V.I.

We shall assume that

$$a(v,v) \geq c\|v\|^2_{H^1(\sigma)}, \quad c > 0, \quad \forall \; v \in H^1(\sigma) . \tag{2.5}$$

We show below the

**Theorem 2.1.**

*Let f be given in $L^\infty(\sigma)$, $f \geq 0$. Then there exists u in $H^1(\sigma) \cap L^\infty(\sigma)$, satisfying (2.3) and such that*

$$u \geq 0 . \tag{2.6}$$

*Moreover if w is any non-negative o solution of (2.3) then* $u \geq w$ (u is "maximal").

(See Remark 2.2 hereafter for the uniqueness.)

Proof

The technique is very similar to the one used in Chapter 3.

We denote by $u^o$ the solution of the Neumann problem.

$$a(u^o,v) = (f,v) \ , \quad \forall \ v \in H^1(\sigma) \ . \tag{2.7}$$

Since f is positive and bounded we have

$$0 \leq u^o \leq m(<\infty) \ . \tag{2.8}$$

We define $u^1$ as the solution of the V.I.

$$\begin{cases} a(u^1,v-u^1) \geq (f,v-u^1) \quad \forall \quad v \leq M(u^o) \ , \\ u^1 \leq M(u^o) \ . \end{cases} \tag{2.9}$$

Since a is not (in general) symmetric, (2.9) is not equivalent to a minimization problem; in order to know that (2.9) uniquely defines $u^1$ one has to rely on Stampacchia [1], Lions-Stampacchia [1].

We have

$$0 \leq u^1 \leq u^o \ . \tag{2.10}$$

Indeed if we take in (2.9) v given by $v-u^1 = -(u^o-u^1)^-$ i.e. $v = \inf(u^o,u^1)$ (so that $v \leq M(u^o)$), and if we take in (2.7) $v = (u^o-u^1)^-$, we obtain:

$$a(u^0-u^1,\ (u^0-u^1)^-) \geq 0$$

so that

$$a((u^0-u^1)^-,\ (u^0-u^1)^-) \leq 0$$

and therefore $(u^0-u^1)^- = 0$ , so that $u^1 \leq u^0$ .
If we take now in (2.9) $v$ given by $v-u^1 = (u^1)^-$ , i.e.
$v = (u^1)^+$ (and $u^1 \leq M(u^0) \Rightarrow (u^1)^+ \leq M(u^0)$ since $u^0 \geq 0$
$\Rightarrow M(u^0) \geq 0$) , we obtain:

$$a(u^1,(u^1)^-) \geq (f,(u^1)^-)$$

i.e.

$$a((u^1)^-,(u^1)^-) + (f,(u^1)^-) \leq 0$$

hence $(u^1)^- = 0$ (since $f \geq 0$) .
We define now recursively $u^n$ by

$$\text{(2.11)} \qquad \begin{cases} a(u^n,v-u^n) \geq (f,v-u^n) \quad \forall \quad v \leq M(u^{n-1}) , \\ u^n \leq M(u^{n-1}) . \end{cases}$$

We admit by induction that

$$\text{(2.12)} \qquad 0 \leq u^{n-1} \leq u^{n-2}$$

and we show that

$$0 \leq u^n \leq u^{n-1} \quad . \tag{2.13}$$

We take $v$ given by $v-u^n = -(u^{n-1}-u^n)^-$ in (2.11) and $v$ given by $v-u^{n-1} = (u^{n-1}-u^n)^-$ in the inequality defining $u^{n-1}$ (1); it follows that $a((u^{n-1}-u^n)^-, (u^{n-1}-u^n)^-) \leq 0$ , hence $u^{n-1} \geq u^n$ ; for the proof of $u^n \geq 0$ , we have only to define $v$ by $v-u^n = (u^n)^-$ in (2.11). []

Summing up, we have proven that

$$u^0 \geq u^1 \geq \ldots \geq u^{n-1} \geq u^n \geq \ldots \geq 0 \; . \tag{2.14}$$

Since $u^{n-1} \geq 0$ , we can choose $v = 0$ in (2.11); it follows that

$$\|u^n\|_{H^1(\sigma)} \leq C \; . \tag{2.15}$$

From (2.14) (2.15) it follows that

$$\begin{cases} u^n \to u \text{ in } L^\rho(\sigma) \text{ strongly, } \forall \; \rho < \infty , \\ u^n \to u \text{ in } H^1(\sigma) \text{ weakly} \\ 0 \leq u \leq u^0 \; . \end{cases} \tag{2.16}$$

From the inequality

(1) The main point here is that (2.12) => $M(u^{n-1}) \leq M(u^{n-2})$ .

$$u^n(x) \leq k + u^{n-1}(x + \xi),\ \xi \geq 0\ , \quad \text{for almost all} \quad x\ ,$$

it follows that

$$u(x\ ) \leq k + u(x + \xi),\ \xi \geq 0$$

i.e.

(2.17) $$u \leq M(u)\ .$$

Let $v$ be given such that

(2.18) $$v \leq M(u)\ .$$

Since $u^n \downarrow u$ , we have $v \leq M(u) \leq M(u^{n-1})$ and we can take this function $v$ in (2.11); therefore

$$a(u^n,v) = (f,v-u^n) \geq a(u^n,u^n)$$

which gives at the limit

$$a(u,v) = (f,v-u) \geq \underline{\lim}\ a(u^n,u^n) \geq a(u,u)$$

and therefore $u$ satisfies the Q.V.I. (2.3). []

Let us show now that the solution $u$ we have just constructed is <u>maximal</u>. Let $w$ be any solution of

$$(2.19)\qquad \begin{cases} a(w,v-w) \geq (f,v-w), & \forall\ v \leq M(w) , \\ w \leq M(w) , & 0 \leq w . \end{cases}$$

<u>Then</u>

$$(2.20)\qquad u \geq w .$$

Indeed let us take in (2.19) $v-w = -(u^0-w)^-$ , i.e. $v = \inf (u^0,w) \leq M(w)$ and in (2.7) $v = (u^0-w)^-$ ; then $a(u^0-w, (u^0-w)^-) \geq 0$ , hence $u^0 \geq w$ . Let us show that $u^1 \geq w$ ; we take in (2.19) $v-w = -(u^1-w)^-$ i.e. $v = \inf (u^1,w) \leq M(w)$ , and we take in (2.9) $v-u^1 = (u^1-w)^-$ i.e. $v = \sup (u^1,w) \leq M(u^0)$ (since $M(u^0) \geq M(w)$) ; we obtain:

$$a(u^1-w, (u^1-w)^-) \geq 0 , \text{ hence } u^1 \geq w .$$

We show inductively that $u^n \geq w$ and (2.20) follows.

<u>Remark 2.2</u>.

The uniqueness is not true if $k = 0$ in $M(u)$ .

We conjecture that there is <u>uniqueness</u> when $k > 0$ . But this is proved only in particular cases (n = 1, or special $f$ ).

Remark 2.3

The (maximal) solution increases with $k$ .

For other comparison results, cf. Bensoussan-Lions, loc.cit.

Remark 2.4

The preceeding proof is constructive, and has indeed been applied by M. Goursat [1] in actual computations.

The main difficulty in applications is the fact that $n$ can be large; sub-optimality and decomposition methods are then to be used; work along these lines is in progress at Iria-Laboria.

Remark 2.5

The problems considered here are closely related to the so-called s-S policy introduced in Management; we refer to Arrow-Harris-Marschak [1], Dvoretzky-Kiefer-Wolfowitz [1], Iglehart and Karlin [1], Scarf [1], Veinott [1], Veinott-Wagner [1].

# CHAPTER 9

# NUMERICAL METHODS

## 1. Parabolic Problem

### 1.1 Position of the Problem.

The state is given in $\Omega = (0,1)\times(0,T)$ by

(1.1)
$$\frac{\partial y}{\partial t} - \frac{\partial^2 y}{\partial x^2} + y = f$$

subject to the boundary and initial conditions:

(1.2)
$$\begin{cases} -\dfrac{\partial y}{\partial x}(0,t) = v(t) \;, & \dfrac{\partial y}{\partial x}(1,t) = g(t) \;, \\ y(x,0) = y_0(x) \;. \end{cases}$$

We suppose that $v \in U = L^2(0,T)$ without constraints; $f,g,y_0$ are given. The cost function is given by

(1.3)
$$J(v) = \int_Q |y-z_d|^2 dx\, dt + N\int_0^T v^2\, dt \;.$$

Let $u$ be the optimal solution:

(1.4)
$$J(u) = \inf J(v),\; v \in U \;.$$

The optimality system is given by:

$$(1.5)\qquad \begin{cases} -\dfrac{\partial p}{\partial t} - \dfrac{\partial^2 p}{\partial x^2} + p = y - z_d \quad \text{(where } y = y(u)) \\[2ex] \dfrac{\partial p}{\partial x}(0,t) = \dfrac{\partial p}{\partial x}(1,t) = 0 \ , \\[2ex] p(x,T) = 0 \ , \end{cases}$$

and

$$(1.6)\qquad p(0,t) + Nu(t) = 0 \ .$$

Remark 1.1

This very simple example was chosen in order to be able--without excessive consumption of computing time--to compare "all" possible methods on this example.

In order to have exact solutions, one uses the optimality system for defining $f, z_d, y_0$ .

We choose, with D. Leroy [1],

$$(1.7)\qquad y(x,t) = (ax^2 - 2ax + c)(T-t) + d$$

$$a, c, d = \text{constants.}$$

Then $u$ is defined by $u(t) = -\dfrac{\partial y}{\partial x}(0,t)$ , i.e.,

$$(1.8)\qquad u(t) = 2a(T-t) \ .$$

We <u>choose</u> next $p$ (subject to boundary conditions in (1.5)):

(1.9) $$p(x,t) = (-2a\,\mu \cos \pi x + \nu)(T-t)$$

where $\mu$ and $\nu$ are constants; (1.6) is satisfied if

(1.10) $$(-2a\mu+\nu) + 2aN = 0 \quad \text{(this will give } \nu \text{ ).}$$

There $f$, $y_0$ and $z_d$ are <u>defined</u> by (1.1)(1.2)(1.5):

$$f = (ax^2-2ax+c)(T-t-1)-2a(T-t)+d ,$$

$$y_0 = (ax^2-2ax+c)T + d ,$$

$$z_d = (ax^2-2ax+c+2a\mu \cos \pi x-\nu)(T-t)+(2a\mu\pi^2 \cos \pi x)(T-t)$$

$$+ 2a\mu \cos \pi x-\nu . \qquad []$$

Constants are chosen as follows:

$$T = 1,\ a = 0.25,\ c = 1.25,\ d = 1,\ \mu = 0.01 .$$

<u>Discretization scheme for the partial differential operator</u>

$$\Delta x = h = \frac{1}{20} ,\quad \Delta t = k = \frac{1}{40} ,$$

Crank-Nicholson scheme.

We emphasize that <u>this is a choice</u> among many possible others; the comparisons made in what follows are <u>all</u> based on this type of discretization. It could be that the conclusions obtained depend on this choice; for instance, in problems of higher space dimension, one has to check if the use of finite element methods changes the conclusions of this section. (Work along these lines is in progress at IRIA - Laboria.)

## 1.2 Numerical Methods

### 1.2.1 Direct solution (by Gauss-Seidel Method) of the Optimality system.

This method has been used by Bossavit [1] and by Miellou [1] [2]. Starting with $u^0$ arbitrary (0), we solve inductively

$$
(1.11)\quad
\begin{cases}
\dfrac{\partial y^n}{\partial t} - \dfrac{\partial^2 y^n}{\partial x^2} + y^n = f , \\[2ex]
-\dfrac{\partial p^n}{\partial t} - \dfrac{\partial^2 p^n}{\partial x^2} + p^n = y^n - z_d , \\[2ex]
-\dfrac{\partial y^n}{\partial x}(0,t) = u^n(t) , \ \dfrac{\partial y^n}{\partial x}(1,t) = g(t) , \ \dfrac{\partial p^n}{\partial x}(1,t) = 0 \\[2ex]
y^n(x,0) = y_o(x), \ p^n(x,T) = 0 ,
\end{cases}
$$

((1.11) is discretized as was said at the end of Section 1.1), and we define $u^{n+1}$ at time $k\Delta t$ by

$$
(1.12)\qquad u^{n+1}(k\Delta t) = - \frac{1}{N} p^n(0,k\Delta t) .
$$

Conditions of convergence of the method are studied in particular in Miellou and are quite strong (in particular N cannot be too small in this method).

The method converges in 10 to 20 iterations if N is of order $\frac{1}{2}$ to 1 .

If N is of the order of 0.2, 0.25 , the method is unstable.

If N is of the order of 0.1, 0.125 , the method diverges.

1.2.2 Gradient method.

This is the classical method: we define inductively

$$u^{n+1} = u^n - \rho^n J'(u^n) \tag{1.13}$$

where $J'$ is computed using the adjoint state.

Computations are made (cf. D. Leroy, loc.cit.) using the following rules for the choice of $\rho^n$ :

> change $\rho^n$ to $\rho^{n/2}$ if $J(u^n) > J(u^{n-1})$ , and in some cases, change $\rho^n$ to $2\,\rho^n$ .

Computations with optimal $\rho$ and by conjugate gradients have also been made.

Iterations are stopped when

$$|J(u^n) - J(u^{n+1})|/J(u^n) \leq \varepsilon \ , \tag{1.14}$$

and $\varepsilon$ was taken to be $10^{-3}$ or $10^{-4}$ .

By the first method, the number of iterations was of the order of 10 , with a computing time about twice the one needed for method 1.2.1.

By the second method (optimal $\rho$ ) and the third one (conjugate gradient), the number of iterations was of the order of 3 , with a computing time about $\frac{1}{2}$ the one of method 1.2.1. [ ]

### 1.2.3 Galerkin Method

One can approximate $U = L^2(0,T)$ by finite dimensional spaces.

D. Leroy has taken spaces of polynomials.

When $U$ is replaced by $U_m$ , the problem is to minimize $J(v)$ over a finite dimensional subspace $U_m$ ; a conjugate gradient method was chosen for doing that.

The efficiency of this method can be compared to the one of 1.2.2, but the organization of the computation is much more cumbersome here. [ ]

### 1.2.4 Penalty method.

One can consider the state equation as a constraint and then "apply a penalty" to this constraint; we are thus led to minimize

$$(1.15) \qquad J_\varepsilon(v) = \int_Q |y - z_d|^2 dx\, dt + N \int_0^T v^2 dt + \\ + \frac{1}{\varepsilon} \int \left|\frac{\partial y}{\partial t} - \frac{\partial^2 y}{\partial x^2} + y - f\right|^2 dx\, dt$$

where $y$ and $v$ are now independent variables, subject to the constraints (1.2).

## BIBLIOGRAPHY

M. Amouroux [1] Sur la représentation des systèmes dynamiques. Report LAAS - Toulouse - June 1973.

K. Arrow, T. Harris, and J. Marschak [1] Optimal inventory policy Econometrica, 19 (1951), 250-272.

C. Baiocchi [1] C.R. Acad. Sc. Paris, 1971.

C. Baiocchi and E. Magenes [1] Rome, Coll. Accad. Naz. Lincei,1972.

A.V. Balakrishnan [1] On a new computing technique in Optimal Control. SIAM J. on Control, G (1968), 169-173.

A. Bamberger [1] Report Laboria (1974).

A. Bamberger and J.P. Yvon [1] Report Laboria, 1974 idem

Banks [1] These proceedings.

J. Baranger [1] Quelques résultats en Optimisation non convexe. Thesis, Grenoble, March 1973.

A. Bensoussan, M. Goursat and J.L. Lions [1] Note C.R. Acad. Sc. Paris, June 1973.

A. Bensoussan, J.L. Lions [1] Notes in the C.R. Acad. Sc., idem 1) May 1973 2) 3) June 1973.

[2] Problèmes de temps d'arrêt Optimal et inéquations variationnelles paraboliques. Applicable Analysis. 1973.

[3] Paper in preparation.

[4] Books in preparation.

[5] On some topics in Optimal Control. (in Russian; in memory of Prof. Petrowski). Ouspechi Mat. Nauk. 1973.

A. Bensoussan, J.L. Lions and R. Temam [1] Report on decomposition Methods. Cahier IRIA, vol. II, (2), 1972; and Chapter 2 of the book Lions-Marchuk, ed. [1].

L.D. Berkowitz [1] to appear.

F. Bidaut [1] Thesis, Paris, June 1973.

A. Bossavit [1] Report, Univ. of Paris, 1969.

C.M. Brauner and P. Penel [1] Un problème de contrôle Optimal non linéaire en Biomathématique. Annali de l'Univ. Ferrera, XVIII (1972), 1-44.

[2] Thesis 3[d] Cycle, Univ. Paris. Sud, 1973.

H. Brézis [1] Équations et inéquations dans les espaces vectoriels en dualité. Ann. Inst. Fourier, 18 (1968), 115-175.

[2] Problèmes Unilatéraux. J. Math. Pures et Appl. 51 (1972), 1-168.

H. Brézis and G. Stampacchia [1] Note C.R. Acad. Sci. Paris, 1973.

F. Browder [1] On the unification of the Calculus of variations and the theory of monotone non linear operators in Banach spaces. Proc.Nat.Acad.Sc. USA, 56, (1966), 1080-1086.

J. Céa, A. Gioan and J. Michel [1] Quelques résultats sur l'identification de domaines. To appear.

G. Chavent [1] Thesis, Paris, 1971.

[2] REPORTS Laboria, 1973, 1974.

M. Delfour and S.K. Mitter [1] Reports Univ. of Montreal, 1972, 1973.

A.S. Demidov [1] On skin effect and asymptotic behaviour of some elliptic pseudo differential operators. (In Russian), Ouspechi Mat. Nank., 97 (1972), 245-246.

G. Duff [1] Report Laboria, 1973.

G. Duvaut [1] Note on the Stefan's problem. C.R. Acad. Sc. Paris, June 1973.

G. Duvaut and J.L. Lions [1] Sur les inéquations en Mécanique et en Physique. Dunod, Paris, 1972. English translation, Springer, to appear.

A. Dvoretzky, J. Kiefer and J. Wolfowitz [1] The inventory problem. Econometrica, 20 (1952), 187-222.

I. Ekeland and R. Temam [1] Analyse Convexe et problèmes variationnels. Paris, Dunod-Gauthier Villars. 1973.

W. Fleming [1] Optimal continuous parameter stochastic control. SIAM Rev. 11 (4), 1969.

A. Friedman [1] Stochastic games and variational inequalities. Archive Rat. Mech. Anal. 1973.

[2] Regularity theorems for variational inequalities. Archive Rat. Mech. Anal. 1973. idem.

E. de Giorgi and S. Spagnolo [1] Sulla convergenza degli integrali dell'energia. To appear.

R. Glowinski, J.L. Lions and R. Trémolières [1] Méthodes de résolution numérique des inéquations variationnelles. Paris, Dunod, 1974.

M. Goursat [1] Reports Laboria, 1974.

J. Hadamard [1].

G. Henry [1] Report Laboria. 1974.

D. Huet [1] Perturbations singulières d'inégalités variationnelles. C.R. Acad. Sc. Paris, 267 (1968), 932-934.

Hullet [1] Ph.D. Thesis, UCLA, 1973.

D. Iglehart and S. Karlin [1] Optimality of (s,S) policies in the infinite horizon dynamic inventory problem. Management Sc. 9 (1963), 259-267.

Jaffré [1] Report Laboria, 1974.

T. Johnson [1] Thesis, MIT, 1973.

J.P. Kernevez [1] Thesis, Paris, 1972.

[2] Control of the flux of substrate entering an enzymatic membrane by an inhibitor at the boundary. J. Opt. theory and Appl. 1973.

[3] IFIP. Optimization Conference. Rome. May 1973.

J.P. Kernevez and D. Thomas [1] Book to appear, Dunod, 1975.

D. Leroy [1] Thesis 3$^{d}$ cycle, Paris, 1973 and Reports Laboria.

D. Leroy and J.P. Yvon [1] Report Laboria, to appear.

J.L. Lions [1] Contrôle optimal des systemes gouvernés pour des équations aux dérivées partielles. Paris, Dunod, 1968. English translation by S.K. Mitter, Springer, 1971.

[2] Some aspects of the optimal control of distributed parameter systems. Reg. Conf. Series in Appl. Math., August 1971. SIAM, Vol. 6, 1972.

[3] Sur le contrôle optimal des systemes distribués. Conferences UMI. L'Enseignement Math. 1973.

[4] <u>Perturbations singulières dans les problèmes aux limites et en contrôle optimal</u>. Lecture Notes in Math., 323, 1973. Springer.

[5] <u>Quelques méthodes de résolution des problèmes aux limites non linéaires</u>. Dunod, Gauthier Villars, 1969.

[6] Contrôle optimal des systèmes distribués. Propriétés de comparaison et perturbations singulières. Rome, Acad. dei lincei, 1972.

[7] <u>Equations différentielles et problèmes aux limites</u>. Springer, III (1961).

[8] Problèmes aux limites en théorie des distributions. Acta. Math. 94 (1955), 13-153.

J.L. Lions and E. Magenes [1] <u>Problèmes aux limites non homogènes et applications</u>. Vol. 1,2,3. Dunod, 1968, 1970. English translation by P. Kenneth, Springer, 1,2, 1972; 3, 1973.

J.L. Lions and G. Marchuk, ed. [1] Book, Dunod, 1974.

J.L. Lions and G. Stampacchia [1] Variational inequalities. Comm. Pure Appl. Math., XX (1967), 493-519.

K.A. Luré [1] Optimum control of Conductivity of a fluid moving in a Channel in a magnetic field. P.M.M. 28 (1966), 258-267.

A. Marino and S. Spagnolo [1] Un tipo di approxximazione dell' operatore. Annali Scuola N. Pisa, XXIII (1969), 657-673.

M. Mercier and R. Temam [1] To appear.

J.C. Miellou [1] Thesis, Grenoble, 1970.

F. Mignot [1] To appear.

G. Minty [1] Monotone (non linear) operators in Hilbert space. Duke Math. J. 29 (1962), 341-346.

S.K. Mitter [1] Lectures IRIA. 1973.

J. Mossino [1] Une application de la dualité au contrôle optimal; contrainte portant sur le contrôle et sur l'état. To appear.

F. Murat [1] C.R. Acad. Sc. Paris, 273 (1971), 708-711; 274 (1972), 395-398.

F. Murat and J. Simon [1] to appear. idem.

J.C. Nédelec [1] Thesis, Paris, 1970.

M.N. Oguztörelli [1] A class of non linear integro-differential equations. Bul. Inst. Pol. Din. Iasi, 14 (1968), 43-49.

[2] Un problema misto. Rend. di Mat. 2 (1969), 245-294.

R.E. O'Malley [1] The singularly perturbed linear state regulator problem. SIAM J., Control, 10 (1972).

[2] Singular perturbation of the time-invariant linear state regulator problem. J. Diff. Eq. 12 (1972), 117-128.

[3] These proceedings.

O. Pironneau [1] On optimum profiles in Stokes flow. J. Fluid Mech. (1973), 59, 117-128.

[2] On optimum drag in fluid mechanics. To appear.

L.L. Pokrovski [1] Doklady Akad. Nauk., 188 (1969), 528-531.

Pritchard [1] To appear.

D.L. Russell [1] Control theory of hyperbolic equations related to certain questions in harmonic analysis and spectral theory.

J. Math. Anal. Appl. 40 (1972), 336-368.

Saguez [1] Report Laboria, 1974.

Saint-Jean-Paulin [1] To appear.

H. Scarf [1] The optimality of (S,s) policies in the Dynamic Inventory oroblem. Chap. 13 in Arrow, Karlin, Suppes (ed.), Math. Methods in the Social Sciences, Stanford U. Press, 1960.

L. Schwartz [1] La transformée de Laplace des distributions, 1952.

S. Spagnolo [1] Sulla convergenza di soluzioni di equazioni paraboliche. ed ellitiche. Annal. Sc. Norm. Sup. Pisa, XXII (1968), 574-597.

[2] Sul limite delle soluzioni. Id. XXI (1967), 657-699.

G. Stampacchia [1] Formes bilinéaires coercitives sur les ensembles convexes. C.R. Acad. Sc. Paris, 258 (1964), 4613-4416.

L. Tartar [1] To appear in the Archive Rat. Mech. Anal.

Van de Vielle [1] To appear.

A. Veinott Jr. [1] The status of Mathematical inventory theory. Management Science 12 (1966), 745-777.

A. Veinott Jr. and H. Wagner [1] Computing Optimal (s,S) inventory policies. Management Science. 11 (1965), 525-552.

I.M. Visik and L.A. Liousternik. [1] Uspechi Mat. Nauk., 12 (1957), 1-121 (Amer.Math.Soc.Translations, (2),20,1962, 239-364).

P.K.C. Wang [1] Paper on optimal fusion.

J.P. Yvon [1] Contrôle optimal d'un problème de fusion. Calcolo. 1974.

[2] Report Laboria, 1973.

[3] Thesis 3$^{d}$ Cycle, Paris, 1969.

# NOTES ADDED IN PROOF , SEPTEMBER 1974.

Free boundary problems arising in infiltration theory for general shapes of dams have been reduced by C. BAIOCCHI to Q.V.I.. Cf :

C. BAIOCCHI, C.R. Acad. Sc. Paris, 178 (1974).

Along the lines of Chapters 2 and 3, interesting results have been obtained by J.P. YVON and his team. Cf. :

A. BERMUDEZ, Report LABORIA (1974)
A. BAMBERGER, C. SAGUEZ, J.P. YVON, Report LABORIA (1974)
A. BERMUDEZ, M. SORINE, J.P. YVON, Report LABORIA (1974)
C. SAGUEZ, Report LABORIA (1974).

New results have been obtained by TARTAR and MURAT for the problems of Chapter 6, Section 2 (control in the coefficients) and by C. SBORDONE ; cf.

C. SBORDONE, Report 43, Istituto di Matematica dell'Università di Napoli (1974),
L. TARTAR, Lecture at IRIA Symposium, June 1974.

For what concerns Remark 2.2. of Chapter 8, uniqueness has been proved by :

L. TARTAR, C.R. Acad. Sc., Paris, 278 (1974)

and by :

Th. LAETSCH, J. of Functional Analysis, 1974.

Other situations related to impulse controls have been studied by A. BENSOUSSAN and the A., in a number of notes in the C.R. Acad. Sc., Paris, 1974.

Problems of impulse games are studied by A. BENSOUSSAN and the A., Meeting of I.E.E.E., Phoenix, November 1974.

The questions of Chapter 9 are being studied in a much more comprehensive manner in a book by J.P. YVON and the A., in preparation.

One can also consult the reports of A. BENSOUSSAN, A. FRIEDMAN and the A. at the International Congress of Mathematicians, Vancouver, 1974.

# GAME THEORY AND SOME INTERFACES WITH CONTROL THEORY

R. M. Thrall

Department of Mathematical Sciences
Rice University

## CONTENTS

The order of topics in these written notes does not follow that of the lectures; in particular, the number of subdivisions is not equal to the number of lectures. Also, the final Section 8 on the "ideal linear weights" is reproduced in the form in which it appeared in the Naval Research Logistics Quarterly (Dec. 1973, Vol. 20, pp. 645-659) rather than just the portion that was covered in the lectures.

## Section 1. Characteristics of Optimization Problems.

Mathematical models are used in an increasing variety of situations. There are three important classes of models

A. Descriptive

B. Predictive

C. Prescriptive.

Descriptive models have long been used in science and engineering and underlie most of classical applied (or physical) mathematics. They are concerned with understanding some phenomenon such as behavior of a falling body, of a space vehicle, or of a chemical reaction. Predictive models resemble descriptive except for an element of extrapolation involving future time or introduction of new parameters. After the fact a predictive model becomes deseriptive (perhaps invalid). Much of the motivation for development of control theory in the past two decades has been to develop descriptive and predictive models related to paths of space vehicles.

Prescriptive or decision models are designed to aid a decision maker in selection of parameters or other variables so as to achieve some objective at minimal cost or to maximize some return. Decision models underlie much of what is called operations research, or management science, or systems analysis. These models have stimulated a discipline now known as mathematical programming as well as the optimization part of control theory.

It seems appropriate to consider some of the characteristics of optimization problems and to provide a general framework which includes many important current areas of optimization research.

I am indebted to some as yet unpublished lectures of Y.C. Ho[1] for much of the content of this section.

TABLE 1. CHARACTERIZATION OF OPTIMIZATION PROBLEMS

| | General | Special |
|---|---|---|
| 1. | dynamic | static |
| 2. | stochastic | deterministic |
| 3. | many persons | one person |
| 4. | decentralized | centralized |
| 5. | multi-payoffs | single payoff |
| 6. | many variables | few variables |
| 7. | non-linear | linear |

TABLE 2. SOME IMPORTANT CASES

| | | |
|---|---|---|
| 1. | s s s s s g s | linear programming |
| 2. | g s s s e s e | control theory |
| 3. | g g s s e s e | stochastic control theory |
| 4. | g s g e e e e | differential games |
| 5. | s s s s s e g | non-linear programming |
| 6. | g s s s s s g | classical optimization |
| 7. | g g g g s g e | games in extensive form |
| 8. | s g g s s e e | games in normal form |
| 9. | s g s s g e e | decision analysis |
| 10. | s s g g g e e | games in characteristic function form |

---

[1] Ho, Y.C., Decisions, Control and Extensive Games, unpublished (1973)

... , Generalized Control Theory, lectures at Navy workshop on Differential Games, July-Aug. 1973, Annapolis, Maryland.

Table 1 lists some of these characteristics in the column labelled "general" with an accompanying column listing "special" instances. Table 2 lists a few important cases obtained under various combinations of specialization. The sequences of symbols in the first column indicate alternatives from the two column of Table 1. The symbol "e" indicates that some characteristic is more or less irrelevant for the case under discussion. For example s s s s s g s indicates the linear programming deals with problems which are static, deterministic, one person (i.e., involve a single decision maker), centralized, single payoff, many variable, linear, and s s g s s e e states that games in characteristic function form are static, deterministic, many person, decentralized, multipayoff, and that the number of variables and linearity are irrelevant.

In what follows I will be concerned with many person optimization, usually with multipayoffs in the sense that each contestant has his own scalar payoff (in contrast with vector payoffs).

We will assume an n-person encounter where player $i$ selects a control $u_i$ . This control may be a vector, a function, or some combination of both and we assume once and for all that $u_i$ is restricted to some domain of feasibility $D_i$ whose precise nature is not of immediate concern. The payoff function $F_i$ for the ith player is assumed to be a function of the composite strategy $u = (u_1,\ldots,u_n)$ of all the players. We now introduce some useful concepts.

An n-tuple $u^* = (u_1^*,\ldots,u_n^*)$ is said to be an <u>equilibrium strategy</u> if for all $u = (u_1,\ldots,u_n)$ we have

$$F_i(u_1^*,\ldots,u_i^*,\ldots,u_n^*) \geqq F_i(u_1^*,\ldots,u_i,\ldots,u_n^*) \quad (i = 1,\ldots,n) ,$$

i.e. if no single player can improve his payoff by deviating from his strategy $u_i^*$ .

An n-tuple $u^P = (u_1^P,\ldots,u_n^P)$ is said to be a Pareto Strategy if for no $u$ is

$$F_i(u_1,\ldots,u_n) > F_i(u_1^P,\ldots,u_m^P) , \quad (i = 1,\ldots,n) .$$

Let $u^i$ denote the vector $u$ with its i-th component deleted. We set

$$R_i(u^i) = \max_{u_i} F_i(u)$$

and call $R_i$ the response function for player $i$ . It represents the best action player $i$ can take if he is permitted to select his strategy knowing what the other players have chosen (we assume existence of the $R_i$ ).

Whereas $R_i$ gives an optimistic picture of what player $i$ can expect, his security base

$$s_i = \max_{u_i} \min_{u^i} F_i(u_1,\ldots,u_n)$$

represents the worst that can happen to him. The vector $S = (s_1,\ldots,s_n)$ is called the security payoff. There is, of course, no assurance that there is any single strategy vector $u$ for which $s_i = F_i(u_1,\ldots,u_n)$ , $(i = 1,\ldots,n)$ .

There is considerable debate as to the relative merits of equilibria in comparison with Pareto strategies. Most work on differential games has been based on equilibrium concepts whereas in ordinary game theory both types are important.

Figure 1 illustrates some of these concepts for the case where $n = 2$ and each $u_i$ is a scalar. The presentation is in the $(u_1, u_2)$ plane and some level curves for $F_1$ and $F_2$ are shown. We assume that the level curves for $F_i$ are ovals about a point $B_1$ and that ovals represent decreasing values of $F_1$ as they progress from $B_1$ ; $B_2$ plays a similar role for $F_2$ .

By definition a point $(u_1, u_2)$ is on $R_2$ if for fixed $u_1, u_2$ is selected so that $(u_1, u_2)$ lies on the level curve for $F_2$ with highest possible value, i.e., closest to $B_2$ . Clearly if the vertical line $u_1$ = constant intersects a level curve twice, then thoro is a better level curve inside. Hence, we conclude that the curve $R_2 : u_2 = R_2(u_1)$ passes through $B_2$ and is a locus of points on the level curves with vertical tangents. Similarly, $R_1$ is a locus of points on the $F_1$ level curves having horizontal tangents. The intersection A of $R_1$ and $R_2$ is seen to be an equilibrium point since departure from it by either contestant decreases his payoff.

Points where level curves of the two classes are tangent externally are clearly Pareto points, since any improvement for one contestant means going inside his level curve and hence outside that of his opponent. It is easy to see that these are the only Pareto points and thus that the Pareto set is (as shown in the figure) a curve extending from $B_1$ to $B_2$ with each contestant preferring his own end of the curve.

The shaded (eye-shaped) region with A in the upper left corner represents strategies which lead to outcomes preferred to A by both players. Note that neither player can penetrate this favored region without the cooperation of the other. Note also, that the portion $P_A$ of the Pareto curve between $C_1$ and $C_2$ which lies in this favored region is of particular interest since if bargaining begins at the equilibrium point and the players move jointly so as to achieve mutual improvement they will end somewhere on $P_A$. What is known as the <u>bargaining problem</u> relates to procedures for selecting a point on $P_A$.

The existence of regions of mutually preferred points shows why equilibria are not universally accepted as reasonable solutions in competitive optimization.

An alternate presentation of two contestant optimization will be given in Section 2 by plotting points $(y_1, y_2)$ where $y_1 = F_1(u_1, u_2)$ and $y_2 = F_2(u_1, u_2)$. In this presentation $u_1, u_2$ need no longer be scalars.

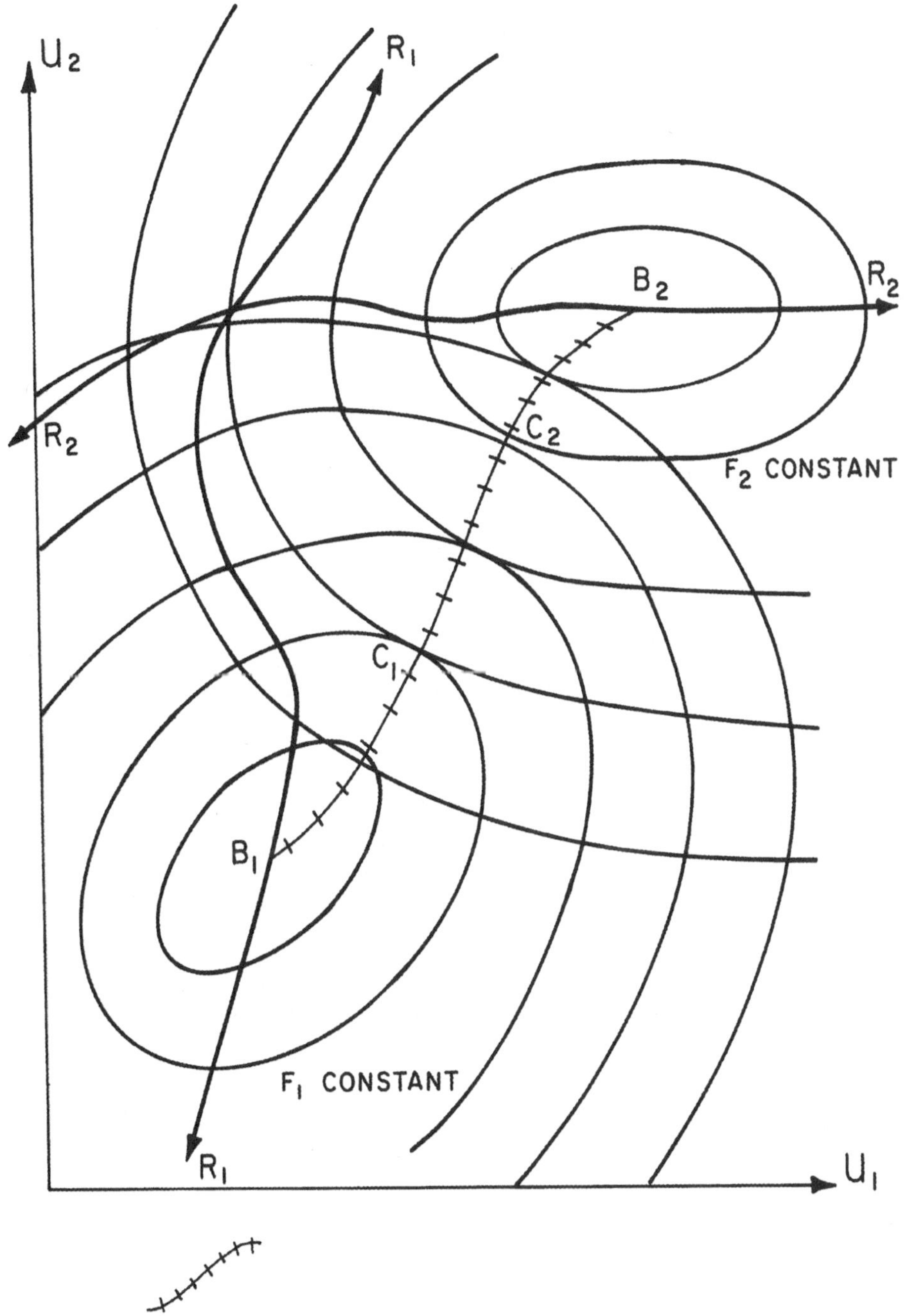
U2
R1
B2
R2
R2
C2
F2 CONSTANT
C1
B1
F1 CONSTANT
R1
U1

## Section 2. Two-Person Games; Nash Bargaining Solutions.

A two-erson nonzero sum game is characterized by a p-by-q matrix A of ordered pairs

$$A = [u_{ij}, v_{ij}]$$

Here p is the number of pure strategies for player 1 , and q the same for player 2. The scalars $u_{ij}$, $v_{ij}$ are real numbers measuring a cardinal utility on outcomes, and we let $U = [u_{ij}]$ and $V = [v_{ij}]$ be the individual payoff matrices. Mixed strategies are permitted for both players and the game is assumed to be non-cooperative, i.e., no side payments or correlation of mixed strategies are permitted. An interesting approach to this situation is due to John Nash [see Owen, Ref. 29, p. 136ff].

Let X denote the convex subset of $R_2$ consisting of payoffs $(x_1,x_2)$ obtainable from (all possible) mixed strategies, and let $s = (s_1,s_2)$ denote the security base (note that s need not be in X ).

Let P and Q respectively denote mixed strategies for players 1 and 2 . Then we have

$$s_1 = \max_P \min_Q P^T UQ ,$$

$$s_2 = \max_Q \min_P P^T VQ .$$

We wish to determine a function $\varphi$ which, given X and s , selects a point $s^* = (s_1^*,s_2^*)$ of X which we can regard as a solution of the game. Thus we write

$$\varphi \; : \; X, s \rightarrow \varphi(X, s) = s^*$$

and search for properties that will make $\varphi$ acceptable or reasonable as a decision model.

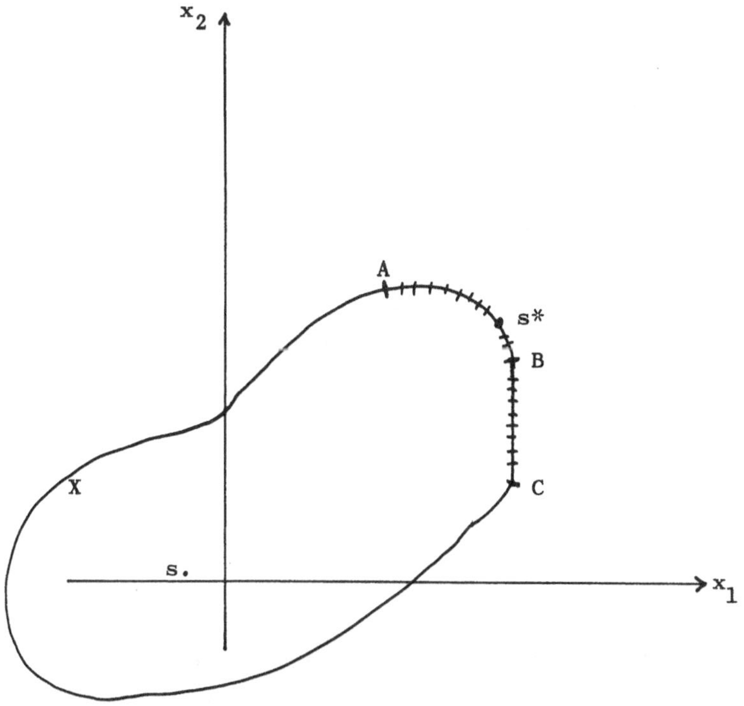


Figure 2.1 Feasible set X

Figure 2.1 pictures a typical feasible set X , security base s , and solution s* . The Pareto set in X containing the undominated elements of X here consists of the curve ABC provided that only strong (every component) domination is required. If weak domination is permitted then the Pareto set is reduced to the arc

AB (since B weakly dominates all other vectors on the vertical segment BC ).

Nash sets forth six axioms and then proves that they characterize a unique function $\varphi$ . These are

N1 (Individual rationality) $s^* \geqq s$

N2 (Feasibility) $s^* \in X$

N3 (Pareto optimality) If $x \in X$ and $x \geqq s^*$ , then $x = s^*$

N4 (Independence of irrelevant alternatives)

If $x \in Y \subset X$ and $x = \varphi(X, s)$ , then $x = \varphi(T, s)$ .

N5 (Independence of linear utility transformations)

Let Y be obtained from X by the linear transformation $x \rightarrow y = \sigma x$ defined by

$$y_1 = \alpha_1 x_1 + \beta_1 \qquad (\alpha_1 > 0)$$

$$y_2 = \alpha_2 x_2 + \beta_2 \qquad (\alpha_2 > 0)$$

Then

$$\varphi(Y, \sigma s) = \sigma s^*$$

N6 (Symmetry) If X is symmetric, i.e., $(x_1, x_2) \in X \Rightarrow (x_2, x_1) \in X$ , then $s_1^* = s_2^*$ .

The first three axioms N1, N2, and N3 have been generally accepted as reasonable requirements for $\varphi$ . On the other hand, N4 has been subject to considerable controversy with objections to its suitability either as a descriptive or as a normative axiom. Axiom N5 is subject to general objections to a linear utility theory. The arguments against N5 seem stronger on descriptive than on normative grounds. The final axiom, N6 , states that there is no discrimination based on whether a contestant is called player 1 or player 2. A stronger formulation would be:

N6' (Strong symmetry) Let $Y$ be the set of all $y = (x_2, x_1)$ where $x = (x_1, x_2) \in X$ and let $t = (s_2, s_1)$ , then

$$\varphi(Y, t) = (s_2^*, s_1^*) .$$

Nash then proves that [see Ref. 29, pp. 142ff] $s^*$ maximizes the function $g : g(x) = (x_1 - s_1)(x_2 - s_2)$ over all elements of $X$ which are individually rational. Moreover, if $h(x) = (s^* - s)^T x$ , then $h(x) \leqq h(s^*)$ for all $x \in X$ and hence the line $L : h(x) = h(s^*)$ supports $X$ (and is the tangent to $X$ at $s^*$ if one exists). Figure 2.2 shows two typical situations.

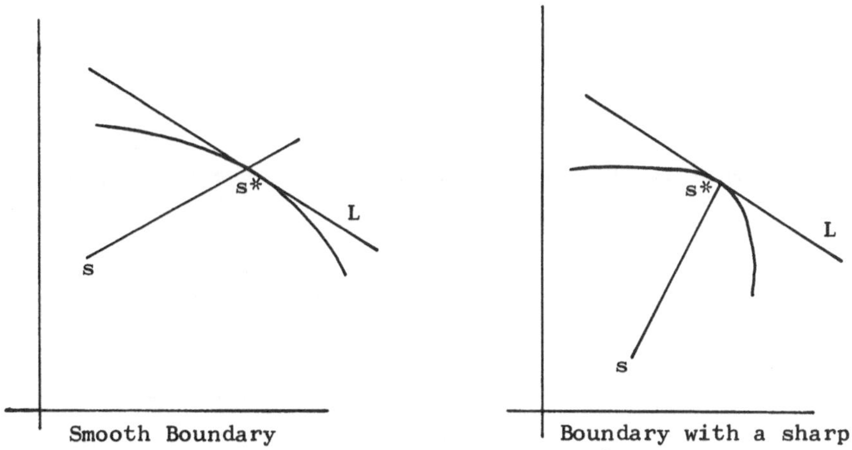


Smooth Boundary

Boundary with a sharp turn

Figure 2.2 Nash solutions

Example 1. Let $X$ be the portion of the first quadrant cut off by the line $x_1 + x_2 = 10$ , and let $s = (1,3)$ . Then $g(x) = (x_1 - 1)(x_2 - 3)$ is clearly maximized at $s^* = (4,6)$ . Then $h(x) = 3x_1 + 3x_2$ and $h(s^*) = 30$ so that $L$ is just the bounding line. If the Pareto curve is linear or piecewise linear, use of $h$ is simpler than that of $g$ in locating $s^*$ since the slope of the segment $s, s^*$ must be the negative of that of the boundary segment at $s^*$ unless $s^*$ is at an angle.

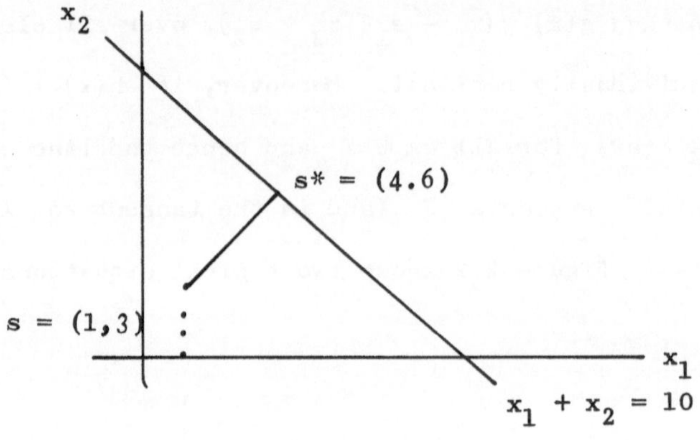


Figure 2.3

Example 2. Let $X$ be the portion of the first quadrant bounded by the arc

$$x_2 = \ln \left( \frac{(200 - x)}{100} \right)$$

and let $s = (0,0)$ . Then $g(x) = x_1, x_2$ is maximized at (54.4, 45.6).

For further discussion of two person games see appropriate sections in references 26, 29, and 33.

## Section 3. N-Person Games in Characteristic Function Form.

Whereas the theory of two-person, zero-sum games is both reasonably complete and fairly well accepted, such is not the case for general games. The classical theory begins with a game in normalized form and passes to what is called characteristic function form; it is this formulation which is central in much of the treatment of games with three or more players.

Denote by $N = \{1,\ldots,n\}$ the set of players in a game, and let $M$ be any subset of $N$. Let $v(M)$ denote the maximum expected total payoff that the members of $M$ can guarantee themselves. The function $v$ thus defined maps the set $2^N$ of all subsets of $N$ into the real numbers and is called the characteristic function of the game.

The function $v$ has two properties

$$v(\emptyset) = 0 \quad \text{where } \emptyset \text{ is the empty set,} \tag{3.1}$$

and

$$v(M_1 \cup M_2) \geqq v(M_1) + v(M_2) \quad \text{if} \quad M_1 \cap M_2 = \emptyset \ . \tag{3.2}$$

This second property is called superadditivity. These two properties provide a direct definition of a characteristic function, since given any function $v$ which satisfies (3.1) and (3.2) there is a game (in normalized form) which has $v$ as its characteristic function.

A vector $a = (a_1,\ldots,a_n)$ is said to be an imputation for a game with characteristic function $v$ if

$$(3.3) \qquad a_i \geq v(\{i\}) \quad (i = 1,\ldots,n) ,$$

and

$$(3.4) \qquad \Sigma a_i = v(N) .$$

The imputations are regarded as possible settlements at the end of a game. Formula (3.3) is referred to as the condition of <u>individual rationality</u> since it limits imputations to those vectors which give each player at least as much as he could assure himself if all the remaining players were allied against him. Formula (3.4), called <u>Pareto</u> (or group) <u>rationality</u>, is based on the argument that if $b$ is any vector with $\Sigma b_i = b_o < v(N)$ then the vector $a$ having $a_i = b_i + (v(N) - b_o)/n$ would be preferred to $b$ by every member of $N$ and therefore that $b$ should not be accepted as a final outcome. This argument is certainly not universally applicable for a descriptive theory since it rules out strikes and wars; however, it does lead to the interesting theory developed by von Neumann and Morgenstern.

Let $a$ and $b$ be two imputations and let $M$ be a subset of $N$. We say that $a$ <u>dominates</u> $b$ <u>relative</u> to $M$, written

$$(3.5) \qquad a \underset{M}{\succ} b$$

if

$$(3.6) \qquad a_i > b_i \quad \text{for all} \quad i \in M ,$$

and

(3.7) $$\Sigma_{i \in M} a_i \leqq v(M) .$$

The first of these conditions is called M-preferability; note that the inequalities are strict, use of $\geqq$ leads to difficulties. The second condition is called M-effectiveness since it does not ask more for members of $M$ than they can collectively assure themselves.

More generally we say that $a$ dominates $b$, written

(3.8) $$a \succ b$$

if there exists a subset $M$ relative to which $a$ dominates $b$. Relative domination is asymmetric and transitive; domination is neither.

A set $S$ of imputations is said to be a solution or a stable set, if

(3.9) $a \succ b$ does not hold for any $a,b$ in $S$,

and

(3.10) for each $b$ not in $S$ there exists an $a$ in $S$ with $a \succ b$.

These conditions can be stated more concisely in terms of the following concept. Let $S$ be a set of imputations; by the dominion of $S$, written dom $S$, we mean the set of all imputations dominated by members of $S$, i.e., $b$ is in dom $S$ if there exists an $a$ in $S$ with $a \succ b$. Now conditions (3.9) and (3.10) become, respectively

$$(3.9') \qquad S \cap \operatorname{dom} S = \emptyset$$

and

$$(3.10') \qquad S \cup \operatorname{dom} S = \{\text{all imputations}\} .$$

The theory of solutions has the following status (see [24] for a more complete review). All solutions for three person games were listed by Von Neumann and Morgenstern [41] and at least one solution has been found for each 4-person zero-sum game. Lucas [21,23] found a 10-person game with no solution. No one yet knows the smallest $n$ for which no solution exists, but Lucas' result has changed the direction of research in game theory, and has given new emphasis to earlier alternatives to the classical solution concept.

One alternative approach was suggested by this author in 1961 and further developed by W.F. Lucas [38,20]. This formulation replaces the characteristic function $v$ by a partition function which for each partition of $N$ assigns a real number to each coalition of the partition. Thus the payoff assigned to a given subset $M$ of $N$ depends on the manner in which the remaining players are grouped.

Let $P = \{P_1,\ldots,P_r\}$ be a partition of $N$ into coalitions (cosets) $P_1,\ldots,P_r$ . The set of all partitions of $N$ is denoted by

$$\Pi = \{P\}$$

Then for each partition $P$ assume there is an <u>outcome</u> <u>function</u> $F_P$ which assigns a real number $F_P(P_i)$ to the coalition $P_i$ when the partition $P$ forms. The function $F$ which maps $P$ in $\Pi$ into function $F_P$ is called the <u>payoff function</u> or <u>partition function</u> for the game. The ordered pair $\Gamma = (N,F)$ is called an n-<u>person game in partition function form</u>.

For each non-empty subset $M$ of $N$ define the <u>value</u> of $M$ as

$$(3.11) \qquad v(M) = \min F_P(M)$$

where the minimum is taken over all partitions having $M$ as a coset, and define $v(\emptyset) = 0$ . This function $v$ is not necessarily superadditive; indeed, there exists a game $\Gamma$ having any preassigned set of numbers $v(M)$ as its set of values.

However, if in (3.11) we took the minimum over all partitions $P$ that have $M$ as union of cosets we would obtain a superadditive function; this would correspond to permitting secret coalitions, whereas the $v$ given by (3.11) is based on a prohibition of secret agreements between the players.

A vector $a = (a_1,\ldots,a_n)$ is called an <u>imputation</u> if (3.3) holds and if

$$(3.4') \qquad \Sigma_{i=1}^{n} a_i = \Sigma_{j=1}^{r} F_P(P_j) \quad \text{for some } P \text{ in } \Pi .$$

Replacing the equality in (3.4') by $\leq$ would allow for a disposal of wealth.

The definition of relative domination now requires in addition to M-preferability (3.6) and M-effectiveness (3.7) also the following condition, called M-<u>realizability</u>.

(3.12) (3.4') holds for a $P$ which has $M$ as a coset.

The concepts of domination and solution are introduced exactly as in the classical case by (3.8) and by (3.9) and (3.10). Here again W.L. Lucas [22] has found a game, this time with 11 persons, which has no solution.

A major feature of games in partition function form is that the absence of superadditivity makes the theory applicable to cases where certain coalitions would reduce rather than enhance the total strength. For example, a coalition of Walter Reuther and Senator Goldwater might reduce the effectiveness of each.

## Section 4. Shapley Value and Owen's Generalization.

A very attractive and relatively early alternative to the stable set solutions was developed by L.S. Shapley [19, pp. 307-317; 26, pp. 245-252]. He proposed a single vector $\varphi(v) = u = (u_1,\ldots,u_n)$, now called the Shapley value, as a solution. The allocation to player $i$ is determined by his average contribution to all coalitions of which he is a member. The formula is

$$(4.1) \qquad \varphi_i(v) = u_i = \sum_{S \subset N} \gamma_n(s)(v(S) - v(S - \{i\}))$$

where $s$ is the number of members in $S$ and

$$(4.2) \qquad \gamma_n(s) = (s - 1)!(n - s)!/n!$$

is a weighting function. Shapley gives an axiomatic characterization of his value and cites a number of examples where it approximates the payoffs actually observed in society. Thus the Shapley value has both a normative justification in terms of providing for a "fair division of the spoils" and a descriptive verification.

We need two definitions before stating Shapley's axioms. A coalition $T$ is called a carrier for the game if for every other coalition $S$ we have $v(S) = v(S \cap T)$. Next, let $\pi$ be any permutation of the player set $N$. Then by $u = \pi v$ we mean the new characteristic function defined by $u(\pi S) = v(S)$. The three axioms are

S1. $\sum_{i \in S} \varphi_i(v) = v(S)$ if $S$ is a carrier

S2. $\varphi_{\pi i}[\pi v] = \varphi_i(v)$ for $i \in N$

S3. $\varphi_i(u + v) = \varphi_i(w + \varphi_i(v))$

As an illustration consider a corporation with 100 shares of stock divided 10, 20, 30, 40 among 4 owners. Any coalition $S$ wins $v(S) = 1$ if it controls a majority of the shares, otherwise $v(S) = 0$. Then the winning coalitions are $\{2,2\}$, $\{3,4\}$, $\{1,2,3\}$, $\{1,2,4\}$, $\{1,3,4\}$, $\{2,3,4\}$, and $\{1,2,3,4\}$. The Shapley value is $\varphi(v) = (\frac{1}{12}, \frac{1}{4}, \frac{1}{4}, \frac{5}{12})$.

We can rewrite (4.1) in the form

$$\varphi_i(v) = \mathrm{Exp}\,[v(S(i,\nabla) \cup \{i\}) - v(S(i,\nabla)] \tag{4.3}$$

where (i) $\nabla$ ranges over the $n!$ orderings of $N$ (ii) $S(i,\nabla) = \{j \mid j$ precedes $i$ in $\nabla\}$, and (iii) each $\nabla$ has probability $1/n!$.

If we change the probability distribution on the orderings, the Shapley is, of course, replaced by some new vector. G. Owen [30] has pursued this possibility as a means for taking into account possible affinities or antagonisms among the players. He assumes:

O1: Any ordering and its reverse have equal probability.

02: Removal of a coalition S should not affect the probabilities assigned to the relative orderings of the remaining set $N - S$.

He shows that these axioms and a few other simple assumptions characterize a new value $\varphi_i^*$ which satisfies (4.3) with the new probabilities.

As an illustrative example consider the four-person weighted majority game with weights (2,1,1,1). For it the Shapley value is

$$\varphi(v) = \left( \frac{1}{2} , \frac{1}{6} , \frac{1}{6} , \frac{1}{6} \right) .$$

If, however, as proposed by M. Maschler, player 1 is the aunt of player 2 so that they tend to be together in coalitions, then Owen uses a geometric representation to get the value

$$\varphi_\alpha^*(v) = (2/3 - \alpha/2\pi,\ 1/3 - \alpha/2\pi,\ 2/\pi,\ 0)$$

depending on an angular parameter which measures the closeness of aunt and nephew. In the limit we have

$$\varphi^*(v) = (2/3,\ 1/3,\ 0,\ 0) .$$

As another example Owen applied his theory to the organization of the 1965 Knesset in Israel. Probabilities of orderings are related to left-right political spectrum, and the organization is regarded as being a weighted majority game. His results given in Table 1 were not far from the proportion of cabinet positions assigned to parties. By contrast, the Gahal has the second largest Shapley value of all the parties and yet received no Cabinet positions.

TABLE 1

| Party | Ordinate | Seats | Value Ø |
|---|---|---|---|
| New Communists | 0.00 | 3 | 0.000 |
| Communists | 0.05 | 1 | 0.000 |
| Poaelei Aguda | 0.20 | 2 | 0.000 |
| Agudat Israel | 0.20 | 4 | 0.063 |
| National Religious | 0.20 | 11 | 0.063 |
| Mapam | 0.30 | 8 | 0.075 |
| The Alignment (Mapai-Ahdut) | 0.40 | 49 | 0.700 |
| Haolan Haze | 0.50 | 1 | 0.000 |
| Independent Liberals | 0.65 | 5 | 0.050 |
| Rafi | 0.80 | 10 | 0.050 |
| Gahal | 1.00 | 26 | 0.000 |

## Section 5. Bargaining Sets, Kernel, Nucleolus.

For annotated discussions of recent developments in solution concepts for n-person games see References [24], [29], and [36]. We give here brief sketches of a few of these concepts.

A coalition structure $T = \{T_1, \ldots, T_n\}$ is defined to be a partition of the player set $N$. A payoff configuration is a pair $(x;T)$ where $T$ is a coalition structure and $x = (x_1, \ldots, x_n)$ is an n-vector for which

$$(5.1) \qquad \Sigma_{i \in T_k} x_i = v(T_k) \text{ , } i = 1, \ldots, m \text{ .}$$

A payoff configuration is said to be individually rational (i.r.p.c.) if also

$$(5.2) \qquad x_i \geqq v(\{i\}) \quad \text{for all} \quad i \in N \text{ ,}$$

and is said to be coalitionally rational (c.r.p.c.) if $x$ satisfies the stronger condition

$$(5.3) \qquad \Sigma_{i \in S} x_i \geqq v(S)$$

if $S$ is a subset of any one of the coalitions $T_k$.

We are interested in stability properties of payoff configurations. It was recognized early that we cannot expect existence of a simple vector so strongly stable as to be undominated. However,

Aumann and Maschler investigated the possibility of a weaker "second order" stability which relates to the ability of these players whose returns are reduced by a dominating strategy to retaliate. To develop this concept we need some further definitions.

Let $T$ be a coalition structure and let $K$ be a coalition. Then by the partners of $K$ in $T$, we mean the set $P(K;T)$ which is the union of all the sets $T_i$ in $T$ which contain members of $K$. Next, let $(x;T)$ be a c.r.p.c. and let $K$ and $L$ be non-empty disjoint subsets of some $T_k \in T$. Then an objection of $K$ against $L$ is a c.r.p.c. $(y;U)$ for which

(5.4) $$P(K;U) \cap L = \emptyset$$

(5.5) $$y_i > x_i \quad \text{for all} \quad i \in K ,$$

(5.6) $$y_i \geqq x_i \quad \text{for all} \quad i \in P(K,U) .$$

Furthermore, a counter-objection of $L$ against $K$ is a c.r.p.c. $(z;V)$ for which

(5.7) $$K \text{ is not a subset of } P(L;V)$$

(5.8) $$z_i \geqq x_i \quad \text{for all} \quad i \in P(L;V)$$

(5.9) $$z_i \geqq y_i \quad \text{for all} \quad i \in P(L;V) \cap P(K;U) .$$

Now, a c.r.p.c. $(x;T)$ is said to be stable if for every

objection against it there is a counterobjection. The bargaining set $M$ is the set of all stable c.r.p.c.'s. There are several significant possible modifications for $M$ obtained by restricting $K$ or $L$ or both to one-element sets and by considering i.r.p.c.'s in place of c.r.p.c.'s. None of these bargaining sets is empty; however, there are no practical algorithms for determining their numbers.

One of these variants $M_1^{(i)}$ is defined as the set of all i.r.p.c.'s $(x,T)$ such that whenever any set $K$ has an objection against $L$ then at least one member of $L$ has a counterobjection. For $M_1^{(i)}$ we have the strong existence theorem that for every coalition structure $T$ there is at least one vector $x$ such that $(x;T) \in M_1^{(i)}$. Indeed, $M_1^{(i)}$ has a subset $K$ called the <u>kernel</u> which has this same property. If $x$ is an imputation and $C$ is any set, then the <u>excess</u> of $C$ is

(5.10) $$e_x(C) = v(C) - \Sigma_{i\in C}\, x_i \ .$$

For two players $i$ and $j$ we define the <u>surplus</u> $s_{ij}$ of $i$ against $j$ as the maximum excess $e(D)$ of all sets containing $i$ but not $j$. Player $i$ is said to <u>outweigh</u> $j$ if $s_{ij} > s_{ji}$ and $x_j > v(\{j\})$. Two players are said to be in equilibrium if neither outweighs the other. The kernel $K$ is the set of all i.r.p.c.'s $(x;T)$ such that any two players belonging to the same set in $T$ are in equilibrium.

We close with a brief discussion of Schmeidler's nucleolus

[35,15]. We define $e_x(\emptyset) = 0$ and let $g(x)$ be the vector consisting of the $2^n$ numbers $e_x(C)$ arranged in descending order. An imputation $y$ is said to belong to the nucleolus if

$$(5.11) \qquad g(y) \lesssim g(x)$$

for all imputations $a$ ; here $\lesssim$ designates lexicographic order (a vector $w = (w_1,\ldots,w_m)$ is said to be lexicographically positive if its first non-vanishing component is positive and $w = w' - w''$ is lexicographically positive). Lexicographic order is a chain order so that a set $W$ of vectors $w$ cannot have more than one least member under this order, and if $W$ is compact it will have a unique least member. The set of all imputations is compact and it is not difficult to show that the same is true for the set $\{g(x) | x \text{ an imputation}\}$. This implies that there is a unique least vector $g(y)$ ; Schmeidler [35] and Kohlberg [15] prove, moreover, that there is a unique imputation $y$ which yields this least vector. Thus for every game the nucleolus consists of exactly one imputation $y$ .

In his dissertation Richard D. Spinetto [Solution Concepts of N-Person Cooperative games as Points in the Game Space, Technical Report No. 138, Dept. of Operations Research, Cornell University, 1971] has introduced a modified nucleolus in the following manner. He considers the excess for each of the $m = 2^n - n - 2$ sets $S$ of $N$ which have more than $1$ and less than $n$ members, and then relates each imputation $a$ to the vector $q^*(a)$ in m-space consisting of the $m$ excesses in descending order. He proves existence and uniqueness for this modified nucleolus. Dropping the two zero components $e_a(\emptyset)$ and $e_a(N)$ is clearly unimportant. However, he observes that [cf. (3.3)] since nonnegativity of the $n$ excesses $e_a(\{i\})$ is a defining condition for imputations that inclusion of these $n$ numbers as components of $g(a)$ places too much weight on the one element coalitions in determining a final outcome for the game.

## Section 6. Howard's Metagame Theory.

This section appears in two parts. The first is a review of Howard's Treatise Paradoxes of Rationality: Theory of Metagames and Political Behavior and the second is a series of supplements expanding certain topics in the review.

### Section 6.1 Review of PARADOXES OF RATIONALITY: THEORY OF METAGAMES AND POLITICAL BEHAVIOR, Nigel Howard, MIT Press, Cambridge, Mass., 1971.

This book clearly belongs to a small select list of major works on game theory. Its contents are highly controversial, but Howard has based his conclusions on arguments which will require careful consideration by game theorists whether they agree or disagree with the general thrust of the author.

The three paradoxes which provide the book with its title can all be explained in the context of two-person, non-zero-sum games. We suppose that player 1 has strategies $s_1,\ldots,s_p$ ; that player 2 has strategies $t_1,\ldots,t_q$ ; and that if player 1 chooses $s_i$ and player 2 chooses $t_j$ , then the outcome is $0_{ij}$ . We suppose further that $u_{ij}$, $v_{ij}$ provide ordinal preference measures of $0_{ij}$ for players 1 and 2, respectively. The matrix of ordered pairs $[(u_{ij},v_{ij})]$ is called the <u>game matrix</u>. (See 6.2.1)

The decision underlying choice of a strategy is said to be (1) <u>subjectively rational</u> if it is the best that can be made based on information available to the decision maker; (2) objectively rational if it is the best that can be made based on the true state of affairs (including the opponents decision). An objectively rational decision is obviously stable in the sense that it will not be changed by any additional information.

The three breakdowns of rationality can now be stated:

First Breakdown: In some games it is impossible for both players to be objectively rational [p. 10].

Second Breakdown: In some games both players do better if irrational than if rational [p.45].

Third Breakdown: In some games following a sure-thing (objectively rational) strategy capitulates entirely to the opponent.

In Howard's words [p. 181], "If, however, there is any possibility of conflict between stable outcomes; that is, if there are two or more possible compromises, of which the one most favored by player 1 is not the one most favored by player 2; then to choose a sure-thing strategy is to be a "sucker" that capitulates entirely to the other side.".

The three breakdowns are illustrated, respectively, by the games known as Matching Pennies, The Prisoner's Dilemma, and Chicken. Game matrices for these are given in Table 1.

| | H | T | | C | D | | A | C |
|---|---|---|---|---|---|---|---|---|
| H | (2,1) | (1,2) | C | (3,3) | (1,4) | A | (3,3) | (2,4) |
| T | (1,2) | (2,1) | D | (4,1) | (2,2) | C | (4,2) | (1,1) |
| Matching Pennies | | | Prisoner's Dilemma | | | Chicken | | |

TABLE I

The first breakdown has been well known for many years and, indeed, was an important barrier to the initiation of a theory of games. The celebrated minimax theorem of von Neumann in the late 1920's provided a way of dealing with this problem and thus marked the true beginning of game theory.

The root of the breakdown is, of course, the logical contradiction involved in having simultaneous moves with each player knowing the other's strategy before he selects his own.

In Matching Pennies "1" means "lose" and "2" means "win"; the row player wishes to match and the column player wishes to differ.

In the Prisoner's Dilemma the strategies are C for "cooperate" and "D" for "defect". For each player, D is a sure-thing strategy, i.e., D is better than C regardless of which strategy the opponent chooses. Thus rational players end at (D,D) with payoff (2,2) whereas if both players select the irrational strategy C they end up at (3,3) , which both prefer.

We defer discussion of the third breakdown.

Howard's main contribution is his concept of <u>metagame</u> which he presents as a remedy to the second breakdown. He associates with each game a hierarchy of games called metagames; metarationality is defined to be rationality in some metagame.

Let G be a game with matrix $A = [(u_{ij}, v_{ij})]$ ; then Howard defines a new game 1G called the 1-metagame (of G ) by changing the rules of G so that player 1 moves after being told the choice of player 2. (See 6.2.2) The metagame 2G and iterates $k_s k_{s-1} \ldots k_1 G$ are defined analogously. For example, the game 2G for the Prisoner's Dilemma is

| | C/C | D/D | C/D | D/C |
|---|---|---|---|---|
| C | (3,3) | (1,4) | (3,3) | (1,4) |
| D | (4,1) | (2,2) | (2,2) | (4,1) |

Here the symbol X/Y stands for the metagame strategy (for player 2) of choosing X against C and Y against D . The game 1,2G would have 16 strategies for player 1 and 4 for player 2; the symbol W/X/Y/Z would represent the strategy W against C/D , X against D/D , Y against C/D , and Z against D/C . (See 6.2.3)

In a game G with r players each permutation $k_r, \ldots, k_1$ of $1, \ldots, r$ defines a metagame $k_r \ldots k_1 G$ . The r! metagames thus obtained are the <u>complete</u>, (because each index appears at least once) <u>primitive</u> (because no index is repeated) <u>metagame</u> descendants of G . Howard shows (1) that no equilibrium is lost in passing from a game to one of its metagames (called descendants), and (2) that any equilibrium obtained in a metagame of G also exists in some primitive descendant of G . These results justify focussing attention on the set of complete primitive metagames.

It is the possibility of obtaining new equilibria that brought the metagame concept widespread attention. For example, in the Prisoner's Dilemma both 12G and 21G have equilibria with payoffs (3,3) . In G itself this is not an equilibrium outcome although it has been widely regarded as representing a desirable social outcome, so in a sense the metagame concept provides a resolution to the Prisoner's Dilemma.

There are four important classes of equilibria: basic (E p.28), symmetric ($\Sigma$ p.122), metagame (T p.120), and general ($\Gamma$ p.120). These are nested in the sequence

$$E \subseteq \Sigma \subseteq T \subseteq \Gamma$$

Howard shows (1) that every metagame has an equilibrium (p. 27), (2) that in every complete metagame (of an ordinal game G ) each Pareto optimum is a metagame equilibrium (p. 154). In a sense the results, respectively, "rescue" rationality from its first two breakdowns. They can be restated in the respective forms " E may be empty", and "elements of E may not be Pareto". But proving that T is non-empty and that its elements are Pareto optima still leaves open the question of symmetry, i.e., the non-emptiness of $\Sigma$ . In the Prisoner's Dilemma the outcome (3,3) is symmetric as well as being a Pareto optimum and for the reviewer a major contribution of metagame theory is that it provides a transition from the ordinary Prisoner's Dilemma to complete metagames whose outcomes have these two properties. (See 6.2.4)

We now return to the game of Chicken and the third breakdown.

In the game 12G the only sure thing strategy yields the metarational outcome (2,4) which is most favored by player 2 and is next to the worst for player 1; in 21G the corresponding sure thing outcome is (4,2) . The lack of concordance between these two complete metagames is the basis for the third breakdown. (See 6.2.5)

Thus the social problem presented by games like Chicken is more severe than those resembling the Prisoner's Dilemma.

Since Howard proposes the metagame structure as a descriptive rather than as a normative theory, it must be judged by its empirical success and not by its logical structure. One positive feature of his approach is that he is willing, yea eager, to put his model to experimental testing. He reports some experimental results in Appendix B and has since gone much further. However, the OR practitioner will understandably yearn for a normative version of the theory.

In my opinion the weakest feature of the theory is its limitation to ordinal utilities. In defense of this he claims that cardinal utilities cannot be practically constructed; this places Nigel Howard in sharp opposition to the claims and practice of Ronald Howard (and other decision theorists). Of course, a corollary to rejection of cardinal utility is the rejection of mixed strategies.

As now developed, metagame theory requires the use of normalized as opposed to characteristic function form. This may be justified in the two person case but the lack of adequate treatment of coalitions and side payments seems a major flaw if metagames are to be taken seriously beyond the two person case (which is precisely where its major triumphs lie).

I found the book stimulating, but the lack of formal structure made reading unnecessarily difficult. Many important theorems and definitions (indeed, even precise statements of the breakdowns of rationality) were not identified by name or number. I also feel that the impact of the book would be greater if less were claimed (e.g., in the n-person case) so that the reader would concentrate more clearly on the major contributions.

In summary, I recommend Paradoxes of Rationality... for every serious student of decision theory; whether or not he agrees with many of its arguments he should find reading it and having it available for reference valuable.

## Section 6.2. Supplementary Comments

### Section 6.2.1 Outcomes and Equilibria.

An outome $O_{ij}$ is to be regarded as including the strategies as well as the preference measures. Thus we write $O_{ij} = (s_i, t_j, (u_{ij}, v_{ij}))$ . We interpret the response functions and equilibria (see Section 1) of a game $G$ as being sets of outcomes. Thus

$$R_1(G) = \{O_{ij} | u_{ij} \geqq u_{hj} \ \forall \ h\} ,$$

$$R_2(G) = \{O_{ij} | v_{ij} \geqq v_{ik} \ \forall \ h\} .$$

Then the equilibria of $G$ are

$$E(G) = R_1(G) \cap R_2(G) \ .$$

The reason for not just using the preference measures is provided by the following example.

| | $t_1$ | $t_2$ | $t_3$ | $R_2(G)$ |
|---|---|---|---|---|
| $s_1$ | 1,1 | 1,3 | 2,2 | $(s_1, t_2, (1,3))$ |
| $s_2$ | 3,1 | 1,1 | 1,3 | $(s_2, t_3, (1,3))$ |
| $s_3$ | 2,2 | 3,1 | 1,1 | $(s_3, t_1, (2,2))$ |
| $R_1(G)$ | $(s_2, t_1, (3,1))$ | $(s_3, t_2, (3,1))$ | $(s_1, t_3, (2,2))$ | |

Observe that the elements of $R_2(G)$ are obtained by selecting in each row the column with preferred preference measure for player 2; similarly, $R_1(G)$ is obtained by maximizing in each column for player 1. Clearly $E(G) = R_1(G) \cap R_2(G)$ is empty. However, without the inclusion of the strategies as part of each outcome we might have been tempted to regard (2,2) as an equilibrium since it appears in the preference parts of both response functions.

## Section 6.2.2 Metages and Sure-Thing Strategies.

A more precise definition of metagame is as follows. The metagame $H = 2G$ has the same strategies for player 1 as does $G$ ; the strategies for player 2 are the $p^q$ functions $f$ of $S_1$ into $S_2$ . Let $f$ be one of these functions for which $f(s_i) = t_j$ . Then the pair $(s_i, f)$ for $H$ is associated with the outcome $O_{ij} = (s_i, t_j, (u_{ij}, v_{ij}))$ of $G$.

Player 2 has a <u>sure thing</u> strategy in $H = 2G$ given by a function $f$ for which $f(s_i) = t_j$ where for each $i$ a corresponding $j$ is determined such that $O_{ij} \in R_2(G)$.

For example, for the matching penny game $G$ we have

$H = 2G =$

| | H/H | H/T | T/H | T/T |
|---|---|---|---|---|
| H | 2,1 | 2,1 | (1,2) | 1,2 |
| T | 1,2 | 2,1 | (1,2) | 2,1 |

The circled positions are equilibria in $H$ and the strategy $f$ denoted by $T/H$ is a sure thing strategy for player 2 in $H$.

## Section 6.2.3 Metaequilibria in the Prisoner's Dilemma.

We list game matrices for the metagames $1G$, $1,2G$, and the significant parts of $2.1G$ with metaequilibria indicated by encircling.

$1G =$

| | C | D |
|---|---|---|
| C/C | 3,3 | 1,4 |
| C/D | 3,3 | 2,2 |
| D/C | 4,1 | 1,4 |
| D/D | 4,1 | (2,2) |

| | C/C | D/D | C/D | D/C |
|---|---|---|---|---|
| C/C/C/C | 3,3 | 1,4 | 3,3 | 1,4 |
| D/D/D/D | 4,1 | (2,2) | 2,2 | 4,1 |
| D/D/D/C | 4,1 | 2,2 | 2,2 | 1,4 |
| D/D/C/D | 4,1 | 2,2 | (3,3) | 4,1 |
| D/D/C/C | 4,1 | 2,2 | 3,3 | 1,4 |
| D/C/D/D | 4,1 | 1,4 | 2,2 | 4,1 |
| D/C/D/C | 4,1 | 1,4 | 2,2 | 1,4 |
| D/C/C/D | 4,1 | 1,4 | 3,3 | 4,1 |
| D/C/C/C | 4,1 | 1,4 | 3,3 | 1,4 |
| C/D/D/D | 3,3 | 2,2 | 2,2 | 4,1 |
| C/D/D/C | 3,3 | 2,2 | 2,2 | 1,4 |
| C/D/C/D | 3,3 | 2,2 | (3,3) | 4,1 |
| C/D/C/C | 3,3 | 2,2 | 3,3 | 1,4 |
| C/C/D/D | 3,3 | 1,4 | 2,2 | 4,1 |
| C/C/D/C | 3,3 | 1,4 | 2,2 | 1,4 |
| C/C/C/D | 3,3 | 1,4 | 3,3 | 4,1 |

The 1-2-Metagame of Prisoner's Dilemma. The symbol "W/X/Y/Z" represents the policy "W against C/C,X against D/D,Y against C/D,Z against D/C." Equilibria are circled.

| | | D/D/D/D | D/C/D/D | C/C/D/D | (13 columns) |
|---|---|---|---|---|---|
| | C/C | 1,4 | 1,4 | 3,3 | 1,4 or 3,3 |
| | C/D | 2,2 | (3,3) | (3,3) | 2,2 or 3,3 |
| 2,1G = | D/C | 1,4 | 1,4 | 1,4 | 1,4 or 4,1 |
| | D/D | (2,2) | 2,2 | 2,2 | 2,2 or 4,1 |

Observe that $(C,C,(3,3))$ is the common representative outcome at G-level of the quilibria with respective strategy pairs $((C/D), (D/C/D/D), (D/D/C/D), (C/D))$ in $2,1G$ and $1,2G$. This verifies the statement that the set $\Sigma$ of symmetric equilibria for the Prisoner's Dilemma is not only non-empty but that it is larger than $E$ and that the added payoff $(3,3)$ is the one desired by many game theorists.

Section 6.2.4 Some properties of equilibria.

Let $G$ be the basic game, let $H = kG$ for some player $k$, and let $K$ denote any metagame descendant of $G$. Then

(1) For every $K$, $E(G) \subset E(K)$ (conservation of equilibria)

(2) Let $K$ be complete and let $K^*$ be a descendant of $K$, Then $E(K) = E(K^*)$

(3) $E(H)$ is not empty.

(4) $T(G)$ is not empty.

(5) $E(G) \subset \Sigma(G) \subset T(G) \subset \Gamma(G)$ where

$$E(G) = \bigcap_{i \in N} R_i(G) \text{, basic equilibria}$$

$$\Gamma(G) = \bigcap_{i \in N} \bigcup_{K} R_i(K) \text{, general equilibria}$$

$$T(G) = \bigcup_{K} E(K) \text{, metagame equilibria}$$

$$\Sigma(G) = \bigcap_{K \text{ complete}} E(K) \text{, symmetric equilibria .}$$

For the Prisoner's Dilemma

$$E(G) = \{D, D, (2,2)\}$$

$$\Sigma(G) = T(G) = \Gamma(G)$$

$$= \{(C, C, (3,3)), (D, D, (2,2))\}$$

Recall that outcomes in metagames are mapped down to G-level before being compared. For example, in 2,1G the equilibrium outcome ((C/D), (D/C/C/D), (3,3)) is mapped first to ((C/D), C, (3,3)) in 1G and then to (C, C, (3,3)) in G .

## Section 6.2.5. Chicken and the Third Breakdown.

Howard discusses "chicken" in terms of its realization in the Cuban Missile Crisis. We show game matrices for G and 2G , and part of the one for 1,2G . Here U.S. is player 1 and U.S.S.R. is player 2.

Inducement: The Third Breakdown of Rationality

| | | Soviet Strategy | |
|---|---|---|---|
| | | Withdraw 'W' | Maintain 'M' |
| U.S. INVASION PLANS: | ABANDON 'A' | 33 COMPROMISE | (24) SOV. VICTORY |
| | CONTINUE 'C' | (42) U.S. VICTORY | 11 NUCLEAR WAR |

"G" Cuban Missile Crisis, Equilibria are circled.

SOVIET POLICIES:

| U.S. STRATEGES | | W/W | M/M | W/M | M/W |
|---|---|---|---|---|---|
| | A | 33 | (24) | 33 | 24 |
| | C | (42) | 11 | 11 | (42) ↑ SOVIET SURE-THING POLICY |

"2G" The Soviet-Metagame. The symbol "X/Y" stands for the Soviet policy "A against A,Y against C". Equilibria are circled.

SOVIET POLICIES:

| | | W/W | M/M | W/M | M/W |
|---|---|---|---|---|---|
| U.S. COUNTER-POLICIES: | A/A/A/A | 33 | (24) | 33 | 24 |
| | C/C/C/C | (42) | 11 | 11 | (42) |
| SURE-THING POLICY | C/A/A/C | 42 | (24) | 33 | 42 |
| 'RETALIATORY' POLICY | C/C/A/C | 42 | 11 | (33) | 42 |

"1,2G" Excerpt from the U.S.-Soviet-Metagame. The symbol "V/X/Y/Z" stands for "V against W/W, X against M/M, Y against W/M, Z against M/W".

Fortunately, in the actual play of the game neither contestant followed his sure-thing policy and the game seems to have been settled at the outcome (A, W, (3,3)) which does not belong to E but which is in $\Sigma$ . This provides an important example of the role of metagame theory as a descriptive model and also points to the need for a normative version.

## Section 7. A Differential Game Analogue of the Prisoner's Dilemma.

This section is based on results in John J. Lawser's University of Michigan dissertation: Properties of Dynamic Games, November 1970, prepared under NSF Grant GK-1925 and U.S. Army Contract DA-ARO-D-31-124-0767.

Dr. Lawser's interest in overlapping areas of ordinary and differential game theory led him to consider the possibility of a continuous differential game analogue of the Prisoner's Dilemma. He wished to investigate the hypothesis that the possibility of instant retaliation might encourage a cooperative strategy. He defined a scalar differential game, adapted it to a hybrid analogue-digital computer, and carried out experiments with "live" players. The experimental results tended to negate the hypothesis but were not regarded as having been sufficiently closely controlled to be conclusive.

I reproduce here some relevant portions of the dissertation.

"Example 3.2. Consider the following scalar differential game.

$$x = u + v \qquad 0 \leqq u \leqq 1 \qquad 0 \leqq v \leqq 1 \qquad x(0) = 0$$

$$J_1(u,v) = \frac{1}{t_f} \{\int_0^{t_f} -u^2(t)dt + x(t_f)\}$$

$$J_2(u,v) = \frac{1}{t_f} \{\int_0^{t_f} -v^2(t)dt + x(t_f)\}$$

While the goal of each player is to make the terminal payoff as large as possible, each is penalized for his control effort in achieving this. Of course, we do not claim that this game models any real situation. However, it does seem that it is representative of the situation where two individuals (or corporations) are jointly working on a project for which there is a terminal payoff. A player receives a payoff for the final product but the net payoff must account for his efforts made during the game. The Hamiltonians for the problem are

$$\mathcal{H}_1 = \langle p_1, u + v\rangle - \frac{1}{t_f} u^2 \tag{3.5}$$

$$\mathcal{H}_2 = \langle p_2, u + v\rangle - \frac{1}{t_f} v^2$$

Note since $\mathcal{H}_1$ and $\mathcal{H}_2$ are independent of $x$ the "open loop" adjoint variables must satisfy

$$\dot{p}_1 = -\frac{\partial \mathcal{H}_1}{\partial x} = 0$$

$$\dot{p}_2 = -\frac{\partial \mathcal{H}_2}{\partial x} = 0$$

Thus, $p_1$ and $p_2$ are constant. The terminal conditions require

$$p_1(t_f) = \frac{1}{t_f}\frac{\partial x(t_f)}{\partial x} = p_2(t_f) = \frac{1}{t_f} \tag{3.6}$$

Substituting eq. (3.6) in eq. (3.5) and choosing $u(v)$ to maximize $\mathcal{K}_1(\mathcal{K}_2)$ gives the equilibrium controls

$$u^e = \frac{1}{2} \qquad v^e = \frac{1}{2}$$

which gives the payoffs

$$J_1^e = \frac{3}{4} \qquad J_2^e = \frac{3}{4} \text{ ."}$$ [1]

"SIMULATION OF A NONZERO-SUM DIFFERENTIAL GAME

## 5.1 Introduction

In this chapter we discuss a hybrid computer simulation of the game in Example 3.2. Some limited experimental results are also presented. Because of the difficulties in defining a solution to non-zero-sum games this author feels such work is important to the development of (dynamic) game theory. The experiments, however, would most likely be best carried out by psychologists and behavioral scientists; therefore no attempt was made to carry out an extensive experimental program. The results given here are of interest, though, because they illustrate the ideas on dynamic bargaining and signaling discussed in Section 2.8. Also, while there have been experiments on repeated plays of discrete games [43] and some work on oscilloscope games [21], this is possibly the first experiment on a differential game. The results described here should at least provide insight into what further work might be done with such simulations.

## 5.2 Discussion of Example 3.2.

In this section we discuss Example 3.2 in terms of the ideas introduced in Section 2.8. Figure 5.1 shows the attainable payoff set, the various point solutions, and negotiation sets as found earlier. The dashed lines represent the payoffs available to the players if one player uses his equilibrium control for the entire game and the other player fixes his control at the beginning of the game. For example, suppose player 2 uses $u_2(t) = 1/2$ and player 1 chooses $u(t) = \alpha$. Then, $\alpha$ parameterizes the vertical parabolic dashed line; $\alpha = 0$ corresponds to the lowest point and $\alpha = 1$ to the highest point on the parabola. Of course, $\alpha = 1/2$ is the equilibrium control. If player 1 uses a control slightly greater than 1/2 then the payoff is slightly above and to the left of the equilibrium payoff. Note that this would hurt his payoff (the definition of equilibrium) but would help the other player. A control slightly less than 1/2 leads to a lower payoff for both players. Note also that by deviating from equilibrium player 1 either hurts or helps the other player more than himself. Because of the symmetry in the game the roles of players 1 and 2 can be interchanged in the discussion above. In this case $\alpha$ would paramaterize the horizontal parabola.

It is also of interest to describe the conditional payoff set. Given that the game has been played to time $t$, $0 \leq t \leq t_f$, the payoffs at the end of the game can be written

$$J_i(t_f/t) = \frac{1}{t_f} \{\int_0^t -u_i^2(\tau)d\tau + x(t)\}$$

$$+ \frac{1}{t_f} \{\int_t^{t_f} -u_i^2(\tau)d\tau + x(t_f) - x(t)\}$$

where i = 1,2. The first term in this equation depends only on $u_i(\tau)$ for $0 \leq \tau \leq t$ and represents the payoff accumulated at time $t$. The set of attainable payoffs for the second term in the equation is the same shape as the original payoff set but is reduced in size by the factor $\frac{t_f - t}{t_f}$. (This follows from the procedures in Example 3.2.) The conditional payoff set for the game in Example 3.2, then, is just the original payoff set reduced in size by the factor $\frac{t_f - t}{t_f}$ and translated by the first term in the equation. Thus, as the game is played the set of attainable payoffs shrinks at constant rate and translates according to how the payoff builds up. Figure 5.2 shows a typical conditional payoff set.

Since the conditional payoff set always has the same shape and the same extremal controls (i.e. those that give payoffs on the boundary) the discussion about the opportunities available at the beginning of the game holds as the game progresses. In other words, at each point in time during play of the game either player can suggest cooperation or threaten for the remainder of the game. Of course, a player's response to such moves would most likely depend on previous play in the game. But this brings up questions that are not answered by mathematics alone. Surely individual personalities would cause wide variation in the game's outcome. In the next two sections some preliminary experiments are discussed.

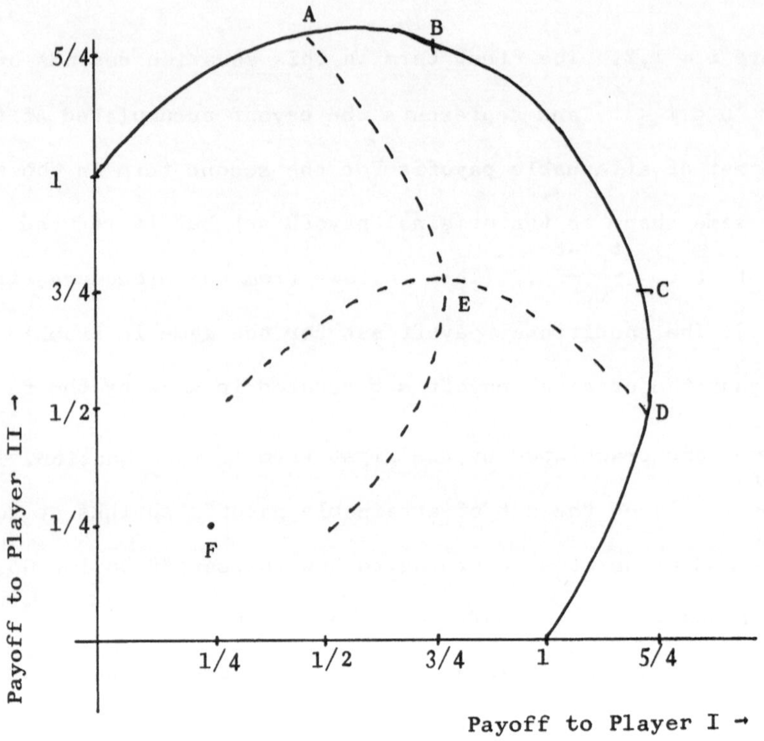


| | |
|---|---|
| ABCD | Pareto Optimal Set |
| ABCD | Negotiation Set |
| BC | Dominant Negotiation Set |
| E | Equilibrium Point |
| F | Security Payoff |

Figure 5.1 Payoff set of example game.

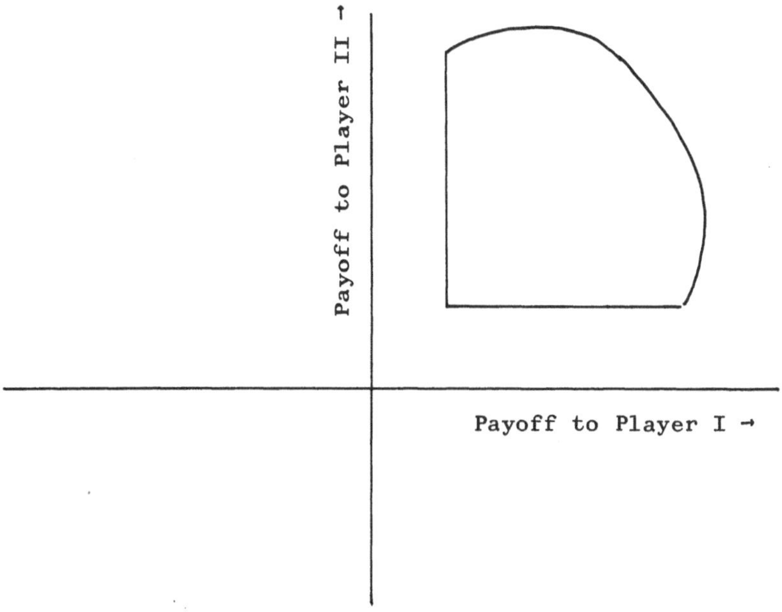


Figure 5.2 A conditional payoff set.

5.3 The Experimental Setup.

While the details of the hybrid computer circuits are given in Appendix C, it is appropriate here to describe the experimental setup. The players viewed an oscilloscope screen which displayed the conditional attainable payoff set and a pair of reference axes. A point representing the projected payoff at the end of the game if the players continued using their current controls to the end of the game was also displayed. In this way the current control of each player was known to the other. The controls were selected by setting a potentiometer between zero and one. While this does not allow discontinuous change in control level, it is not felt that this limitation significantly changed the strategies a player would use; one can change the potentiometer from zero to one in a very small fraction of the game's 200 second duration. The players did not communicate verbally during play of the game and were separated by a screen. Figure 5.3 shows the arrangements.

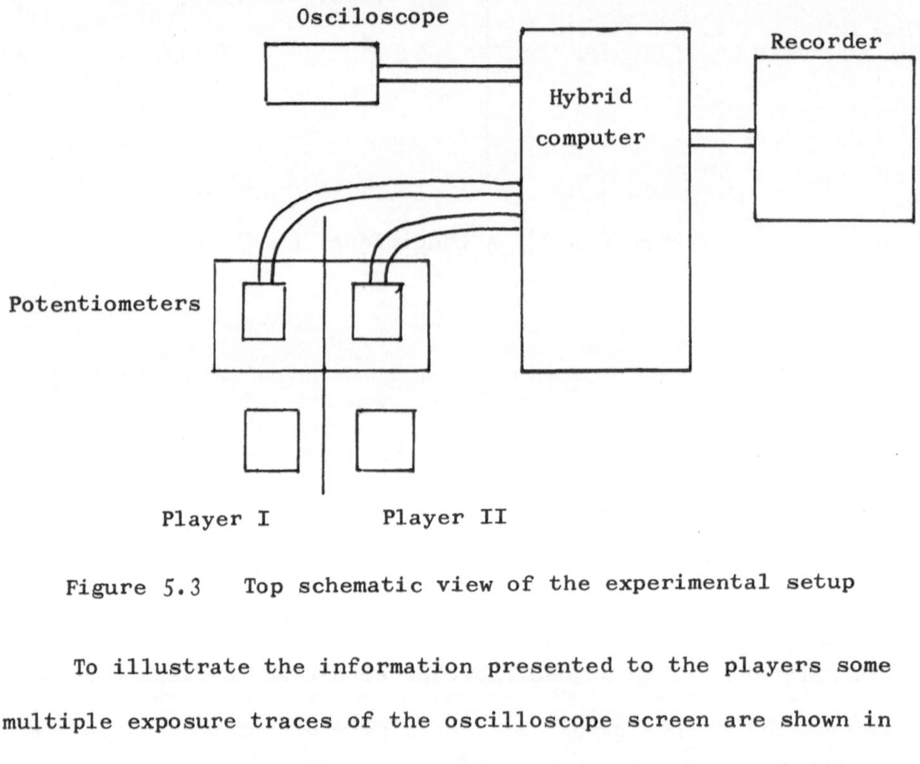


Figure 5.3 Top schematic view of the experimental setup

To illustrate the information presented to the players some multiple exposure traces of the oscilloscope screen are shown in Figure 5.4. The center photograph in Figure 5.4 is a multiple exposure of the oscilloscope trace, taken at $t = 0, .2t_f, \ldots, t_f$, with both controls set to the equilibrium control $1/2$. This leads to a payoff of $3/4$ each (the dot). The left trace is a similar multiple exposure with both controls set to 1 . This leads to a payoff of 1 each. The trace on the right was obtained by setting one player's control to 1/2 (equilibrium) and making exposures with the other player's control set to 0.,.2,...,1.. (restarting the game for each new setting). Then this process was repeated with the other player's control set to 1/2 . The sequence of dots in this photograph clearly illustrates the meaning of an equilibrium solution. Unfortunately, though, these photographs do not quite adequately describe the continuously evolving nature of the game. It should be remembered that the payoff set shrinks continuously and that at each point in time during play of the game each player must decide whether or not to continue using his current control.

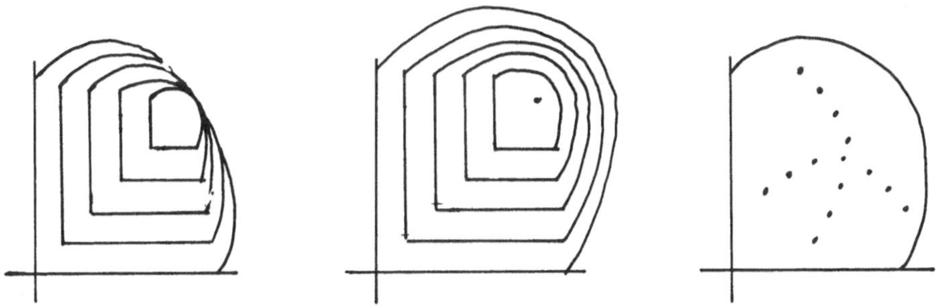

Figure 5.4 Oscilloscope traces of simulated game."[2]

FOOTNOTES

[1]John J. Lawser. Properties of Dynamic Games, Ph.D. dissertation, The University of Michigan, November 1970, pp. 63-64.

[2]Ibid., pp. 165-172.

## Section 8. Theory of Ideal Linear Weights for Heterogeneous Combat Forces.*

### Introduction

In conducting military operations research, analysts frequently make use of indices of force effectiveness which are intended to measure the contribution of some force component to the overall power of a military force in some hypothetical military conflict.

---

*Reproduced from Naval Research Logistics Quarterly, Vol. 20, 1973 and co-authored with David R. Howes, U.S. Army Concepts Analysis Agency. The original version of this paper appears as Part B of chapter 2 of [26] and was presented at the 11th U.S. Army Operations Research Symposium, May 1972. The authors wish to thank the referee for many helpful suggestions.

An example of such an index is the "Firepower Potential" which has been used in a number of U.S. Army analyses as a measure of force strength.† In the alternative considered here, indices are derived from inter-weapon effective matrices (tables) such as might emerge from a detailed combat simulation or from other sources (see for example, [1], [2], [3], [4], [15], [21], [22]).

When such tables are given it is possible to construct from them a system of weapon weights each of which is a weighted average of the effects of a given weapon against each of the enemy's weapons. This paper will describe the construction of such weights.

## 1. Effectiveness Matrices

Weapon effectiveness may be considered a function of casualty-production which lies in depriving the enemy of the value of weapons lost (cf. [13]). Therefore, it is appropriate to consider numbers which measure the killing power of each weapon against each opposing weapon. An effectiveness matrix may be regarded as a table whose entries are these killing powers or relative effectivenesses.

More precisely, consider a combat situation between two opponents, Blue and Red. We suppose that Blue has $m$ classes of weapons and consider the Blue force vector

†Several references on current procedures are cited below; some others (e.g., [11], [16], [28]) are included among the references without having been cited in the text. A full exposition of past efforts at constructing indices of effectiveness would require access to many classified or otherwise unavailable sources and would go far beyond the scope and purpose of the present paper.

(1.1)
$$U_B = \begin{bmatrix} u_{1B} \\ \cdot \\ \cdot \\ \cdot \\ u_{mB} \end{bmatrix}$$

where $u_{1B}$ is the number of Blue weapons of class $1,\ldots,u_{mB}$ is the number of Blue weapons of Blue class $m$. Similarly, suppose that Red has $n$ classes of weapons and that

(1.2)
$$U_R = \begin{bmatrix} u_{1R} \\ \cdot \\ \cdot \\ \cdot \\ u_{nR} \end{bmatrix}$$

is the Red force vector.

In the discussion which follows, it is assumed that the Blue and Red vectors of weights are to be derived in some way from certain interweapon effectiveness matrices; however, there are precedents constructing weight vectors based directly on other considerations. For example, various military operations research organizations (i.e., CORG, RAC, STAG) have from time to time constructed weight vectors based on a consensus of military judgement, individuals being asked to score lists of weapons of types of military unit. Other systems of weights have been based on such considerations as World War II casualties to personnel materiel or on the average damage radii observed during proving ground tests of ammunition (see, for example, [6] and [20]).

The effectiveness matrix concept is connected with Lanchester-type theory of combat in section 5.

We wish to find Blue and Red <u>weight vectors</u>

(1.3) $$W_B = \begin{bmatrix} w_{1B} \\ \cdot \\ \cdot \\ \cdot \\ w_{mB} \end{bmatrix} \cdot \qquad W_R = \begin{bmatrix} w_{1R} \\ \cdot \\ \cdot \\ \cdot \\ w_{nR} \end{bmatrix}$$

such that the linear combinations

(1.4) $$S(B) = w_{1B}u_{1B} + \ldots + w_{mB}u_{mB} = W_B^T U_B$$

and

(1.5) $$S(R) = w_{1R}u_{1R} + \ldots + w_{nR}u_{nR} = W_R^T U_R$$

are <u>good</u> measures of the respective overall strengths of Blue and Red. Then the fraction

(1.6) $$T = S(B)/S(R)$$

can be used as an index of the relative strengths.

A Blue-vs-Red effectiveness matrix $M_{BR}$ is a matrix (table) having $m$ rows and $n$ columns where the element $m_{BR}(i,j)$ measures the effectiveness (killing power) of a single weapon of Blue class $i$

against Red weapon class $j$ . Similarly a Red-vs-Blue effectiveness matrix

$$M_{RB} = [m_{RB}(j,i)] \tag{1.7}$$

has $n$ rows and $m$ columns and, inversely, $m_{RB}(j,i)$ measures the effectiveness of a single Red weapon of class $j$ against Blue weapon class $i$ . The numbers $m_{BR}(i,j)$ and $m_{RB}(j,i)$ may be positive or zero, but, by definition, cannot be negative.

For example, suppose that $m = n = 2$ , that both Red and Blue weapon class one are infantry weapons and that both Red and Blue weapon class two are artillery weapons. Then the effectiveness matrices

$$M^1_{BR} = \begin{bmatrix} 0.5 & 0 \\ 0.7 & 0.2 \end{bmatrix}, \quad M^1_{RB} = \begin{bmatrix} 0.6 & 0 \\ 0.6 & 0.1 \end{bmatrix} \tag{1.8}$$

would describe a situation in which (1) in infantry combat Red was more effective than Blue (0.6 vs 0.5), (2) neither infantry could harm the enemy artillery, and (3) the Blue artillery is superior to the Red artillery, and (4) each artillery battery has a positive effectiveness against its counterpart.

The effectiveness matrices

$$M^2_{BR} = \begin{bmatrix} 0.5 & 0.1 \\ 0.7 & 0.2 \end{bmatrix}, \quad M^2_{RB} = \begin{bmatrix} 0.6 & 0.2 \\ 0.6 & 0.1 \end{bmatrix} \tag{1.9}$$

would describe a change which gave each infantry capability against the opposing artillery.

The matrices

$$(1.10)\qquad M_{BR}^{3} = \begin{bmatrix} 0.5 & 0 \\ 0.7 & 0.8 \end{bmatrix}, \quad M_{RB}^{3} = \begin{bmatrix} 0.6 & 0 \\ 0.6 & 0.5 \end{bmatrix}$$

would describe a different type of change in which the artillery attritions are substantially increased.

If we assume that the artillery units are either concealed or out of each other's range then we could have effectiveness matrices

$$(1.11)\qquad M_{BR}^{4} = \begin{bmatrix} 0.5 & 0 \\ 0.7 & 0 \end{bmatrix}, \quad M_{RB}^{4} = \begin{bmatrix} 0.6 & 0 \\ 0.6 & 0 \end{bmatrix}$$

## 2. Ideal Linear Weights

We turn next to consideration of suitable weight vectors, $W_B$ and $W_R$. These should be derived in some reasonable way from the corresponding effectiveness matrices, $M_{BR}$ and $M_{RB}$.

For example, one could simply let $W_B$ be the average of the columns of $M_{BR}$. Using $M_{BR}^{1}$ and $M_{RB}^{1}$ this would give

$$(2.1)\qquad W_B^{1} = \frac{1}{2}\begin{bmatrix} 0.5 + 0 \\ 0.7 + 0.2 \end{bmatrix} = \begin{bmatrix} 0.25 \\ 0.45 \end{bmatrix}, \text{ and } W_R^{1} = \begin{bmatrix} 0.3 \\ 0.35 \end{bmatrix};$$

similarly from $M_{BR}^{2}$ and $M_{RB}^{2}$ we would obtain

$$(2.2) \qquad W_B^2 = \begin{bmatrix} 0.3 \\ 0.45 \end{bmatrix}, \quad W_R^2 = \begin{bmatrix} 0.4 \\ 0.35 \end{bmatrix}.$$

This naive approach has the advantage of simplicity, but lacks credibility since it places equal emphasis on effectiveness against enemy infantry and artillery whereas one of these might be considered much more dangerous than the other.

The naive approach places equal weight on each column. A more general procedure is to select as weights nonnegative numbers which add to one. Thus in example 2, if we consider enemy artillery to be twice as important a target as enemy infantry, we would choose weights 1/3, 2/3 and get

$$W_B^2 = \begin{bmatrix} 1/3(0.5) + 2/3(0.1) \\ 1/3(0.7) + 2/3(0.2) \end{bmatrix} = 1/3 \begin{bmatrix} 0.7 \\ 1.1 \end{bmatrix}.$$

A vector with nonnegative elements that sum to one is called a <u>probability</u> <u>vector</u>. Then the more general procedure would consist of selecting two probability vectors

$$(2.3) \qquad Z_B = \begin{bmatrix} z_{B1} \\ \cdot \\ \cdot \\ \cdot \\ z_{Bm} \end{bmatrix}, \qquad Z_R = \begin{bmatrix} z_{R_1} \\ \cdot \\ \cdot \\ \cdot \\ z_{Rn} \end{bmatrix}$$

and then defining the linear weights by

$$(2.4) \qquad W_B = M_{BR}Z_R \,, \quad W_R = M_{RB}Z_B \,.$$

We observe (i) that (2.4) gives each weighting factor $w_{iB}$ as a weighted average (probability combination) of the effectiveness numbers corresponding to the ith Blue weapon type, and (ii) that the same weighted average is used for all $i$. A still more general procedure would be to permit a different weighted average for each $i$; this would replace (2.4) by

$$(2.5) \qquad w_{iB} = \sum_j m_{BR}(i,j)z_R(i,j) \,, \quad w_{jR} = \sum_i m_{RB}(j,i)z_B(j,i) \,,$$

where all columns of the matrices $Z_R$ and $Z_B$ are probability vectors.

Returning now to (2.4) the next step is selection of $Z_B$ and $Z_R$. In the naive approach we took

$$(2.6) \qquad Z_B = \frac{1}{m} E_m = \frac{1}{m} \begin{bmatrix} 1 \\ \cdot \\ \cdot \\ \cdot \\ 1 \end{bmatrix} , \quad Z_R = \frac{1}{n} E_n = \frac{1}{n} \begin{bmatrix} 1 \\ \cdot \\ \cdot \\ \cdot \\ 1 \end{bmatrix}$$

Here (and later) we use the symbol $E_p$ to represent the column vector consisting of $p$ ones, e.g.,

$$E_3 = \begin{bmatrix} 1 \\ 1 \\ 1 \end{bmatrix} .$$

A second, somewhat more reasonable selection is

$$(2.7) \qquad Z_B = M_{BR}E_n/\gamma_B \, , \qquad Z_R = M_{RB}E_m/\gamma_R \, ,$$

where

$$\gamma_B = E_n^T M_{RB} E_m = \sum_{i,j} m_{RB}(j,i) \, , \quad \gamma_R = E_m^T M_{BR} E_n = \sum_{i,j} m_{BR}(i,j)$$

then

$$(2.8) \qquad W_B = M_{BR}M_{RB}E_m/\gamma_R \, , \quad W_R = M_{RB}M_{BR}E_n/\gamma_B \, .$$

In Example 2 this gives

$$(2.9) \qquad Z_B^2 = \begin{bmatrix} 0.6 \\ \\ 0.9 \end{bmatrix} \div 1.5, \quad Z_R^2 = \begin{bmatrix} 0.8 \\ \\ 0.7 \end{bmatrix} \div 1.5$$

and

$$(2.10) \qquad W_B^2 = M_{BR}^2 Z_R^2 = \begin{bmatrix} 0.47 \\ \\ 0.70 \end{bmatrix} \div 1.5 \, , \quad W_R^2 = \begin{bmatrix} 0.54 \\ \\ 0.45 \end{bmatrix} \div 1.5 \, .$$

In (2.7) the jth component of the averaging vector $Z_R$ is proportional to the sum of all Red effectiveness numbers corresponding to the jth Red type. This tacitly assumes equal importance for all Blue weapon types. Clearly, we could modify (2.7) by selecting any nonnegative linear combination $M_{RB}V_B$ of the columns of $M_{RB}$ and then taking $Z_R$ as the unique probability vector proportional to $M_{RB}V_B$. The <u>ideal linear weights</u> which we next introduce correspond to the choice $V_B = W_B$, $V_R = W_R$.

To motivate this choice we consider the following argument. Suppose that $W_R$ has been determined; this means that relative values for the Red weapon systems are known. Then it seems reasonable to select as $Z_R$ the unique probability vector proportional to $W_R$ . Similar reasoning would apply in selection of $Z_B$ if $W_B$ is given. This line of argument would lead to

$$(2.11) \qquad Z_B = W_B/\alpha_B , \quad Z_R = W_R/\alpha_R$$

where

$$\alpha_B = E_m^T W_B , \quad \alpha_R = E_n^T W_R ,$$

then, we get

$$(2.12) \qquad W_B = M_{BR} Z_R = M_{BR} W_R/\alpha_R , \quad W_R = M_{RB} Z_B = M_{RB} W_B/\alpha_B ,$$

and by substituting each of these equations in the other we get

$$(2.13) \qquad W_B = M_{BR} M_{RB} W_B/\alpha_B \alpha_R , \quad W_R = M_{RB} M_{BR} W_R/\alpha_R \alpha_B .$$

Now, let

$$(2.14) \qquad P_B = M_{BR} M_{RB} , \quad P_R = M_{BR} \; M_{BR} , \quad \lambda = \alpha_B \alpha_R$$

and we have the equations

$$(2.15) \qquad P_B W_B = \lambda\, W_B\,, \quad P_R W_R = \lambda\, W_R\,.$$

The ideal weights must satisfy these equations and also be nonnegative vectors (and also nonzero). At first glance it might seem that (2.11) and (2.12) involve a circular logic since each of the weights is ultimately (cf. 2.15) defined in terms of itself. However, this is a familir situation in mathematics and is a characteristic of eigenvalue problems which crop up in a wide variety of mathematical models. In particular, Equations (2.15) are well known in linear algebra. First, they require that $\lambda$ be an eigenvalue of each of the square matrices $P_B$ $(m \times m)$ and $P_R (n \times n)$ and that $W_B$, $W_R$ be eigenvectors. Since the effectiveness matrices $M_{BR}, M_{RB}$ have nonnegative elements, the same is true of their products $P_B, P_R$.

The classical Perron-Frobenius theory of eigenvalues and eigenvectors of nonnegative matrices applies to our situation and guarantees solutions to (2.14) with $W_B, W_R$ nonnegative and $\lambda$ positive. Moreover, it follows from the general theory of matrices that $P_B$ and $P_R$ have the same nonzero eigenvalues. The pertinent facts from the classical Perron-Frobenius theory can be found (with proofs) in chapter XIII of Gantmacher, Vol. II [10]. This chapter also has a comprehensive bibliography (see also Varga [29]). The original papers by Perron and Frobenius appear, respectively, as References [19] and [8], (see also [9], pp. 404-414 and 546-567, [14], [27]).

## 3. Examples of Ideal Weights

We return to our four examples to illustrate the theory.

EXAMPLE 1.

$$(3.1)\qquad P_B^1 = \begin{bmatrix} 0.30 & 0 \\ 0.54 & 0.02 \end{bmatrix}, \quad P_R^1 = \begin{bmatrix} 0.30 & 0 \\ 0.37 & 0.02 \end{bmatrix}.$$

The eigenvalues for both $P_B^1$ and $P_R^1$ are $\lambda_1^1 = 0.30$, $\lambda_2^1 = 0.02$. Then

$$(3.2)\qquad Z_B^1 = \begin{bmatrix} 0.34 \\ 0.66 \end{bmatrix}, \quad Z_R^1 = \begin{bmatrix} 0.43 \\ 0.57 \end{bmatrix}$$

are the unique probability eigenvectors corresonding to $\lambda_1^1$. The corresponding weights are

$$(3.3)\qquad W_B^1 = M_{BR}^1 Z_R^1 = \begin{bmatrix} 0.215 \\ 0.415 \end{bmatrix}, \quad W_R^1 = \begin{bmatrix} 0.204 \\ 0.270 \end{bmatrix}$$

$$\alpha_B^1 = 0.63, \quad \alpha_R^1 = 0.474, \quad \alpha_B^1 \alpha_R^1 = \lambda_1^1 = 0.3 .$$

The second eigenvalue $\lambda_2^1$ gives

$$(3.4)\qquad Z_B^{1*} = Z_R^{1*} = \begin{bmatrix} 0 \\ 1 \end{bmatrix}, \quad W_B^{1*} = \begin{bmatrix} 0 \\ 0.2 \end{bmatrix}, \quad W_R^{1*} = \begin{bmatrix} 0 \\ 0.1 \end{bmatrix}.$$

We will see later that this second eigenvalue yields less meaningful weights than the first.

EXAMPLE 2.

$$(3.5) \qquad P_B^2 = \begin{bmatrix} 0.36 & 0.11 \\ 0.54 & 0.16 \end{bmatrix}, \quad P_R^2 = \begin{bmatrix} 0.44 & 0.10 \\ 0.37 & 0.08 \end{bmatrix}.$$

The characteristic equation for both matrices is

$$(3.6) \qquad \lambda^2 - 0.52\lambda - 0.0018 = 0 ,$$

and has as its roots the eigenvalues

$$(3.7) \qquad \lambda_1^2 = 0.5235 , \quad \lambda_2^2 = -0.0035 .$$

From $\lambda_1^2$ we get the unique probability eigenvectors

$$(3.8) \qquad z_B^2 = \begin{bmatrix} 0.40 \\ 0.60 \end{bmatrix}, \quad z_R^2 = \begin{bmatrix} 0.545 \\ 0.455 \end{bmatrix}$$

for $P_B^2$ and $P_R^2$ respectively.

$$(3.9) \qquad W_B^2 = \begin{bmatrix} 0.32 \\ 0.48 \end{bmatrix}, \quad W_R^2 = \begin{bmatrix} 0.36 \\ 0.30 \end{bmatrix}$$

$$\alpha_B^2 = 0.8 , \quad \alpha_R^2 = 0.66 , \quad \alpha_B^2\alpha_R^2 = 0.528 \sim \lambda_1^2 .$$

EXAMPLE 3.

$$(3.10)\qquad P_B^3 = \begin{bmatrix} 0.3 & 0 \\ 0.9 & 0.4 \end{bmatrix}, \quad P_R^3 = \begin{bmatrix} 0.30 & 0 \\ 0.65 & 0.40 \end{bmatrix}, \quad \lambda_1^3 = 0.3,\ \lambda_2^3 = 0.4.$$

This example differs from Example 1 since this time the second eigenvalue is larger than the first and hence $\lambda_1^3$ does not correspond to probability eigenvectors (cf. [10] vol. 2, p. 66). The only probability eigenvectors come from $\lambda_2^3$ and are

$$(3.11)\qquad Z_B^3 = \begin{bmatrix} 0 \\ 1 \end{bmatrix}, \quad Z_R^3 = \begin{bmatrix} 0 \\ 1 \end{bmatrix}.$$

These give

$$(3.12)\qquad W_B^3 = \begin{bmatrix} 0 \\ 0.8 \end{bmatrix}, \quad W_R^3 = \begin{bmatrix} 0 \\ 0.5 \end{bmatrix}, \quad \alpha_B^3 = 0.8,\ \alpha_R^3 = 0.5,$$
$$\alpha_B^3\alpha_R^3 = 0.4 = \lambda_2^3 .$$

EXAMPLE 4.

$$(3.13)\qquad P_B^4 = \begin{bmatrix} 0.30 & 0 \\ 0.42 & 0 \end{bmatrix}, \quad P_R^4 = \begin{bmatrix} 0.30 & 0 \\ 0.30 & 0 \end{bmatrix}, \quad \lambda_1^4 = 0.3,\quad \lambda_2^4 = 0 .$$

This example resembles Example 1 in that the first eigenvalue is larger than the second.

From the first eigenvalue we get

$$(3.14) \qquad Z_B^4 = \begin{bmatrix} 0.42 \\ 0.58 \end{bmatrix}, \quad Z_R^4 = \begin{bmatrix} 0.5 \\ 0.5 \end{bmatrix}$$

$$W_B^4 = \begin{bmatrix} 0.25 \\ 0.35 \end{bmatrix}, \quad W_R^4 = \begin{bmatrix} 0.25 \\ 0.25 \end{bmatrix}, \quad \alpha_B^4 = 0.6, \; \alpha_R^4 = 0.5, \; \alpha_B^4\alpha_R^4 = 0.3 = \lambda_1^4 .$$

The second eigenvalue gives

$$(3.15) \qquad Z_B^{4*} = Z_R^{4*} = \begin{bmatrix} 0 \\ 1 \end{bmatrix}$$

$$W_B^4 = W_R^4 = \begin{bmatrix} 0 \\ 0 \end{bmatrix}, \quad \alpha_B^4 = \alpha_R^4 = 0 ,$$

and thus does not provide useful weighting vectors.

Example 2 illustrates a general class of situations where each Blue weapon system is (at least minimally) effective against each Red one and vice versa. If a square matrix $P$ has positive (not merely nonnegative) elements then it has a unique probability eigenvector $Z$ and the corresponding eigenvalue $\lambda_1$ (called the Perron eigenvalue) is not only positive, but has the largest absolute value of all the eigenvalues of $P$. It is then easy to calculate $Z$ and $\lambda_1$ by the following sequential process (see [14], pp. 151-152 or [27], p. 250). Let $V_o = E_m$ (where $P$ is $m \times m$), let $\alpha(V_o) = E_m^T V_o = m$, let $Z_o = V_o/\alpha(V_o)$, and proceeding inductively let $V_{i+1} = PZ_i$, let $Z_{i+1} = V_{i+1}/\alpha(V_{i+1})$, $i = 1,2 \ldots$ . Then

$$(3.16) \qquad Z = \lim_{i\to\infty} Z_i , \quad \lambda_1 = \lim_{i\to\infty} \alpha(V_{i+1}) .$$

These results still hold (see [10], vol. 2, p. 80) even if P has some, but not too many, zero elements (i.e., as long as P remains irreducible and primitive).

Indeed, when $P_B$ and $P_R$ are positive, we can use a limiting process to define the ideal weights $W_B$, $W_R$ .

We can begin with $W_R^0$ any positive vector (e.g., $W_R^0 = E_n$) then in turn set $Z_R^0 = W_R^0/\alpha(W_R^0)$, $W_B^0 = M_{BR}Z_R^0$, $Z_B^0 = W_B^0/\alpha(W_B^0)$ , and proceeding inductively

(3.17) $$W_R^i = M_{RB}Z_B^{i-1} \ , \ Z_R^i = W_R^i/\alpha(W_R^i)$$

$$W_B^i = M_{BR}Z_R^i \ , \quad Z_B^i = W_B^i/\alpha(W_B^i) \ , \qquad i = 1,2,\ldots \ .$$

Then the six sequences

(3.18) $$W_R^i,\ Z_R^i,\ W_B^i,\ Z_B^i,\ \alpha(W_R^i),\ \alpha(W_B^i)$$

converge, respectively, to

(3.19) $$W_R, Z_R, W_B, Z_B, \alpha_R, \alpha_B$$

where $Z_B, Z_R$ are the unique Perron probability eigenvectors of $P_B, P_R$ , respectively; $W_R$, $W_B$ are the ideal weights for R, B , respectively; $W_R = \alpha_R Z_R$, $W_B = \alpha_B Z_B$ , and $\lambda_1 = \alpha_R \alpha_B$ is the Perron eigenvalue for both $P_B$ and $P_R$ .

This approach provides a computationally convenient algorithm for calculating the ideal weights. When m and n exceed two, this approach is clearly preferable to calculating and solving the characteristic equation for $P_B$ or $P_R$. There are other more refined computational algorithms which are, in general, more efficient than this one. However, a computer program written for this iterative process gave quite satisfactory numerical results for moderate values of m and n. An example involving 40 weapon types converged in nine iterations to an accuracy of 0.0001.

## 4. Interpretation of Reducibility

Examples 1,3,4 illustrate some of the possible effects of zeros in $P_B$, $P_R$. All of the P's in these examples are what is called reducible. A non-negative square matrix P is said to be reducible if it has the form

$$P = \begin{bmatrix} P_1 & 0 \\ P_{21} & P_2 \end{bmatrix}, \tag{4.1}$$

where $P_1$ and $P_2$ are square, or more generally, if this form can be obtained by a reordering of the rows followed by the same reordering of the columns.

In our combat context, we encounter reducible matrices when as in Examples 1,3,4 there are two classes of weapons on each side and the first class of Blue is totally ineffective against the second class of Red and vice versa.

Let us assume that both $P_B$ and $P_R$ are reducible with $P_{B1}, P_{B2}, P_{R1}, P_{R2}$ all positive, that $P_{B1}$ , $P_{R1}$ have the Perron eigenvalue $\lambda_1$ , and that $P_{B2}, P_{R2}$ have the Perron eigenvalue $\lambda_2$ . [These assumptions all hold for Examples 1 and 3.] Then, if we apply our computational algorithm beginning with $W_R^0 = E_n$ , the limiting eigenvectors obtained will correspond to the larger eigenvalue.

Thus, in Example 1 we would get $W_B^1, W_R^1$ and not $W_B^{1*}, W_R^{1*}$ . In Example 3 we would, of course, get $W_B^3, W_R^3$ and in this case there is no possibility of positive ideal weights.

Moreover, in Example 1 the only way to get the starred vectors would be to start with $W_R^0$ of the form $\begin{bmatrix} 0 \\ a \end{bmatrix}$ , i.e., almost all starting vectors $W_R^0$ will yield $W_B^1, W_R^1$ . For this reason we choose to limit the term "ideal" to $W_B^1, W_R^1$ .

There is a possible interpretation for the different types of weights found in Examples 1 and 3. In Example 1 the attrition of infantry is so much greater than that of artillery that we visualize one phase of the battle ending when one side has lost all of its infantry even though both sides still have artillery left. However, at that time the starred weights do become relevant for the ensuing artillery duel.

On the other hand, in Example 3 the artillery attrition is more rapid than that of infantry. Moreover, when one side runs out of artillery the remaining infantry forces will ultimately be annihilated by the surviving artillery. Hence a zero weight for infantry is not inappropriate.

Example 4 is much like Example 1 for even though $P_{B2} = P_{R2} = 0$ the larger eigenvalue $\lambda_1^4$ still gives a viable ideal weight.

## 5. Calculation of Effectiveness Matrices and an Application to Lanchester Theory

There are several possible approaches to calculation of the effectiveness matrices. Only one of these will be discussed in the present paper.

A sufficiently detailed combat simulation can be expected to produce loss matrices

$$(5.1) \qquad L_{BR} = [\ell_{BR}(i,j)] \; , \quad L_{BR} = [\ell_{BR}(j,i)] \; ,$$

where $\ell_{BR}(i,j)$ is the number of Red weapons of class $j$ lost by action of Blue weapons of class $i$, etc. Then we may define effectiveness matrices $M_{BR}$, $M_{RB}$ whose elements are the effectiveness numbers;

$$(5.2) \qquad m_{BR}(i,j) = \ell_{BR}(i,j)/u_{iB} \; , \quad m_{RB}(j,i) = \ell_{RB}(j,i)/u_{jR}$$

where $U_B$ and $U_R$ are as in section 1 (formulas (1.1) and (1.2)).

The $u_{iB}$ and $u_{jR}$ might refer either to the initial Blue and Red strengths, or to certain average strengths during the battle. The choice of an appropriate average would relate to questions not considered here; however, a simple case of such an average might be

$[u_{iR(t=0)} + u_{iR(t=t_1)}]/2$ where $t_1$ is an arbitrary time chosen as a unit of measurement. The interval $(0,t_1)$ must, of course, not exceed the battle length and should be small enough so that combat losses have not yet changed the character of the encounter.

This procedure has as its main drawbacks (1) that the validity of the results obtained depends on the simulation scenario, on the simulation model, and on the extent of sampling error, (2) that it fails to consider military appurtenances which, although affecting the combat action, do not cause attributable casualties to opposing weapon systems, and (3) that it does not take into account scale factors (i.e. it tacitly assumes that the losses are strictly proportional to the number of weapons in a class).

Effectiveness matrices calculated as above are closely related to the Lanchester parameters appropriate to a heterogeneous Lanchester linear system and could be interpreted as estimates of such parameters. Such a system represents an extension of the formula which F. W. Lanchester [17] used to describe the attrition inflicted on each other by two hostile forces to the case where each force is composed of various subelements. In such cases, each force can be represented as a vector of elements and the (scalar) Lanchester attrition coefficients have as counterparts matrices whose elements describe the interacting effects between the elements. These attrition matrices, if known, could serve as examples of effectiveness matrices as discussed in this paper. Conversely, effectiveness matrices, when based on data from real or simulated combat, might be interpreted as Lanchester parameters as noted above.

The generalization of Lanchester equations to the heterogenous case was explored by Snow [23], then by Kolansky [7], and by Bonder and Farrell [5]. It should be noted that effectiveness matrices may be derived in other ways and also that the statistical problem of parameter estimation from sample data is far more complex than might be suggested by the discussion given here.

Dare and James, in Defense Operational Analysis Establishment Memorandum M7120 have made an analysis based on a Lanchester interpretation with results parallel to those which follow next. In Tab E, Appendix II to Annex L of the TATAWS III study, BAARINC Inc. has based a similar analysis on another interpretation.

More specifically, if we have the Lanchester systems

$$(5.3) \qquad \dot{U}_B = -C_R U_R , \quad \dot{U}_R = -C_B U_B$$

then the $(i,j)$ elements $c_R(i,j)$ of $C_R$ represents the effectiveness of $R$ weapon $j$ against $B$ weapon $i$, i.e.,

$$c_R(i,j) = m_{RB}(j,i) .$$

Reasoning similarly for $C_B$ we conclude that

$$(5.4) \qquad C_R = M_{RB}^T , \qquad C_B = M_{BR}^T$$

are reasonable choices for the Lanchester coefficient matrices.

Now, differentiating Equation (1.4) with respect to time we get

$$(5.5) \qquad \dot{S}(B) = W_B^T \dot{U}_B = - W_B^T M_{RB}^T U_R$$

$$= - (M_{RB}\, W_B)^T U_R$$

$$= - (M_{RB}\, M_{BR}\, W_R\, /\alpha_B)^T U_R$$

$$= - \frac{\lambda}{\alpha_R} W_R^T U_R = - \alpha_B W_R^T U_R \qquad (\text{since} \quad \lambda = \alpha_R \alpha_B) .$$

Now substituting from (1.5) this gives

$$(5.6) \qquad \dot{S}(B) = -\alpha_B S(R) .$$

Similarly, differentiating (1.5) yields

$$(5.7) \qquad \dot{S}(R) = -\alpha_R S(B) .$$

Equations (5.6) and (5.7) are the ones obtained by Dare and James. A note of caution is appropriate here. The heterogeneous systems (5.3), (5.4) are not valid past the time $t^*$ at which any component of $U_R$ or $U_B$ becomes zero. Although the summarizing homogeneous systems (5.6) and (5.7) will in general yield solutions $S(B)$, $S(R)$ which both remain positive far beyond $t^*$, the attrition-rate coefficients $\alpha_B$ and $\alpha_R$ must be modified whenever the weights $W_B$ and $W_R$ change due to the annihilation of a target type (see the discussion of Example 1 in section 4).

## 6. A Larger Example

An example of extended calculation is given below based on results obtained in a particular detailed war game. No claims are warranted concerning the representativeness of these results, which are dependent on the particular scenario, and the random statistical variation inherent in the game model used. Weapons classes for both sides were the same. They were (following some aggregation of similar type):

1. Small arms
2. Armored personnel carriers
3. Tanks
4. Armed reconnaissance vehicles
5. Anti-tank weapons
6. Mortars
7. Artillery

Red forces were in the attack, Blue in the defense.

7 Red Weapons 7 Blue Weapons

Red Effects

$$M_{RB} = \begin{bmatrix} 0.0145 & 0.0012 & 0.0000 & 0.0229 & 0.0004 & 0.0000 & 0.0000 \\ 0.0510 & 0.0326 & 0.0000 & 0.0638 & 0.0012 & 0.0048 & 0.0000 \\ 0.1060 & 0.4600 & 0.4540 & 0.4900 & 0.0056 & 0.0515 & 0.0000 \\ 0.4440 & 0.2220 & 0.0000 & 0.4440 & 0.0700 & 0.0000 & 0.0000 \\ 0.0000 & 0.1370 & 0.7400 & 0.2740 & 0.0137 & 0.0000 & 0.0000 \\ 6.1500 & 0.0000 & 0.0000 & 0.0000 & 0.0630 & 0.0740 & 0.0000 \\ 21.0000 & 0.2320 & 0.0750 & 0.2770 & 0.1570 & 0.0800 & 0.1960 \end{bmatrix} \qquad (6.1)$$

Blue Effects

$$M_{BR} = \begin{bmatrix} 0.0334 & 0.0028 & 0.0000 & 0.0290 & 0.0004 & 0.0000 & 0.0000 \\ 0.1170 & 0.0940 & 0.0000 & 0.1111 & 0.0045 & 0.0000 & 0.0000 \\ 0.4770 & 2.5300 & 2.0900 & 1.8200 & 0.0730 & 0.0000 & 0.0000 \\ 0.8200 & 0.4730 & 0.0000 & 0.5550 & 0.0008 & 0.0000 & 0.0000 \\ 0.0000 & 2.8300 & 0.5000 & 3.3300 & 0.1860 & 0.1940 & 0.0000 \\ 12.0800 & 0.0000 & 0.0000 & 0.0000 & 0.1580 & 0.1502 & 0.0000 \\ 9.7100 & 0.1220 & 0.1000 & 0.1350 & 0.1180 & 0.0680 & 0.2590 \end{bmatrix} \tag{6.2}$$

$$P_R = \begin{bmatrix} 0.0194 & 0.0121 & 0.0002 & 0.0146 & 0.0001 & 0.0001 & 0.0000 \\ 0.1158 & 0.0368 & 0.0006 & 0.0445 & 0.0012 & 0.0010 & 0.0000 \\ 1.2978 & 1.4398 & 0.9517 & 1.1711 & 0.0448 & 0.0088 & 0.0000 \\ 0.4049 & 0.4302 & 0.0350 & 0.5171 & 0.0146 & 0.0136 & 0.0000 \\ 0.5937 & 2.0535 & 1.5534 & 1.5597 & 0.0574 & 0.0027 & 0.0000 \\ 1.0993 & 0.1955 & 0.0315 & 0.3881 & 0.0259 & 0.0233 & 0.0000 \\ 3.8610 & 0.8696 & 0.2548 & 1.4743 & 0.0801 & 0.0558 & 0.0508 \end{bmatrix} \tag{6.3}$$

$$P_B = \begin{bmatrix} 0.0135 & 0.0066 & 0.0003 & 0.0139 & 0.0021 & 0.0000 & 0.0000 \\ 0.0558 & 0.0285 & 0.0033 & 0.0592 & 0.0080 & 0.0005 & 0.0000 \\ 1.1656 & 1.4585 & 1.0029 & 2.0245 & 0.1433 & 0.1198 & 0.0000 \\ 0.2824 & 0.1397 & 0.0006 & 0.2956 & 0.0398 & 0.0023 & 0.0000 \\ 2.8689 & 1.0870 & 0.3646 & 1.9550 & 0.2541 & 0.0537 & 0.0000 \\ 1.0989 & 0.0361 & 0.1169 & 0.3199 & 0.0165 & 0.0111 & 0.0000 \\ 6.0748 & 0.1679 & 0.1521 & 0.4432 & 0.0606 & 0.0315 & 0.0508 \end{bmatrix} \tag{6.4}$$

Clearly this is a reducible case with one obvious Perron eigenvalue $\lambda_2 = 0.0508$ . Applying seven iterations we find that the other Perron eigenvalue $\lambda_1$ has the positive probability eigenvectors.

$$(6.5) \qquad Z_{1R} = \begin{bmatrix} 0.00052 \\ 0.00198 \\ 0.30482 \\ 0.03033 \\ 0.48015 \\ 0.03087 \\ 0.15134 \end{bmatrix} , \quad Z_{1B} = \begin{bmatrix} 0.00082 \\ 0.00433 \\ 0.54771 \\ 0.01396 \\ 0.26523 \\ 0.06485 \\ 0.10310 \end{bmatrix}$$

where also

$$(6.6) \qquad \alpha_{1R} = 0.85983 , \qquad \alpha_{1B} = 1.33191$$

$$\lambda_1 = \alpha_R \alpha_B = 1.14522$$

$$W_{1R} = \alpha_{1R} Z_{1R} , \qquad W_{1B} = \alpha_{1B} Z_{1B} .$$

Since $\lambda_1$ is much greater than $\lambda_2$ the ideal weights obtained from $\lambda_1$ may be regarded as being more significant than those obtained from $\lambda_2$ as given in (6.7) and (6.8) below:

$$(6.7) \qquad Z_{2R} = Z_{2B} = \begin{bmatrix} 0 \\ 0 \\ 0 \\ 0 \\ 0 \\ 0 \\ 1 \end{bmatrix}$$

$$(6.8) \qquad \alpha_{2R} = 0.1960 , \ \alpha_{2B} = 0.2590 , \ \lambda_2 = 0.0508 ,$$

$$W_{2R} = \alpha_{2R} Z_{2R} , \ W_{2B} = \alpha_{2B} Z_{2B} .$$

# BIBLIOGRAPHY FOR SECTION 8 ONLY


[1] Barfoot, C.B., "The Attrition-Rate Coefficient, Some Comments on Seth Bonder's Paper and a Suggested Alternative Method", Operations Research 17, 888-894 (1969).

[2] Bonder, Seth, "A Theory for Weapon System Analysis", Proc. U.S. Army Operations Research Symposium, 111-128 (1965).

[3] Bonder, Seth, "The Lanchester Attrition-Rate Coefficient", Operations Research 15, 221-232 (1967).

[4] Bonder, Seth, "The Mean Lanchester Attrition Rate", Operations Research 18, 179-181 (1970).

[5] Bonder, S. and R. Farrell, "Development of Models for Defense Systems Planning", SRL 2147, Systems Research Laboratory, University of Michigan, Ann Arbor, Michigan (1970).

[6] Corg, "Measuring Combat Effectiveness", Vol. II, Technical Operations Incorporated Inc. Combat Operations Research Group, Alexandria, Va. (Jan. 1970).

[7] Dolansky, L., "Present State of the Lanchester Theory of Combat", Operations Research 12, 344-358 (1964).

[8] Frobenius, Georg, "Uber Matrizen aus nicht negativen Elementen", Sitzungsberichte der Kgl Preussischen Akademie der Wissenschaften zu Berlin (1912), Berlin, pp. 456-477.

[9] Frobenius, Georg, <u>Gesammelte Abhandlungen, Band III</u> (Edited by J.P. Serre), Springer-Verlag, Berlin (1968).

[10] Gantmacher, F.R., <u>The Theory of Matrices</u> (Chelsea, 1959), 2 vols.

[11] Grubbs, Frank E. and John H. Shuford, "A New Formulation of Lanchester Combat Theory", Operations Research 21, 926-941 (1973).

[12] Hayward, P., "The Measurement of Combat Effectiveness", Operations Research 16, 314-323 (1968).

[13] Hero, "Comparative Analyses of Historical Studies", Historical Evaluation and Research Office, 2223 Wisconsin Avenue, Washington, D.C. (15 Oct. 1964), Annex III-H.

[14] Householder, A., Principles of Numerical Analysis (McGraw-Hill, New York, 1953).

[15] Kimbleton, S., "Attrition Rates for Weapons with Markov-Dependent Fire", Operations Research 19, 698-706 (1971).

[16] Koopman, B.O., "A Study of the Logical Basis of Combat Simulation", Operations Research 18, 855-882 (1970).

[17] Lanchester, F.W., Aircraft in Warfare, the Dawn of the Fourth Arm (Constable, London, 1916).

[18] Morse, Philip M., and George E. Kimball, Methods of Operations Research (John Wiley, New York, 1951).

[19] Perron, Oskar, "Zur Theorie der Matrices", Mathematische Annalen, Vol. 64 (1907).

[20] RAC-TP-III, "Tacspiel War Game Procedures and Rules of Play", Research Analysis Corp. McLean, Va. (Nov. 1963) (Secret).

[21] Rustagi, J. and R. Laitinen, "Moment Estimation in a Markov-Dependent Firing Distribution", Operations Research 18, 918-923 (1970).

[22] Rustagi, J. and R. Srivastava, "Parameter Estimation in a Markov Dependent Firing Distribution", Operations Research 16, 1222-1227 (1968).

[23] Snow, R.N., "Contributions to Lanchester Attrition Theory", Project RAND RA-15078 Douglas Aircraft Co., Santa Monica, Cal. (Apr. 1942).

[24] Shuford, John H., "A New Probability Model for Lanchester's Equations of Combat", Masters Thesis submitted to the George Washington University (Dec. 1971).

[25] Taylor, James G., "A Note on the Solution to Lanchester Type Equations with Variable Coefficients", Operations Research 19, 709-712 (1971).

[26] Thrall, R.M. and Associates, Final Report to U.S. Army Strategy and Tactics Analysis Group, RMT-200-R4-33 (1 May 1972).

[27] Todd, J. (Editor), Survey of Numerical Analysis (McGraw-Hill, New York, 1962).

[28] United States Army Combat Developments Command Report, Measuring Combat Effectiveness, by Technical Operations Incorporated, Combat Operations Research Group, Vol. I "Firepower Potential Methodology (U)" (Confidential-NO FORN).

[29] Varga, R., Matrix Iterative Analysis (Prentice-Hall, Englewood Cliffs, New Jersey, 1962).

[30] Weiss, H.K., "Lanchester-Type Models of Warfare", Proc. First International Conference on Operations Research (Dec. 1957), pp. 82-89.

# BIBLIOGRAPHY


1. Blackwell, David and Girshick, M.A. Theory of Games and Statistical Decisions. John Wiley and Sons, New York, 1954.

2. Burger, Ewald., Introduction to the Theory of Games. Prentice-Hall, Inc., Englewood Cliffs, New Jersey, 1963.

3. Case, J.N., Toward a Theory of Many Player Differential Games, SIAM J. Control, Vol. 7 (1969), pp. 179-197.

4. Copeland, A.H., Review: Theory of Games and Economic Behavior (John von Neumann and Oskar Morgenstern), Bulletin Amer. Math. Soc., Vol. 51 (1945), pp. 498-504.

5. Davis, M.D., Game Theory: A Nontechnical Introduction, Basic Books, Inc., New York, 1970.

6. Dresher, M., Tucker, A.W. and Wolfe, P. Contributions to the Theory of Games, Vol. III, Ann. Math. Studies, 39, Princeton University Press, Princeton, New Jersey,1957.

7. Dresher, M., Shapley, L.S. and Tucker, A.W., EDS., Advances in Game Theory, Annals of Math. Studies, No. 52, Princeton University Press, Princeton, 1964.

8. Dresher, Melvin, Games of Strategy: Theory and Applications, Prentice-Hall, Inc., Englewood Cliffs, New Jersey, 1961.

9. Eisenman, R.L., Alliance Games of N-persons, Nav. Res. Logist. Quarterly, 13 (1966), pp. 403-411.

10. Glicksman, A.M., An Introduction to Linear Programming and the Theory of Games, J. Wiley and Sons, Inc., New York, 1963.

11. Grotte, J.H., Computation of and Observations on the Nucleolus, the Normalized Nucleolus, and the Central Games, Master's Thesis, Applied Mathematics Dept., Cornell University, Ithaca, New York, 1970.

12. Howard, Nigel, Paradoxes of Rationality: Theory of Metagames and Political Behavior, MIT Press, Cambridge, Mass. (1971).

13. Isaacs, R., Differential Games: A Mathematical Theory with Applications to Warfare and Pursuit, Control and Optimization, J. Wiley and Sons, Inc., New York, 1965.

14. Karlin, Samuel, Mathematical Methods and Theory in Games, Programming, and Economics, Vols. I and II, Addison-Wesley Publishing Company, Inc., Reading, Mass., 1959.

15. Kohlberg, E., On the Nucleolus of a Characteristic Function Game, SIAM J. Appl. Math., 20 (1971), pp. 62-66.

16. Kohlberg, E., The Nucleolus as a Solution of a Minimization Problem, SIAM J. Appl. Math., to appear.

17. Kopelowitz, A., Computation of the Kernels of Simple Games and the Nucleolus of N-person Games, RPGTME RM 31, Dept. of Mathematics, Hebrew Univ., Jerusalem, Sept., 1967.

18. Kuhn, Harold W. and Tucker, Albert W., Contributions to the Theory of Games, Ann. Math. Studies, 24, Princeton University Press, Princeton, New Jersey, 1950.

19. Kuhn, Harold W. and Tucker, Albert W., Contributions to the Theory of Games, Vol. II, Ann. Math. Studies, 28, Princeton University Press, Princeton, New Jersey, 1953.

20. Lucas, W.F., Solutions for Four-person Games in Partition Function Form, SIAM J. Appl. Math., 13 (1965), pp.118-128.

21. Lucas, W.F., A Game with No Solution, Bull. Amer. Soc., Vol. 74 (1968), pp. 237-239.

22. Lucas, W.F., A Game in Partition Function Form with No Solution, J. SIAM, Vol. 16 (1968), pp. 582-585.

23. Lucas, W.F., The Proof That a Game May Not Have a Solution, Trans. Amer. Math. Soc., Vol. 137 (1969), pp. 219-229.

24. Lucas, W.F., Some Recent Developments in N-Person Game Theory, SIAM Review, Vol. 13 (1971).

25. Lucas, W.F., An Overview of the Mathematical Theory of Games, Management Science 18 (1972), pp. P3-19.

26. Luce, R. Duncan and Raiffa, Howard, Games and Decisions: Introduction and Critical Survey, John Wiley and Sons, New York, 1957.

27. McKinsey, J.C.C., Introduction to the Theory of Games, McGraw-Hill Book Company, New York, 1952.

28. Megiddo, N., The Kernel and the Nucleolus of a Product of Simple Games, RPGTME RM 45, Dept. of Mathematics, Hebrew Univ., Jerusalem, April, 1969.

29. Owen, G., Game Theory, W.B. Saunders Co., Philadelphia, 1968.

30. Owen, G., Political Games, Nav. Res. Logist. Quart., 18 (1971), pp. 345-355.

31. Owen, G., Optimal Threat Strategies of Bimatrix Games, Int. J. Game Th., 1, (1971), pp. 1-9.

32. Owen, G., Multilinear Extensions of Games, Management Sci., 18, (1971), pp. P64-P79.

33. Rapoport, A., Two-Person Game Theory: The Essential Ideas, The University of Michigan Press, Ann Arbor, 1966.

34. Rapoport, A., N-Person Game Theory: Concepts and Applications, University of Michigan Press, Ann Arbor, 1970.

35. Schmeidler, D., The Nucleolus of a Characteristic Function Game, SIAM J. Appl. Math., Vol. 17 (1969), pp. 1163-1170.

36. Schwodiauer, G., Glossary of Game Theoretical Terms, Working Paper No. 1, Dept. of Ec., New York Univ., (1971), pp.88.

37. Shubik, M., ED., Game Theory and Related Approaches to Social Behavior, John Wiley and Sons, Inc., 1964.

38. Thrall, R.M. and Lucas, W.F., N-Person Games in Partition Function Form, Nav.Res.Logist.Quart., Vol.10, (1963),pp.281-298.

39. Tucker, Albert W., and Luce, R. Duncan, Contributions to the Theory of Games, Vol. IV, Ann. Math. Studies, 40, Princeton University Press, Princeton, New Jersey, 1959.

40. Vorob'ev, N.N., The Development of Game Theory, (Translated by E. Schwodiauer) Working Paper No. 2, Dept. of Ec., New York Univ., (1971), 124 + 18 pp.

41. von Neumann, John and Morgenstern, Oskar, Theory of Games and Economic Behavior, Princeton University Press, Princeton, New Jersey, 1st ed. 1944, 2nd ed. 1947.

42. Williams, John D., The Compleat Strategist: Being a Primer on the Theory of Games of Strategy, McGraw-Hill Book Comoany, New York, 1954.

# LINEAR MULTIVARIABLE CONTROL

by

W. M. Wonham

Department of Electrical Engineering
University of Toronto

These lectures are devoted to qualitative aspects of the design of linear time-invariant multivariable control systems of finite dynamic order. A "deterministic" setting is adopted, starting from a more-or-less classical point of view. On the other hand, we consider problems of structure and synthesis which have been solved only quite recently. At this level no role, as yet, is played by notions of optimality.

The lecture material is based on the following references. Unpublished articles are available from the author in report form, on request.

## 1. INTRODUCTION

Controllability, pole assignment, stabilizability; observability, observers, detectability.

1 W. M. Wonham, On pole assignment in multi-input controllable linear systems, I.E.E.E. Trans. Aut. Control AC-12(6), 1967, pp. 660-665.

2 W. M. Wonham, On a matrix Riccati equation of stochastic control, SIAM J. Control 6(4), 1968, pp. 681-697.

3 W. M. Wonham, Dynamic observers: geometric theory, I.E.E.E. Trans. Aut. Control AC-15(2), 1970, pp. 258-259.

4 W. M. Wonham, Algebraic methods in linear multivariable control, in System Structure (ed. A. S. Morse), Control Systems Society, I.E.E.E. Catalog No. 71C61-CSS, August, 1971.

5 W. M. Wonham, and A. S. Morse, Feedback invariants of linear multivariable systems, Automatica 8, 1972, pp. 93-100.

## 2. NONINTERACTING CONTROL

(A,B)-invariant subspaces, (A,B)-controllability subspaces, decoupling.

6 W. M. Wonham and A. S. Morse, Decoupling and pole assignment in linear multivariable systems: a geometric approach, SIAM J. Control 8(1), 1970, pp. 1-18.

7 A. S. Morse and W. M. Wonham, Decoupling and pole assignment by dynamic compensation, SIAM J. Control 8(3), 1970, pp. 317-337.

8 A. S. Morse and W. M. Wonham, Triangular decoupling of linear multivariable systems, I.E.E.E. Trans. Aut. Control, AC-15 (4), 1970, pp. 447-449.

9 A. S. Morse and W. M. Wonham, Status of noninteracting control, I.E.E.E. Trans. Aut. Control AC-16 (6), 1971, pp. 568-580.

10 E. Fabian and W. M. Wonham, Generic solvability of the decoupling problem, Control System Report No. 7301, Department of Electrical Engineering, University of Toronto, January, 1973; to appear, SIAM J. Control, 1974.

11 E. Fabian and W. M. Wonham, Decoupling, disturbance rejection and sensitivity, Control System Report No. 7309, Department of Electrical Engineering, University of oronto, June, 1973; submitted for publication, I.E.E.E. Trans. Aut. Control.

## 3. TRACKING AND REGULATION

The general regulator-servomechanism problem with the requirements of internal stability and output regulation. Symthesis with the requirement of parametric insensitivity.

12 S. P. Bhattacharyya, J. B. Pearson and W. M. Wonham, On zeroing the output of a linear system, Information and Control 20(2), 1972, pp. 135-142.

13 W. M. Wonham, Tracking and regulation in linear multivariable systems, Control Systems Report No. 7202 (revised), Department of Electrical Engineering, University of To onto, May, 1972; SIAM J. Control 11(3), 1973, pp. 424-437.

14 W. M. Wonham and J. B. Pearson, Regulation and internal stabilization in linear multivariable systems, Control System Report No. 7212, Department of Electrical Engineering, University of Toronto, August, 1972; SIAM J. Control 12(1), 1974, to appear.

15 B. Francis, O. A. Sebakhy and W. M. Wonham, Symethesis of multivariable regulators, Proc. Eleventh Annual Allerton Conference on Circuit and System. Theory, University of Illinois, October, 1973, to appear.

Notation

System $\dot{x} = Ax + Bu$

State space $\mathcal{X}$ , $d(\mathcal{X}) = n$

Control space $\mathcal{U}$ , $d(\mathcal{U}) = m$

Measured outputs: $y = Cx$

Measured output space $\mathcal{Y}$, $d(\mathcal{Y}) = p$

Regulated outputs: $z = Dx$

Regulated output space $\mathcal{Z}$

All maps $A: \mathcal{X} \to \mathcal{X}$ , $B: U \to \mathcal{X}$ , $C: \mathcal{X} \to \mathcal{Y}$ , $D: \mathcal{X} \to \mathcal{Z}$ etc.

are constant (independent of $t$ )

$\mathcal{B} = B\mathcal{U} = \mathrm{Im} B$

$\langle A|\mathcal{B}\rangle = \mathcal{B} + A\mathcal{B} + \ldots + A^{n-1}\mathcal{B}$

$\langle A|\mathcal{B}\rangle$ is the <u>controllable space</u> of $(A,B)$

Thus $\langle A|\mathcal{B}\rangle \subset \mathcal{X}$ is largest subspace of $\mathcal{X}$ reachable from $x = 0$ by appropriate choice of $u(\cdot) =$ (say) $u(t)$, $0 \leq t \leq 1$ .

Definition

$(A,B)$ is <u>controllable</u> if $\langle A|\mathcal{B}\rangle = \mathcal{X}$ .

Prop.

$\langle A|\mathcal{B}\rangle = \mathcal{X} \Rightarrow \langle A + BF|\mathcal{B}\rangle = \mathcal{X}$ for all $F: \mathcal{X} \to \mathcal{U}$ .

Next, if $\mathcal{R} \subset \mathcal{X}$ and $A\mathcal{R} \subset \mathcal{R}$ , we have $(\bar{A},\bar{B})$ defined in $\bar{\mathcal{X}} = \mathcal{X}/\mathcal{R}$, $\mathcal{U}$ :

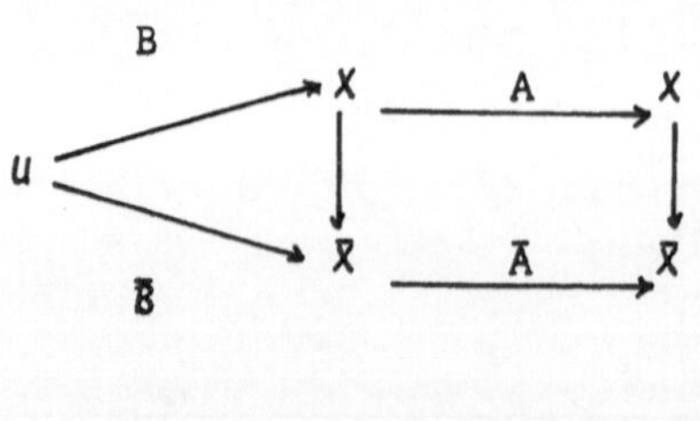

Prop.

$$\langle A \mid \mathcal{B} \rangle = \mathcal{X} \Rightarrow \langle \bar{A} \mid \bar{\mathcal{B}} \rangle = \bar{\mathcal{X}}$$

Prop.

Let $d(\mathcal{B}) = 1$, $\mathcal{B} = b$, $(A,b)$ cont'able. Let ch.p. of $A$ (= min. poly. of $A$ in this case) be $\lambda^n - (a_1 + a_2\lambda + \ldots + a_n\lambda^{n-1})$. Then there is a (unique) basis in which

$$A = \begin{bmatrix} 0 & 1 & 0 & \ldots & 0 \\ 0 & 0 & 1 & \ldots & 0 \\ \cdot & \cdot & \cdot & \ldots & \cdot \\ 0 & \cdot & \cdot & 0 & 1 \\ a_1 & \cdot & \cdot & \ldots & a_n \end{bmatrix}, \quad b = \begin{bmatrix} 0 \\ \cdot \\ \cdot \\ 0 \\ 1 \end{bmatrix}$$

In what follows, all maps $A$ etc. and vector spaces $\mathcal{X}$ etc. are real. But $\sigma(A)$ means complex spectrum; $|\sigma(A)| = n$, and $\sigma(A)$ is symmetric about real axis of $\mathbb{C}$.

Prop.

$(A,b)$ controllable and $\Lambda$ a symmetric set of $n$ complex numbers $\Rightarrow \exists f' = f'(\Lambda) \ni \sigma(A + bf') = \Lambda$.

Proof

Use standard form for $(A,b)$ given above.

Prop. [1]

$(A,B)$ controllable and $A$ cyclic $\Rightarrow \exists b \in \mathcal{B} \ni (A,b)$ controllable. [Almost all $b \in \mathcal{B}$ have this property.]

Prop. [1]

$(A,B)$ controllable $\Rightarrow A + BF$ cyclic for a.a. $F: \mathcal{X} \to \mathcal{U}$ .

Theorem (Pole Assignment). [1]

$(A,B)$ controllable iff $\forall \Lambda \subset \mathbb{C}$ , symmetric, $|\Lambda| = n$ ,

$\exists F = F(\Lambda): \quad \mathcal{X} \to \mathcal{U} \ni \sigma(A + BF) = \Lambda$ .

Proof

(Only if) $(A,B)$ controllable $\Rightarrow \exists F_0: \mathcal{X} \to \mathcal{U}$ and $b = Bu \ni$

$(A + BF_0, b)$ controllable. Then as shown before, $\exists f' \ni$

$\sigma(A + BF_0 + bf') = \Lambda$ . Set $F = F_0 + uf'$ .

Prop.

Almost all pairs $(A,B)$ are controllable (i.e., all except "points" $(A,B)$ lying on a proper variety in $R^{n^2+nm}$) .

Proof

$(A,B)$ controllable $\Leftrightarrow$ Rank $[B, AB, \ldots, A^{n-1}B] = n$ .

Such a property (relative to a suitable $R^N$ of parameter points) will be called generic.

Let $\mathbb{C}_g \cup \mathbb{C}_b$ be a partition of the complex plane $\mathbb{C}$ , with $\mathbb{C}_g$ symmetric about the real axis, and $\mathbb{C}_g \cap \mathbb{R} \neq \emptyset$ . Let

$\alpha(\lambda)$ = min. poly. of $A$

$= \alpha_g(\lambda)\alpha_b(\lambda)$

where roots of $\alpha_g(\alpha_b) \in \mathbb{C}_g(\mathbb{C}_b)$ . Let $\mathcal{X}_g(A) = \text{Ker}\, \alpha_g(A)$ , $\mathcal{X}_b(A) = \text{Ker}\, \alpha_b(A)$ . Then $\mathcal{X} = \mathcal{X}_g(A) \oplus \mathcal{X}_b(A)$ .

Prop. [1]

$\exists F: \quad \mathcal{X} \to \mathcal{U} \ni \sigma(A + BF) \subset \mathbb{C}_g \quad \text{iff} \quad \mathcal{X}_b(A) \subset \langle A|\mathcal{B}\rangle$ .

In particular if $\mathbb{C}_g = \mathbb{C}^- \triangleq \{s: \text{Re } s < 0\}$ , we write $\mathcal{X}_g(A) = \mathcal{X}^-(A)$, $\mathcal{X}_b(A) = \mathcal{X}^+(A)$ .

Definition

$(A,B)$ is stabilizable if $\exists F: \quad \mathcal{X} \to \mathcal{U} \ni A + BF$ is stable, i.e., $\sigma(A + BF) \subset \mathbb{C}^-$ . Thus $(A,B)$ stabilizable iff $\mathcal{X}^+(A) \subset \langle A|\mathcal{B}\rangle$ . Finally

$$(\forall F)\langle A + BF|\mathcal{B}\rangle = \langle A|\mathcal{B}\rangle \ .$$

If $\bar{\mathcal{X}} = \mathcal{X}/\langle A|\mathcal{B}\rangle$ then $\overline{A + BF} = \bar{A}$ for all $F$ , and

$$\sigma(A + BF) = \underbrace{\sigma(A + BF|\langle A|\mathcal{B}\rangle)}_{\substack{\text{freely assignable}\\ \text{by choice of } F}} \mathbin{\dot\cup} \underbrace{\sigma(\bar{A})}_{\text{fixed}}$$

(A,B)-invariant subspaces.

Definition.

$\mathcal{V} \subset \mathcal{X}$ is (A,B)-invariant if $\exists F:\mathcal{X} \to \mathcal{U} \ni (A + BF)\mathcal{V} \subset \mathcal{V}$ .

Prop. [6]

$\mathcal{V}$ is (A,B)-inv. iff $A\mathcal{V} \subset \mathcal{B} + \mathcal{V}$ .

Notation

$\mathcal{V} \in \underline{\mathcal{I}} = \underline{\mathcal{I}}(A,B)$ .

Prop. [6]

$\mathcal{V}_1, \mathcal{V}_2 \in \underline{\mathcal{I}} \Rightarrow \mathcal{V}_1 + \mathcal{V}_2 \in \underline{\mathcal{I}}$ [ $\mathcal{V}_1, \mathcal{V}_2 \ \underline{\mathcal{I}} \not\Rightarrow \mathcal{V}_1 \cap \mathcal{V}_2 \in \underline{\mathcal{I}}$ ]

Prop. [6]

Given $\mathcal{K} \subset \mathcal{X}$ arbitrary, $\exists \mathcal{V}^* \subset \mathcal{K}$, $\mathcal{V}^* \in \underline{\mathcal{I}}$ such that

$$(\forall \mathcal{V})\mathcal{V} \subset \mathcal{K} \text{ and } \mathcal{V} \in \underline{\mathcal{I}} \Rightarrow \mathcal{V} \subset \mathcal{V}^* \ .$$

<u>Notation</u> $\mathcal{V}^* = \sup\{ \mathcal{V} : \mathcal{V} \subset \mathcal{K}, \ \mathcal{V} \in \underline{I} \}$

<u>Notation</u> $A^{-1}\mathcal{K} = \{x : Ax \in \mathcal{K}\}$ .

So: $\mathcal{V}^* = \sup\{ \mathcal{V} : \mathcal{V} \subset \mathcal{K} \cap A^{-1}(\mathcal{V} + \mathcal{B})\}$

<u>Proof</u>

$$\mathcal{V}^* = \lim \mathcal{V}^{\mu} \text{ , where}$$

$$\mathcal{V}^{\mu+1} = \mathcal{K} \cap A^{-1}(\mathcal{V}^{\mu} + \mathcal{B}) , \qquad \mu = 0,1,\ldots .$$

$$\mathcal{V}^0 = \mathcal{K} .$$

[Note: $\mathcal{V}^{\mu}\downarrow$ ] .

<u>Disturbance rejection I</u>.

$$\dot{x} = Ax + Bu + Eq$$

$$z = Dx$$

$q(\cdot)$ = disturbance, unmeasurable

Problem: Find (if possible) $u = Fx$ such that $z(\cdot)$ is decoupled from $q(\cdot)$ ,

i.e., $D(sI - A - BF)^{-1}E \equiv 0$

i.e., $D(A + BF)^{i-1}E = 0, \ i \in \underline{n}$

i.e., $D\langle A + BF|\mathcal{E}\rangle = 0$

i.e., $\langle A + BF|\mathcal{E}\rangle \subset \text{Ker } D$

<u>DRP I</u>.

Given $A: \mathcal{X} \to \mathcal{X}$ , $B: \mathcal{U} \to \mathcal{X}$ , $D: \mathcal{X} \to \mathcal{Z}$ and $\mathcal{E} \subset \mathcal{X}$ , find $F: \mathcal{X} \to \mathcal{U}$ $\langle A + BF|\mathcal{E}\rangle \subset \text{Ker } D$ .

Let $\mathcal{V}^* = \sup\{ \mathcal{V} : \mathcal{V} \subset \text{Ker } D \cap A^{-1}(\mathcal{V} + \mathcal{B})\}$ .

<u>Theorem</u>. [6]

DRP I is solvable iff $\mathcal{V}^* \supset \mathcal{E}$ .

Proof

(Only if) Let $\mathcal{V} \triangleq \langle A + BF | \quad \rangle$ . Then $\mathcal{V} \in \underline{I}$ and $\mathcal{V} \subset \text{Ker } D$ and $\mathcal{V} \supset \mathcal{E}$ . So $\mathcal{E} \subset \mathcal{V} \subset \mathcal{V}^*$ . (If) Let $(A + BF)\, \mathcal{V}^* \subset \mathcal{V}^*$ . Then

$$\begin{aligned} \langle A + BF | \mathcal{E} \rangle &\subset (A + BF | \mathcal{V}^*) \\ &= \mathcal{V}^* \\ &\subset \text{Ker } D . \end{aligned}$$

## Controllability Subspaces [6]

Fix $(A,B)$ .

Definition

$\mathcal{R} \subset \mathcal{X}$ is a controllability subspace (c.s.) if $\exists F : \mathcal{X} \to \mathcal{U}$ and $G : \mathcal{U} \to \mathcal{U} \ni \mathcal{R} = \langle A + BF | \text{Im}(BG) \rangle$

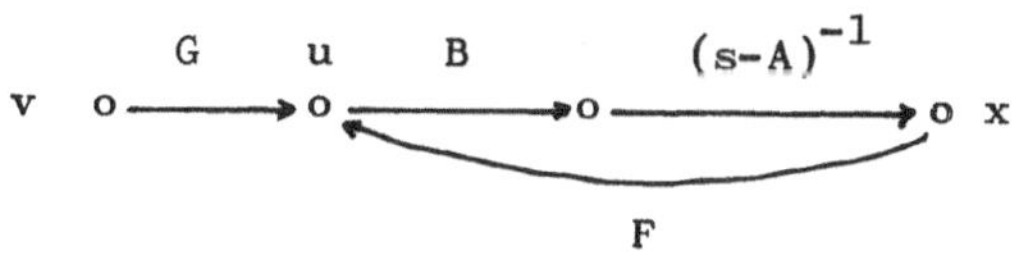


Prop.

$\mathcal{R}$ is a c.s. iff $\exists F : \mathcal{X} \to \mathcal{U} \ni \mathcal{R} = \langle A + BF | \mathcal{B} \cap \mathcal{R} \rangle$

Notation

If $\mathcal{V} \subset \mathcal{X}$ ,

$\underline{F}(\mathcal{V}) \triangleq \{F : (A + BF)\, \mathcal{V} \subset \mathcal{V}\}$ .

$(\underline{F}(\mathcal{V}) \neq \emptyset$ iff $\mathcal{V} \in \underline{I})$ .

Prop.

$\mathcal{R}$ is a c.s. $\Rightarrow \mathcal{R} = \langle A + BF | \mathcal{B} \cap \mathcal{R} \rangle$ , all $F \in \underline{F}(\mathcal{R})$

Prop.

Let $\mathcal{R}$ be a c.s., $d(\mathcal{R}) = \rho$ .

If $\Lambda \subset \mathbb{C}$ symmetric, $|\Lambda| = \rho$, $\exists F \in \underline{F}(\mathcal{R}) \ni \sigma[(A+BF) | \mathcal{R}] = \Lambda$ .

Prop.

Let $\mathcal{R} \subset X$ , $d(\mathcal{R}) = \rho$ . If $\forall \Lambda$ symm. and $|\Lambda| = \rho$ there exists $F \in \underline{F}(\mathcal{R}) \ni \sigma[(A + BF)|\mathcal{R}] = \Lambda$ , then $\mathcal{R}$ is a c.s.

Prop. [7]

Let $\mathcal{R} \subset X$ . $\mathcal{R}$ is a c.s. iff $\forall\, \tilde{x} \in \mathcal{R}\ \exists u(t)$, $0 \leq t \leq 1$ , with the property:

The solution of

$$\dot{x} = Ax + Bu \qquad 0 \leq t \leq 1$$
$$x(0) = 0$$

is such that $x(1) = \tilde{x}$ and $x(t) \in \mathcal{R}$ , $0 \leq t \leq 1$ .

Controllability subspace algorithm.

For $\mathcal{R} \subset X$ let

$$S^{\mu+1} = \mathcal{R} \cap (AS^{\mu} + \mathcal{B}) \qquad \mu = 0,1,\ldots$$
$$S^0 = 0$$

Then $S^{\mu} \uparrow S_* = S_*(\mathcal{R})$ .

Prop. [6]

$\mathcal{R}$ is a c.s. iff

(1) $A\mathcal{R} \subset \mathcal{R} + \mathcal{E}$ (i.e. $\mathcal{R} \in \underline{I}$ )

(2) $\mathcal{R} = S_*(\mathcal{R})$

[One shows $S^{\mu} = \sum_{j=1}^{\mu} (A+BF)^{j-1}(\mathcal{B} \cap \mathcal{R})$ for $F \in \underline{F}(\mathcal{R})$]

Prop. [6]

$\mathcal{R}_1$ and $\mathcal{R}_2$ c.s. => $\mathcal{R}_1 + \mathcal{R}_2$ is a c.s.

Prop. [6]

For any $K \subset X$ there exists

$$\mathcal{R}^* = \mathcal{R}^*(K) = \sup\{\mathcal{R} : \mathcal{R} \text{ a c.s. } \subset K\} .$$

Notation

$\underline{C}(\mathcal{K})$ = set of c.s. contained in $\mathcal{K}$ .

Prop.

Let $\mathcal{V}^* = \sup\{ \mathcal{V} : \mathcal{V} \subset \mathcal{K} \cap A^{-1}(\mathcal{V} + \mathcal{B})\}$ .

Then $\mathcal{R}^*(\mathcal{K}) = \lim \mathcal{R}^\mu$ , where

$$\mathcal{R}^{\mu+1} = \mathcal{V}^* \cap (A\,\mathcal{R}^\mu + \mathcal{B})$$

$$\mu = 0,1,\ldots$$

$$\mathcal{R}^0 = 0 .$$

Prop.

Fix $\mathcal{K}$ . Let $\mathcal{V}^* = \mathcal{V}^*(\mathcal{K})$, $\mathcal{R}^* = \mathcal{R}^*(\mathcal{K})$ . Then

$$\mathcal{R}^* = \langle A+BF \mid \mathcal{B} \cap \mathcal{V}^* \rangle ,$$

$$\text{all } F \in \underline{F}(\mathcal{V}^*) .$$

Prop.

Let $\mathcal{V}^* \in \underline{I}$ , $\mathcal{R}^* = \sup \underline{C}(\mathcal{V})$ .

Let $F \in \underline{F}(\mathcal{V}^*)$, $A_F = A + BF$, $\bar{A}_F$ the map induced in $\mathcal{V}^*/\mathcal{R}^*$ by $A_F$.

Then $\bar{A}_F$ <u>is the same for all</u> $F \in \underline{F}(\mathcal{V}^*)$ .

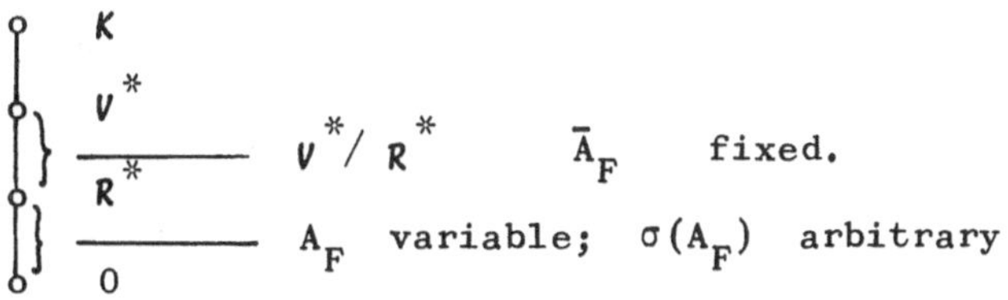


Disturbance rejection problem II [4]

$$\dot{x} = Ax + Bu + Eq$$

$$z = Dx$$

Find $F \ni \langle A+BF | \mathcal{E} \rangle \subset \text{Ker } D$ <u>and</u> $A + BF$ stable.

Assume $(A,B)$ controllable.

<u>Solution</u>

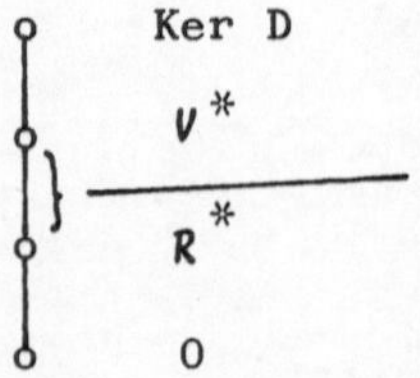


$$\mathcal{V}^*/\mathcal{R}^* = (\mathcal{V}^*/\mathcal{R}^*)^+ \oplus (\mathcal{V}^*/\mathcal{R}^*)^-$$

$$\exists\, \mathcal{V}_g^* \text{ unique}, \ni \mathcal{R}^* \subset \mathcal{V}_g^* \subset \mathcal{V}^*$$

$$\text{and} \quad \mathcal{V}_g^*/\mathcal{R}^* = (\mathcal{V}^*/\mathcal{R}^*)^-$$

<u>Theorem</u> [4]

DRP II is solvable iff $\mathcal{V}_g^* \supset \mathcal{E}$ .

[One shows that

$$\mathcal{V}_g^* = \sup\{\mathcal{V} : \mathcal{V} \subset \text{Ker } D \cap A^{-1}(\mathcal{V} + \mathcal{B}) \text{ and } (A + BF) | \mathcal{V} \text{ stable, some } F \in \underline{F}(\mathcal{V})\} .$$

Finish up by stabilizing the pair $(\widetilde{A+BF}, \tilde{B})$ induced in $\mathcal{X}/\mathcal{V}_g^*$].

<u>Noninteracting Control</u> [6],[7],[9] .

$$\dot{x} = Ax + Bu$$

$$z_i = D_i x , \quad i \in \underline{k} .$$

where

$$A : \mathcal{X} \to \mathcal{X} , \quad B : \mathcal{U} \to \mathcal{X} , \quad D_i : \mathcal{X} \to \mathcal{Z}_i .$$

Allow controls

$$u = Fx + \sum_{i=1}^{k} G_i v_i$$

where

$$F : \mathcal{X} \to \mathcal{U}, \quad G_i : \mathcal{U} \to \mathcal{U} .$$

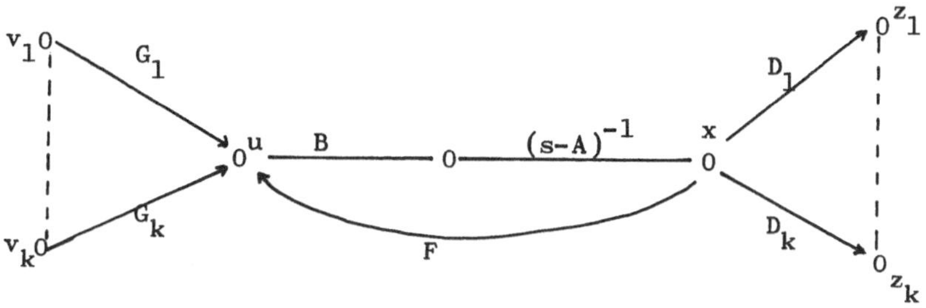


$v_i$ generates the c.s.

$$\mathcal{R}_i = \langle A+BF \mid \mathrm{Im}(BG_i) \rangle$$

Objective:

Choose $F$ and $G_i$, $i \in \underline{k}$, so that $v_i$ controls $z_i$ but doesn't influence $z_j$, $j \neq i$. Equivalently

$$D_i \mathcal{R}_j = 0 , \qquad i, j \in \underline{k} , i \neq j$$

$$D_i \mathcal{R}_i = \mathrm{Im}\, D_i , \quad i \in \underline{k} .$$

<u>Restricted decoupling problem (RDP)</u>:

Given $A$, $B$, $D_i (i \in \underline{k})$ find $F : \mathcal{X} \to \mathcal{U}$ and c.s. $\mathcal{R}_i (i \in \underline{k})$ such that

$$\mathcal{R}_i = \langle A+BF \mid \mathcal{B} \cap \mathcal{R}_i \rangle , \ i \in \underline{k}$$

<u>compatibility</u>

$$\mathcal{R}_i \subset \bigcap_{j \neq i} \mathrm{Ker}\, D_j , \ i \in \underline{k} ,$$ <u>noninteraction</u>

$$\mathcal{R}_i + \mathrm{Ker}\, D_i = \mathcal{X} \quad , \ i \in \underline{k}$$

<u>output controllability</u>

Approach:

Let $\mathcal{R}_i^* = \sup \text{c.s.} \subset \bigcap_{j \neq i} \text{Ker } D_j \quad (i \in \underline{k})$

? $\mathcal{R}_i^* + \text{Ker } S_i = \mathcal{X} \quad (i \in \underline{k})$

If not, RDP unsolvable.

If so,

$$? \exists F: \quad \mathcal{R}_i^* = \langle A + BF \mid \mathcal{B} \cap \mathcal{R}_i^* \rangle \qquad (i \in \underline{k})$$

i.e. $\quad ? \quad \bigcap_i \underline{F}(\mathcal{R}_i^*) \neq \emptyset$

i.e. $\quad ? \exists F : (A + BF) \mathcal{R}_i^* \subset \mathcal{R}_i^* \qquad (i \in \underline{k})$

If so, RDP is solvable and we are done.

If not, the solvability question remains open. The problem must be extended by enlarging the class of controls. To see how, consider a special case of original RDP .

Special case.

Assume $\bigcap_{i=1}^{k} \text{Ker } D_i = 0 \qquad (*)$

Claim: the $\mathcal{R}_i^*$ are independent.

In fact, $\mathcal{R}_i^* \subset \bigcap_{j \neq i} \text{Ker } D_j$

and

$$\left( \bigcap_{j \neq i} \text{Ker } D_j \right) \cap \sum_{i \neq i} \left( \bigcap_{\ell \neq k} \text{Ker } D_\ell \right) = 0 \text{ , by } \quad (*) \text{ .}$$

Next

$$(\forall i)(\exists F_i)\ (A+BF_i)\ \mathcal{R}_i^* \subset \mathcal{R}_i^* \text{ .}$$

Let $$d(\mathcal{R}_i^*) = \rho_i$$

$$\Lambda_i \subset \mathbb{C} \quad , \text{ symm. } \quad |\Lambda_i| = \rho_i \ .$$

Can choose $$F_i = F_i(\Lambda_i)$$

$$\sigma[(A+BF_i)| \ \mathcal{R}_i^*] = \Lambda_i \ , \quad i \in \underline{k}$$

Finally, by independence,

$$\exists \ F : \mathcal{X} \to \mathcal{U} \quad \ni \quad F_i^* = F_i | \ \mathcal{R}_i^* \ , \quad i \in \underline{k} \ .$$

<u>Theorem</u>

Let $(A,B)$ be controllable and $\bigcap_{i=1}^{k} \text{Ker } D_i = 0$ .

Then RDP is solvable iff

$$\mathcal{R}_i^* + \text{Ker } D_i = \mathcal{X} \ , \quad i \in \underline{k} \ .$$

Furthermore, in this case if

$$|\Lambda_i| = d(\ \mathcal{R}_i^*) \ , \quad i \in \underline{k}$$

and $$|\Lambda_o| = n - \sum_{i=1}^{k} d(\ \mathcal{R}_i^*) \ ,$$

then $F \in \bigcap_{i=1}^{k} \underline{F}(\ \mathcal{R}_i^*)$ such that

$$\sigma(A+BF) = \bigcup_{j=0}^{k} \Lambda_j \quad .$$

<u>Extended Decoupling Problem (EDP)</u> [7]

"Add auxiliary dynamics (i.e. extra state space) to split the $\mathcal{R}_i^*$ apart enough to achieve independence".

Crudely:

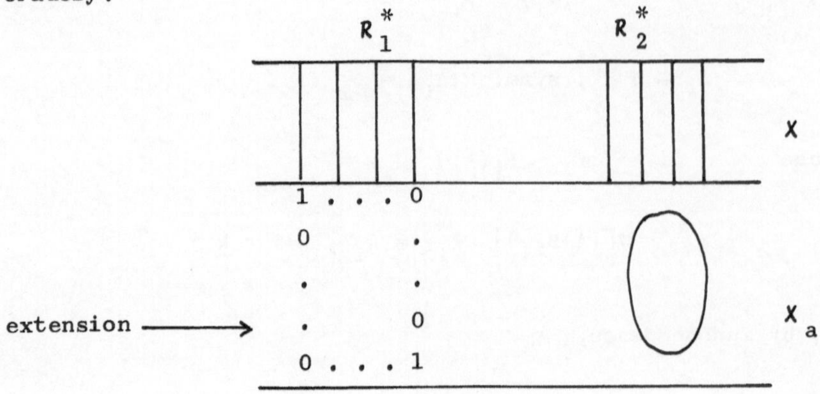


Replace:

$$A \quad \text{by} \quad \begin{bmatrix} A & 0 \\ 0 & 0 \end{bmatrix}$$

$$B \quad \text{by} \quad \begin{bmatrix} B & 0 \\ 0 & B_a \end{bmatrix}$$

$$D_i \quad \text{by} \quad (D_i \;\; 0)$$

$$F \quad \text{by} \quad \begin{bmatrix} F_{11} & F_{12} \\ F_{21} & F_{22} \end{bmatrix}$$

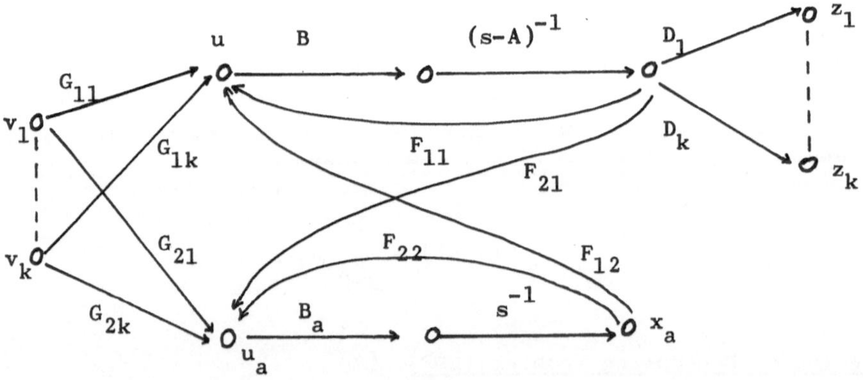

EDP has same formal definition as RDP , but now on $\mathcal{X} \oplus \mathcal{X}_a$ and using extended maps.

Theorem

EDP is solvable iff, for RDP ,

$$\mathcal{R}_i^* + \text{Ker } D_i = \mathcal{X} , \quad i \in \underline{k} .$$

In that case, one can take

$$d(\mathcal{X}_a) = \sum_{i=1}^{k} d(\mathcal{R}_i^*) - d(\sum_{i=1}^{k} \mathcal{R}_i^*) .$$

Freedom of pole assignment, as before.

Generic solvability of EDP [10]

$$A : n \times n , \quad B : n \times m , \quad D_i : q_i \times n .$$

$$N = n^2 + nm + \sum_{i=1}^{k} q_i n .$$

Assume: $1 \leq m \leq n$ , $1 \leq q_i \leq n$ .

Theorem

EDP is solvable for a.a. data sets $(A,B,D_1,\ldots,D_k)$ in $R^N$ , iff

1) $$\sum_{i=1}^{k} q_i \leq n$$

2) $$m \geq 1 + \sum_{i=1}^{k} q_i - \min_{1 \leq i \leq k} q_i .$$

In that case one can (a.a.) take

$$d(\mathcal{X}_a) = \begin{cases} 0 & \text{if} \quad \sum_i q_i = n \\ (k-1)(n - \sum_{i=1}^{k} q_i) & \text{if} \quad \sum_i q_i < n \end{cases} .$$

E.g.

If $n = 15,\ k = 2,\ q_1 = 3,\ q_2 = 4$

then $m \geq 5$ and $d(\mathcal{X}_a) = 8$ .

For applications to adaptive control and sensitivity see [11].

## Tracking and Regulation [12]-[15]

$\dot{x} = Ax + Bu$

$y = Cx$ (measured)

$z = Dx$ (to be regulated).

Coordinatization:

Let $\mathcal{X}_1 = \langle A \mid \mathcal{B} \rangle$

$$\mathcal{X} = \mathcal{X}_1 \oplus \mathcal{X}_2$$

$$\begin{bmatrix} \dot{x}_1 \\ \dot{x}_2 \end{bmatrix} = \begin{bmatrix} A_1 & A_3 \\ 0 & A_2 \end{bmatrix} \begin{bmatrix} x_1 \\ x_2 \end{bmatrix} + \begin{bmatrix} B_1 \\ 0 \end{bmatrix} u$$

$$y = Cx_1 + C_2x_2$$

$$z = D_1x_1 + D_2x_2$$

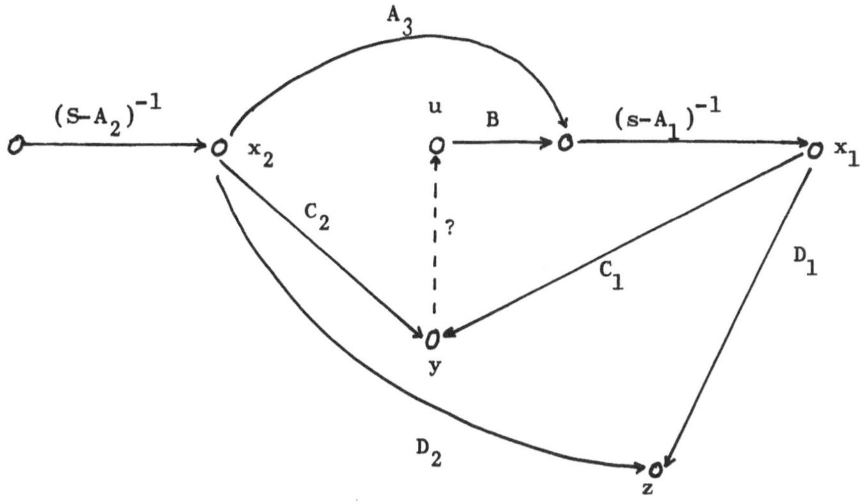


Let $N = \bigcap_{i=1}^{n} \mathrm{Ker}(CA^{i-1}) \quad (\subset X)$

$N$ is the <u>unobservable subspace</u> of $(C,A)$ .

| | | | |
|---|---|---|---|
| $u(\cdot)$ | Dynamic | | |
| | | $\bar{x}(\cdot)$ | (See [3]) |
| $y(\cdot)$ | observer | | |

$$\bar{x}(t) = x(t) \bmod N$$
$$\in X / N$$

$X/N$ is the observable (factor) space.

Controls: $u = Fx$

<u>Observability constraint</u>: $\mathrm{Ker}\ F \supset N$ . (For discussion see [13])

Objective:

1) Output regulation:

$$z(t) \longrightarrow 0 \ (t \longrightarrow \infty) \quad \text{for all} \quad x(0) .$$

i.e. $\mathcal{X}^+(A+BF) \subset \operatorname{Ker} D$ .

2) Internal stability: i.e. stability of controllable, observable "subsystem"; i.e.

$$\frac{\mathcal{X}^+(A+BF) + \mathcal{N}}{\mathcal{N}} \cap \frac{\langle A|\mathcal{B}\rangle + \mathcal{N}}{\mathcal{N}} = 0$$

i.e.

$$\mathcal{X}^+(A+BF) \cap (\langle A|\mathcal{B}\rangle + \mathcal{N}) \subset \mathcal{N} .$$

Regulator problem with internal stability (RPIS) [14]

Given $A: \mathcal{X} \to \mathcal{X}$ , $B: \mathcal{U} \to \mathcal{X}$ , $D: \mathcal{X} \to \mathcal{Z}$ and $\mathcal{N} \subset \mathcal{X}$ with $A\mathcal{N} \subset \mathcal{N}$ , find $F: \mathcal{X} \to \mathcal{U}$ such that

$$\operatorname{Ker} F \supset \mathcal{N}$$

$$\mathcal{X}^+(A+BF) \cap (\langle A|\mathcal{B}\rangle + \mathcal{N}) \subset \mathcal{N}$$

and

$$\mathcal{X}^+(A+BF) \subset \operatorname{Ker} D$$

Theorem 1 [14]

RPIS is solvable iff $\exists$ a subspace $\mathcal{V} \subset \mathcal{X}$ such that

$$\mathcal{V} \subset \operatorname{Ker} D \cap A^{-1}(\mathcal{V} + \mathcal{B})$$

$$\mathcal{X}^+(A) \cap \mathcal{N} + A(\mathcal{V} \cap \mathcal{N}) \subset \mathcal{V}$$

$$\mathcal{V} \cap (\langle A | \mathcal{B} \rangle + \mathcal{N}) \subset \mathcal{V}$$

and

$$\mathcal{X}^+(A) \subset \langle A | \mathcal{B} \rangle + \mathcal{V}$$

Pf. (only if) Identify $\mathcal{V} = \mathcal{X}^+(A+BF)$ .

Theorem 1 is nonconstructive, but can be used to show that dynamic control (beyond ordinary state feedback) won't help. Formally, introduce extended state space, and maps, much as in the decoupling problem. Define "extended regulator problem with internal stability" (ERPIS) in the same way.

Theorem 2 [14]

ERPIS solvable => RPIS solvable.

Pf. Project the extended version of the conditions in Theorem 1.

Constructive solution when $\mathcal{N} = 0$.

For $\mathcal{N} = 0$, Theorem 1 says RPIS is solvable iff $\exists\ \mathcal{V} \subset \mathcal{X} \ni$

$$\mathcal{V} \subset \text{Ker } D \cap A^{-1}(\mathcal{V} + \mathcal{B})$$

$$\mathcal{V} \cap \langle A | \mathcal{B} \rangle = 0$$

$$\mathcal{X}^+(A) \subset \langle A | \mathcal{B} \rangle + \mathcal{V}$$

So we try to construct a $\mathcal{V}$ .

Defn.

Let $A\mathcal{Z} \subset \mathcal{Z}$ and $A\mathcal{R} \subset \mathcal{R} \subset \mathcal{Z}$. $\mathcal{R}$ decomposes $\mathcal{Z}$ rel. $A$ if $\exists\ \mathcal{S} \ni A\mathcal{S} \subset \mathcal{S}$ and $\mathcal{R} \oplus \mathcal{S} = \mathcal{Z}$ .

Let $\mathcal{V}^* = \sup(A,B)$ - inv. in Ker D

$\mathcal{R}^* = \sup$ c.s. in Ker D .

Theorem 3 [14]

Let $\mathcal{N} = 0$ . RPIS is solvable iff

(1) $\mathcal{X}^+(A) \subset \langle A \mid \mathcal{B} \rangle + \mathcal{V}^*$

and

(2) With $F \in \underline{F}(\mathcal{V}^*)$, $A_F \triangleq A + BF$ , the (factor) subspace

$$\frac{\mathcal{V}^* \cap \mathcal{X}^+(A_F) \cap \langle A \mid \mathcal{B} \rangle + \mathcal{R}^*}{\mathcal{R}^*}$$

decomposes the subspace

$$\frac{\mathcal{V}^* \cap \mathcal{X}^+(A_F) + \mathcal{R}^*}{\mathcal{R}^*}$$

relative to the map $\bar{A}_F$ induced by $A_F$ in $\mathcal{V}^*/\mathcal{R}^*$ .

Pf. (If)

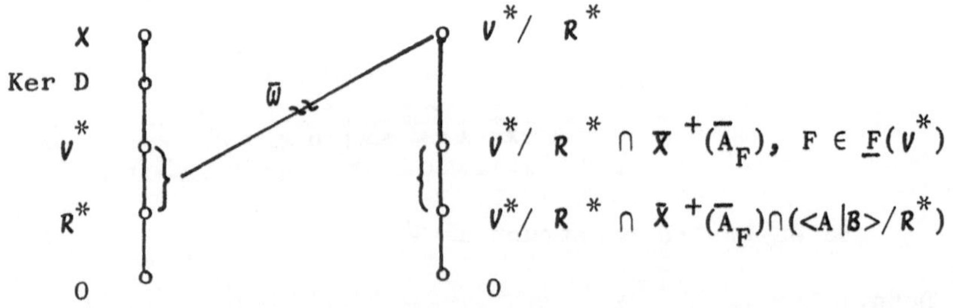


$$\exists \mathcal{W} : \mathcal{R}^* \subset \mathcal{W} \subset \mathcal{V}^* \text{ and } \mathcal{W}/\mathcal{R}^* = \bar{\mathcal{W}} .$$

Choose $A_F \ni$

$$\sigma(A_F | \mathcal{R}^*) \cap \sigma(\bar{A}_F | \mathcal{V}^*/\mathcal{R}^*) = \emptyset .$$

Then $\exists\, \mathcal{V} : A_F \mathcal{V} \subset \mathcal{V}$ and $\mathcal{R}^* \oplus \mathcal{V} = \mathcal{W}$ .

This $\mathcal{V}$ will do.

Example:

Regulation in the presence of step disturbances.

$$\dot{x}_1 = A_1 x_1 + A_3 x_2 + B_1 u$$

$$\dot{x}_2 = 0$$

$$y = (x_1, x_2)$$

$$z = D_1 x_1 + D_2 x_2$$

$(A_1, B_1)$ is controllable.

i.e. $\mathcal{N} = 0$

$$\text{Im } A \subset \langle A | \mathcal{B} \rangle$$

Theorem 4 [14]

RPIS is solvable iff

$$\langle A | \mathcal{B} \rangle + \text{Ker } D \cap A^{-1}\mathcal{B} = \mathcal{X} .$$

Pf.

(If) Instead of $\mathcal{V}^*$ take $\tilde{\mathcal{V}} = \text{Ker } D \cap A^{-1}\mathcal{B}$ .

Then $\mathcal{X}^+(A) \subset \mathcal{X} = \langle A | \mathcal{B} \rangle + \tilde{\mathcal{V}}$ .

Since $A\tilde{\mathcal{V}} \subset \mathcal{B}$ , $\exists F \in \underline{F}(\tilde{\mathcal{V}}) \ni A_F \tilde{\mathcal{V}} = 0$ . Then

$$\tilde{\mathcal{V}} \subset \text{Ker } A_F \subset \mathcal{X}^+(A_F) .$$

So

$$\tilde{V} \cap X^{+}(A_F) \cap \langle A \mid B \rangle = \tilde{V} \cap \langle A \mid B \rangle .$$

Let $\tilde{R}$ = sup c.s. in $\tilde{V}$ . Then

$$\tilde{R} = \langle A_F \mid B \cap \tilde{V} \rangle$$

$$= B \cap \tilde{V} .$$

Does

$$\frac{\tilde{V} \cap \langle A \mid B \rangle}{B \cap \tilde{V}} \quad \text{decompose} \quad \frac{\tilde{V}}{B \cap \tilde{V}} ,$$

relative to $\bar{A}_F$ induced by $A_F$ in $\tilde{V}/(B \cap \tilde{V})$ ?

Yes, trivially, since $A_F \mid \tilde{V} = 0$ .

<u>Extension to the case $N \neq 0$</u>

<u>Theorem 5</u> [14]

RPIS is solvable iff

1) $X^{+}(A) \cap N \subset \operatorname{Ker} D$

and

2) The "reduced problem" defined in $X / X^{+}(A) \cap N$ is solvable.

[After reduction, check condn. 2 using Thm. 3].

Regulation and Tracking - Synthesis [15], [16].

Naive application of RPIS theory may produce engineering nonsense.

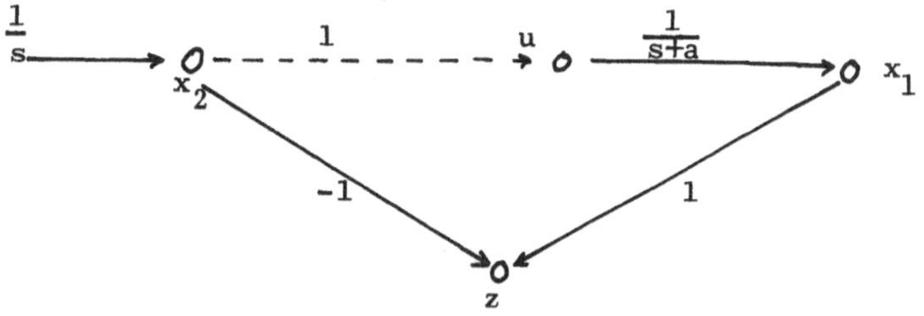


$$\dot{x}_1 = -ax_1 + u$$

$$\dot{x}_2 = 0$$

$$z = x_1 - x_2 \quad \text{(tracking error)}$$

$$y = (x_1, x_2) .$$

Assume $a = 1$ .

A solution of RPIS is $u = x_2$

$$\dot{z} = -z$$

$$z(t) \to 0 \qquad (t \to \infty) .$$

But if $a = 1 + \varepsilon$ , our control gives

$$\dot{z} = -(1 + \varepsilon)z - \varepsilon x_2 , \; t > 0$$

$$z(t) \longrightarrow -\frac{\varepsilon}{1+\varepsilon} x_2(0+) .$$

"offset error"

Moral: Naive implementation may lead to loss of regulation under small parameter changes on plant.

"Correct synthesis:

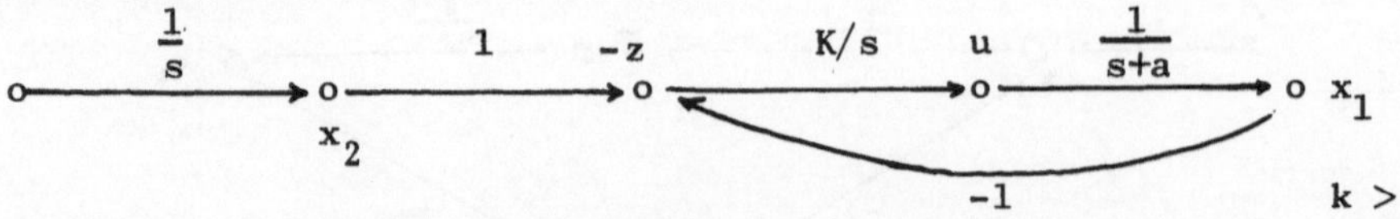


$k > 0$

$$\hat{z}(s) = \frac{s(s+a)}{s^2 + as + K} \hat{x}_2(s)$$

$a \approx 1$

$z(t) \longrightarrow 0$ for all $K > 0$ and $a$ near $1$.

We could arrive at a correct synthesis by redefining $y$:

$$\dot{y} = z .$$

The observer will then supply the loop integration $1/s$ needed for parametric insensitivity.

Well-posed RPIS

Assume:

1) $\bar{A}$ induced in $\mathcal{X}/\langle A|\mathcal{B}\rangle$ is totally unstable: $\sigma(\bar{A}) \subset \mathbb{C}^+$

2) $(C,A)$ is "detectable", i.e.

$$\mathcal{X}^+(A) \cap \mathcal{N}_C = 0 . \qquad \text{(See [17])}$$

3) $D\langle A|\mathcal{B}\rangle = \mathcal{Z}$.

Coordinatize:

$$\mathcal{X}_1 = \langle A \mid \mathcal{B} \rangle , \quad \mathcal{X} = \mathcal{X}_1 \oplus \mathcal{X}_2$$

$$A = \begin{bmatrix} A_1 & A_3 \\ 0 & A_2 \end{bmatrix} , \quad B = \begin{bmatrix} B_1 \\ 0 \end{bmatrix}$$

$$C = (C_1 \quad C_2) , \quad D = (D_1 \quad D_2)$$

$$d(\mathcal{X}_1) = n_1 , \quad d(\mathcal{X}_2) = n_2 , \quad d(Z) = q .$$

Regard $A_2$ , C, D as fixed.

Regard $\underline{p} = (A_1, A_3, B_1)$ as a variable data point in $R^N$ ,

$N = n_1^2 + n_1 n_2 + n_1 m$ .

<u>Defn.</u>

RPIS is <u>well-posed</u> at $\underline{p}$ if it is solvable at all data points in some (open) nbhd of $\underline{p}$ in $R^N$ .

<u>Theorem</u> [15] [18]

Subject to assumptions (1) - (3), RPIS is well-posed at $\underline{p}$ iff, over $\mathbb{C}$ ,

$$(A-\lambda)(\text{Ker } D \cap \langle A \mid \mathcal{B} \rangle) + \mathcal{B} = \langle A \mid \mathcal{B} \rangle$$

for all $\lambda \in \sigma(\bar{A})$ .

{Equivalently

$$\text{Rank} \begin{bmatrix} A_1-\lambda I & B_1 \\ D_1 & 0 \end{bmatrix} = n_1 + q$$

for all $\lambda \in \sigma(A_2)$}

<u>Corollary</u> [15][18]

RPIS is generically solvable iff $m \geq q$ . If $m < q$ , no data point is well-posed.

<u>"Theorem"</u> (For a precise statement see [15]

Under mild <u>a priori</u> assumptions, RPIS admits a "correct" (i.e. parametrically insensitive synthesis) iff RPIS is well-posed.

<u>Remark</u>. This Theorem leads to the "internal model principle", according to which parametrically insensitive regulation is possible <u>only if</u> feedback is utilized, and the feedback loop incorporates a fixed, reduplicated model of the dynamic structure of the exogenous signals which the system is required to process. For details see [15] and [16].

# BIBLIOGRAPHY

[1] W.M. Wonham, On pole assignment in multi-input controllable linear systems, I.E.E.E. Trans. Aut. Control AC-12(6), 1967, pp. 660-665.

[2] W.M. Wonham, On a matrix Riccati equation of stochastic control, SIAM J. Control 6(4), 1968, pp. 681-697.

[3] W.M. Wonham, Dynamic observers: geometric theory, I.E.E.E. Trans. Aut. Control AC-15(2), 1970, pp. 258-259.

[4] W.M. Wonham, Algebraic methods in linear multivariable control, in System Structure (ed. A.S. Morse), Control Systems Society, I.E.E.E. Catalog No. 71C61-CSS, August, 1971.

[5] W.M. Wonham, and A.S. Morse, Feedback invariants of linear multivariable systems, Automatica 8, 1972, pp. 93-100.

[6] W.M. Wonham and A.S. Morse, Decoupling and pole assignment in linear multivariable systems: a geometric approach, SIAM J. Control 8(1), 1970, pp. 1-18.

[7] A.S. Morse and W.M. Wonham, Decoupling and pole assignment by dynamic compensation, SIAM J. Control 8(3), 1970, pp. 317-337.

[8] A.S. Morse and W.M. Wonham, Triangular decoupling of linear multivariable systems, I.E.E.E. Trans. Aut. Control, AC-15(4), 1970, pp. 447-449.

[9] A.S. Morse and W.M. Wonham, Status of noninteracting control, I.E.E.E. Trans. Aut. Control AC-16(6), 1971, pp. 568-580.

[10] E. Fabian and W.M. Wonham, Generic solvability of the decoupling problem, Control System Report No. 7301, Department of Electrical Engineering, University of Toronto, January, 1973; to appear, SIAM J. Control, 1974.

[11] E. Fabian and W.M. Wonham, Decoupling, distrubance rejection and sensitivity, Control System Report No. 7309, Department of Electrical Engineering, University of Toronto, June, 1973; submitted for publication, I.E.E.E. Trans. Aut. Control.

[12] S.P. Bhattacharyya, J.B. Pearson and W.M. Wonham, On zeroing the ouput of a linear system, Information and Control, 20(2), 1972, pp. 135-142.

[13] W.M. Wonham, Tracking and regulation in linear multivariable systems, Control Systems Report No. 7202 (revised), Department of Electrical Engineering, University of Toronto, May, 1972; SIAM J. Control 11(3), 1973, to appear.

[14] W.M. Wonham and J.B. Pearson, Regulation and internal stabilization in linear multivariable systems, Control System Report No. 7212, Department of Electrical Engineering, University of Toronto, August, 1972; SIAM J. Control 12(1), 1974, to appear.

[15] B. Francis, O.A. Sebakhy and W.M. Wonham, Synthesis of multivariable regulators, Proc. Eleventh Annual Allerton Conference on Circuit and System Theory, University of Illinois, October, 1973, to appear.

[16] B. Francis, O.A. Sebakhy and W.M. Wonham, "Synthesis of Multivariable Regulators: The Internal Model Principle", Int. J. Appl. Maths. and Optimization 1(1), 1974, to appear.

# LECTURERS and PARTICIPANTS

Adler, Eric L., Associate Professor, Department of E.E., McGill University.

Adler, Lee S., Assistant Professor, Department of Mathematics, Sir George Williams University.

Alagheband, M. Ali, Graduate Student, Department of Mathematics, University of Utah.

Audley, David R., Research Mathematician, Aerospace Research Lab., Wright-Patterson AFB.

Banks, H. Thomas, Assoc. Professor, Division of Applied Maths., Brown University.

Berman, Ari, Senior Lecturer, Dept. of Mathematics, Institute of Tech., Haifa, Israel.

Bishop, E. Robert, Assoc. Professor, Dept. of Mathematics, Acadia University.

Bodkin, Ronald G., Manager, CANDIDE Project, Economic Council of Canada.

Boyarsky, Abraham, Assist. Professor, Dept. of Mathematics, Sir George Williams University.

Buoncristiani, Martin, Assist. Professor, Dept. of Mathematics, Ohio State University.

Butz, Edward, Post-Doctoral Fellow, Dept. of Math.,University of Alberta.

Bryan, Robert N., Assoc. Professor, Dept. of Mathematics, University of Western Ontario.

Clark, Colin, Professor, Dept. of Mathematics, University of British Columbia.

Clarke,Frank H., Research Assist., Dept. of Math., University of Washington.

Crawford, William S.H., Professor, Dept. of Mathematics, Mount Allison University.

Davis, Jon H., Assist. Professor, Dept. of Mathematics, Queen's University.

Lecturers and Participants Cont'd:

Delfour, Michel, Attache de Recherche, Centre de Recherches, Math. Univ. de Montreal.

Dobell, Ronald, Professor, Inst. for Policy Analysis, Univ. of Toronto.

Faulkner, Frank, Professor, Dept. of Mathematics, U.S. Naval Postgrad. School.

Forster, Bruce A., Assist. Professor, Dept. of Economics, Univ. of Guelph.

Gaerhart, William B., Assoc. Professor, Division of Math. and Systems Design, Univ. of Texas.

Gregory, David A., Assist. Professor, Dept. of Mathematics, Queen's University.

Gruyaert, Frans R., Graduate Student, Dept. of Chem. Eng., McMaster University.

Halkin, H., Professor of Math., Univ. of California.

Hall, Richard L., Assoc. Professor, Dept. of Mathematics, Sir George Williams University.

Haussmann, Ulrich, Assist. Professor, Dept. of Mathematics, University of British Columbia.

Heidel, John W., Assoc. Professor, Dept. of Mathematics, University of Tennessee.

Hill, David R., Senior Teach. Fellow, Dept. of Mathematics, Univ. of Pittsburgh.

Hoffman, William C., Professor, Dept. of Mathematics, Oakland U.

Hsiang, Thomas, Consult.-Mathematical Stats., Bell Canada.

Hsu, Bernadette, Graduate Student, Dept. of Mathematics, Case Western Reserve Univ.

Hughes, Edward, Assist. Professor, Dept. of Mathematics, Carleton Univ.

Hum, Derek, Assist. Professor, Dept. of Economics, Univ. of Manitoba.

Lecturers and Participants Cont'd:

Jacobs, Marc Q., Assoc. Professor, Dept. of Mathematics, Univ. of Missouri.

Jakubow, Roman, Graduate Student, Dept. of E.E., Queen's Univ.

Jurdjevic, Velimir, Assist. Professor, Dept. of Mathematics, Univ. of Toronto.

Kirby, Bruce J., Professor, Dept. of Mathematics, Queen's Univ.

Laub, Alan J., Research Assist., Control Science Dept., Univ. of Minnesota.

Lee, Richard, Assoc. Professor, Dept. of Mathematics, Univ. of New Brunswick.

Lions, Jacques L., Prof. d'Analyse Numérique a l'Ecole, Polytechnique, Paris.

Macchia, Roberto, Graduate Student, Dept. of Mathematics, Stevens Inst. of Tech.

Malik, M.A., Assoc. Professor, Dept. of Mathematics, Sir George Williams Univ.

Mallet-Paret, John, Ph.D. Candidate, School of Math., Univ. of Minnesota.

Manitius, Andrzej, Visiting Prof., Dept. of Computer Information, Univ. of Minnesota.

May, Sherry, Graduate Student, Dept. of Applied Math., Univ. of Waterloo.

McCalla, Clement, Assist. Professor, Dept. of Mathematics, Mass. Inst. of Technology.

McCann, Roger, Assist. Professor, Dept. of Mathematics, Case West. Reserve Univ.

McNamee, John, Executive Director, Canadian Math. Congress.

Moore, Bruce, Assist. Professor, Dept. of Computer Science, Louisiana State Univ.

Mukherjee, Swapan, Graduate Student, Dept. of E.E., McMaster Univ.

Nicolaou, Costas, Assist. Professor, Dept. of Economics, Lakehead U.

Lecturers and Participants Cont'd:

Norman, R. Daniel, Assoc. Professor, Dept. of Mathematics, Queen's University.

O'Malley, Robert E., Professor, Dept. of Mathematics, Univ. of Arizona.

Quinn, John P., Assist. Professor, Dept. of Math., Univ. of Toronto.

Rasmy, Mohamed, Graduate Student, Dept. of E.E., Univ. of Calgary.

Rebhuhn, Deborah, Graduate Student, Dept. of Mathematics, Univ. of Illinois.

Renner, Richard C., TELESAT CANADA, Ottawa.

Ricciardi, Luigi M., Assist. Professor, Dept. of Theoretical Biology, Univ. of Chicago.

Ritcey, Lee, Professor, Dept. of Math., U. of Western Ontario.

Ritchie, Michael, Graduate Student, Dept. of Mathematics, Acadia University.

Sagan, Hans, Professor, Dept. of Mathematics, North Carolina State University.

Sen, Abhijit, Graduate Student, Dept. of E.E., McMaster Univ.

Sethi, Suresh, Assist. Professor, Dept. of Management Studies, Univ. of Toronto.

Showalter, Ralph E., Assoc. Professor, Dept. of Mathematics, Univ. of Texas.

Stenger, Frank, Centre de Recherches, Math., Univ. of Montreal, and Math. Dept., Univ. of Utah.

Svoboda, R., Assistant Professor, Division of Math. Sciences, Purdue Univ.

Talman, James D., Professor, Dept. of Applied Math., University of Western Ontario.

Thrall, Robert M., Chairman, Dept. of Mathematical Sciences, Rice Univ.

Triggiani, R., Instructor, School of Math., Univ. of Minnesota.

Lecturers and Participants Cont'd:

Waltman, Paul, Professor, Dept. of Mathematics, University of Iowa.

Wang, Yuan Chia, Teach. Assist., Dept. of Mathematics, Univ. of Wisconsin.

Wong, Man Wah, Undergraduate, Dept. of Mathematics, Sir George Williams University.

Wonham, W. Murray, Professor, Dept. of Electrical Engineering, Univ. of Toronto.

Yeung, D.S., Graduate Student, Dept. of Math. and Stats., Case Western Reserve University.

Zecca, A.R., Research Engineer, Process Control, Armco Steel Corp.

Vol. 59: J. A. Hanson, Growth in Open Economics. IV, 127 pages. 1971. DM 16,-

Vol. 60: H. Hauptmann, Schätz- und Kontrolltheorie in stetigen dynamischen Wirtschaftsmodellen. V, 104 Seiten. 1971. DM 16,-

Vol. 61: K. H. F. Meyer, Wartesysteme mit variabler Bearbeitungsrate. VII, 314 Seiten. 1971. DM 24,-

Vol. 62: W. Krelle u. G. Gabisch unter Mitarbeit von J. Burgermeister, Wachstumstheorie. VII, 223 Seiten. 1972. DM 20,-

Vol. 63: J. Kohlas, Monte Carlo Simulation im Operations Research. VI, 162 Seiten. 1972. DM 16,-

Vol. 64: P. Gessner u. K. Spremann, Optimierung in Funktionenräumen. IV, 120 Seiten. 1972. DM 16,-

Vol. 65: W. Everling, Exercises in Computer Systems Analysis. VIII, 184 pages. 1972. DM 18,-

Vol. 66: F. Bauer, P. Garabedian and D. Korn, Supercritical Wing Sections. V, 211 pages. 1972. DM 20,-

Vol. 67: I. V. Girsanov, Lectures on Mathematical Theory of Extremum Problems. V, 136 pages. 1972. DM 16,-

Vol. 68: J. Loeckx, Computability and Decidability. An Introduction for Students of Computer Science. VI, 76 pages. 1972. DM 16,-

Vol. 69: S. Ashour, Sequencing Theory. V, 133 pages. 1972. DM 16,-

Vol. 70: J. P. Brown, The Economic Effects of Floods. Investigations of a Stochastic Model of Rational Investment Behavior in the Face of Floods. V, 87 pages. 1972. DM 16,-

Vol. 71: R. Henn und O. Opitz, Konsum- und Produktionstheorie II. V, 134 Seiten. 1972. DM 16,-

Vol. 72: T. P. Bagchi and J. G. C. Templeton, Numerical Methods in Markov Chains and Bulk Queues. XI, 89 pages. 1972. DM 16,-

Vol. 73: H. Kiendl, Suboptimale Regler mit abschnittweise linearer Struktur. VI, 146 Seiten. 1972. DM 16,-

Vol. 74: F. Pokropp, Aggregation von Produktionsfunktionen. VI, 107 Seiten. 1972. DM 16,-

Vol. 75: GI-Gesellschaft für Informatik e.V. Bericht Nr. 3. 1. Fachtagung über Programmiersprachen · München, 9-11, März 1971. Herausgegeben im Auftrag der Gesellschaft für Informatik von H. Langmaack und M. Paul. VII, 280 Seiten. 1972. DM 24,-

Vol. 76: G. Fandel, Optimale Entscheidung bei mehrfacher Zielsetzung. 121 Seiten. 1972. DM 16,-

Vol. 77: A. Auslender, Problemes de Minimax via l'Analyse Convexe et les Inégalités Variationelles: Théorie et Algorithmes. VII, 132 pages. 1972. DM 16,-

Vol. 78: GI-Gesellschaft für Informatik e.V. 2. Jahrestagung, Karlsruhe, 2.-4. Oktober 1972. Herausgegeben im Auftrag der Gesellschaft für Informatik von P. Deussen. XI, 576 Seiten. 1973. DM 36,-

Vol. 79: A. Berman, Cones, Matrices and Mathematical Programming. V, 96 pages. 1973. DM 16,-

Vol. 80: International Seminar on Trends in Mathematical Modelling, Venice, 13-18 December 1971. Edited by N. Hawkes. VI, 288 pages. 1973. DM 24,-

Vol. 81: Advanced Course on Software Engineering. Edited by F. L. Bauer. XII, 545 pages. 1973. DM 32,-

Vol. 82: R. Saeks, Resolution Space, Operators and Systems. X, 267 pages. 1973. DM 22,-

Vol. 83: NTG/GI-Gesellschaft für Informatik, Nachrichtentechnische Gesellschaft. Fachtagung „Cognitive Verfahren und Systeme", Hamburg, 11.-13. April 1973. Herausgegeben im Auftrag der NTG/GI von Th. Einsele, W. Giloi und H.-H. Nagel. VIII, 373 Seiten. 1973. DM 28,-

Vol. 84: A. V. Balakrishnan, Stochastic Differential System I. Filtering and Control. A Function Space Approach. V, 252 pages. 1973. DM 22,-

Vol. 85: T. Page, Economics of Involuntary Transfers: A Unified Approach to Pollution and Congestion Externalities. XI, 159 pages. 1973. DM 18,-

Vol. 86: Symposium on the Theory of Scheduling and Its Applications. Edited by S. E. Elmaghraby. VIII, 437 pages. 1973. DM 32,-

Vol. 87: G. F. Newell, Approximate Stochastic Behavior of n-Server Service Systems with Large n. VIII, 118 pages. 1973. DM 16,-

Vol. 88: H. Steckhan, Güterströme in Netzen. VII, 134 Seiten. 1973. DM 16,-

Vol. 89: J. P. Wallace and A. Sherret, Estimation of Product. Attributes and Their Importances. V, 94 pages. 1973. DM 16,-

Vol. 90: J.-F. Richard, Posterior and Predictive Densities for Simultaneous Equation Models. VI, 226 pages. 1973. DM 20,-

Vol. 91: Th. Marschak and R. Selten, General Equilibrium with Price-Making Firms. XI, 246 pages. 1974. DM 22,-

Vol. 92: E. Dierker, Topological Methods in Walrasian Economics. IV, 130 pages. 1974. DM 16,-

Vol. 93: 4th IFAC/IFIP International Conference on Digital Computer Applications to Process Control, Zürich/Switzerland, March 19-22, 1974. Edited by M. Mansour and W. Schaufelberger. XVIII, 544 pages. 1974. DM 36,-

Vol. 94: 4th IFAC/IFIP International Conference on Digital Computer Applications to Process Control, Zürich/Switzerland, March 19-22, 1974. Edited by M. Mansour and W. Schaufelberger. XVIII, 546 pages. 1974. DM 36,-

Vol. 95: M. Zeleny, Linear Multiobjective Programming. XII, 220 pages. 1974. DM 20,-

Vol. 96: O. Moeschlin, Zur Theorie von Neumannscher Wachstumsmodelle. XI, 115 Seiten. 1974. DM 16,-

Vol. 97: G. Schmidt, Über die Stabilität des einfachen Bedienungskanals. VII, 147 Seiten. 1974. DM 16,-

Vol. 98: Mathematical Methods in Queueing Theory. Proceedings of a Conference at Western Michigan University, May 10-12, 1973. Edited by A. B. Clarke. VII, 374 pages. 1974. DM 28,-

Vol. 99: Production Theory. Edited by W. Eichhorn, R. Henn, O. Opitz, and R. W. Shephard. VIII, 386 pages. 1974. DM 32,-

Vol. 100: B. S. Duran and P. L. Odell, Cluster Analysis. A survey. VI, 137 pages. 1974. DM 18,-

Vol. 101: W. M. Wonham, Linear Multivariable Control. A Geometric Approach. X, 344 pages. 1974. DM 30,-

Vol. 102: Analyse Convexe et Ses Applications. Comptes Rendus, Janvier 1974. Edited by J.-P. Aubin. IV, 244 pages. 1974. DM 25,-

Vol. 103: D. E. Boyce, A. Farhi, R. Weischedel, Optimal Subset Selection. Multiple Regression, Interdependence and Optimal Network Algorithms. XIII, 187 pages. 1974. DM 20,-

Vol. 104: S. Fujino, A Neo-Keynesian Theory of Inflation and Economic Growth. V, 96 pages. 1974. DM 18,-

Vol. 105: Optimal Control Theory and its Applications. Part I. Proceedings of the Fourteenth Biennual Seminar of the Canadian Mathematical Congress. University of Western Ontario, August 12-25, 1973. Edited by B. J. Kirby. VI, 425 pages. 1974. DM 35,-

Vol. 106: Optimal Control Theory and its Applications. Part II. Proceedings of the Fourteenth Biennual Seminar of the Canadian Mathematical Congress. University of Western Ontario, August 12-25, 1973. Edited by B. J. Kirby. VI, 403 pages. 1974. DM 35,-

# Ökonometrie und Unternehmensforschung
# Econometrics and Operations Research